Lecture Notes in Physics

The Editorial Policy for Proceedings

The series Lecture Notes in Physics reports new developments in physical research and teaching – quickly, informally, and at a high level. The proceedings to be considered for publication in this series should be limited to only a few areas of research, and these should be closely related to each other. The contributions should be of a high standard and should avoid lengthy redraftings of papers already published or about to be published elsewhere. As a whole, the proceedings should aim for a balanced presentation of the theme of the conference including a description of the techniques used and enough motivation for a broad readership. It should not be assumed that the published proceedings must reflect the conference in its entirety. (A listing or abstracts of papers presented at the meeting but not included in the proceedings could be added as an appendix.)

When applying for publication in the series Lecture Notes in Physics the volume's editor(s) should submit sufficient material to enable the series editors and their referees to make a fairly accurate evaluation (e.g. a complete list of speakers and titles of papers to be presented and abstracts). If, based on this information, the proceedings are (tentatively) accepted, the volume's editor(s), whose name(s) will appear on the title pages, should select the papers suitable for publication and have them refereed (as for a journal) when appropriate. As a rule discussions will not be accepted. The series editors and Springer-Verlag will normally not interfere with the detailed editing except in fairly obvious cases or on technical matters.

Final acceptance is expressed by the series editor in charge, in consultation with Springer-Verlag only after receiving the complete manuscript. It might help to send a copy of the authors' manuscripts in advance to the editor in charge to discuss possible revisions with him. As a general rule, the series editor will confirm his tentative acceptance if the final manuscript corresponds to the original concept discussed, if the quality of the contribution meets the requirements of the series, and if the final size of the manuscript does not greatly exceed the number of pages originally agreed upon. The manuscript should be forwarded to Springer-Verlag shortly after the meeting. In cases of extreme delay (more than six months after the conference) the series editors will check once more the timeliness of the papers. Therefore, the volume's editor(s) should establish strict deadlines, or collect the articles during the conference and have them revised on the spot. If a delay is unavoidable, one should encourage the authors to update their contributions if appropriate. The editors of proceedings are strongly advised to inform contributors about these points at an early stage.

The final manuscript should contain a table of contents and an informative introduction accessible also to readers not particularly familiar with the topic of the conference. The contributions should be in English. The volume's editor(s) should check the contributions for the correct use of language. At Springer-Verlag only the prefaces will be checked by a copy-editor for language and style. Grave linguistic or technical shortcomings may lead to the rejection of contributions by the series editors. A conference report should not exceed a total of 500 pages. Keeping the size within this bound should be achieved by a stricter selection of articles and not by imposing an upper limit to the length of the individual papers. Editors receive jointly 30 complimentary copies of their book. They are entitled to purchase further copies of their book at a reduced rate. As a rule no reprints of individual contributions can be supplied. No royalty is paid on Lecture Notes in Physics volumes. Commitment to publish is made by letter of interest rather than by signing a formal contract. Springer-Verlag secures the copyright for each volume.

The Production Process

The books are hardbound, and the publisher will select quality paper appropriate to the needs of the author(s). Publication time is about ten weeks. More than twenty years of experience guarantee authors the best possible service. To reach the goal of rapid publication at a low price the technique of photographic reproduction from a camera-ready manuscript was chosen. This process shifts the main responsibility for the technical quality considerably from the publisher to the authors. We therefore urge all authors and editors of proceedings to observe very carefully the essentials for the preparation of camera-ready manuscripts, which we will supply on request. This applies especially to the quality of figures and halftones submitted for publication. In addition, it might be useful to look at some of the volumes already published. As a special service, we offer free of charge LATEX and TEX macro packages to format the text according to Springer-Verlag's quality requirements. We strongly recommend that you make use of this offer, since the result will be a book of considerably improved technical quality. To avoid mistakes and time-consuming correspondence during the production period the conference editors should request special instructions from the publisher well before the beginning of the conference. Manuscripts not meeting the technical standard of the series will have to be returned for improvement.

For further information please contact Springer-Verlag, Physics Editorial Department II, Tiergartenstrasse 17, D-69121 Heidelberg, FRG

A. Alekseev   A. Hietamäki   K. Huitu
A. Morozov   A. Niemi   (Eds.)

Integrable Models and Strings

Proceedings of the 3rd Baltic Rim Student Seminar
Held at Helsinki, Finland, 13-17 September 1993

Springer-Verlag
Berlin Heidelberg GmbH

Editors

Anton Alekseev
Department of Theoretical Physics
Uppsala University, Box 803
S-751 08 Uppsala, Sweden

Antero Hietamäki
Research Institute for Theoretical Physics
University of Helsinki, Siltavuorenpenger 20C
FIN-00170 Helsinki, Finland

Katri Huitu
Research Institute for High Energy Physics
University of Helsinki, Siltavuorenpenger 20C
FIN-00170 Helsinki, Finland

Alexei Morozov
Institute for Theoretical and Experimental Physics
B. Cheremushinskaya 25
117259 Moscow, Russia

Antti Niemi
Department of Theoretical Physics
Uppsala University, Box 803
S-751 08 Uppsala, Sweden
and
Research Institute for Theoretical Physics
University of Helsinki, Siltavuorenpenger 20C
FIN-00170 Helsinki, Finland

ISBN 978-3-662-13968-4   ISBN 978-3-540-48810-1 (eBook)
DOI 10.1007/978-3-540-48810-1

CIP data applied for

Originally published by Springer-Verlag Berlin Heidelberg New York in 1994
Softcover reprint of the hardcover 1st edition 1994

Typesetting: Camera ready by author/editor
SPIN: 10127058   55/3140-543210 - Printed on acid-free paper

Preface

In this volume we present for the first time a collection of lecture notes of a conference series in the field of mathematical physics. This series started three years ago, first under the name Helsinki-St.Petersburg Student Seminars and presently - as a consequence of recent historical developments - under the name of Baltic Rim Student Seminar. The first meeting in this series was organized in September 1991 in Helsinki, and the second meeting took place in St.Petersburg in September 1992. The third meeting - represented by these lecture notes - was held again in Finland, in Espoo just outside of Helsinki in September 1993. In the future, we expect these meetings to also take place in other locations around the Baltic Rim.

The seminars have rapidly grown up from a small, local conference intended mainly for graduate students in Helsinki and St.Petersburg to a representative international meeting attracting many experts in the field. Participants and lecturers of the 1993 meeting came from Austria, Belgium, Estonia, Germany, Finland, France, Russia, Sweden, Switzerland and the United States, and we hope that the material of the scientific talks collected in this volume may attract even wider attention in the international community.

The original idea of these Seminars was to bring together young scientists from East and West. Usually the program consists of two main parts. During the first part the junior participants present their current work and the second part is a school in Modern Mathematical Physics with senior participants as lecturers presenting short courses consisting of 2-3 lectures on recent developments in this field. Until now, the selection of topics has mainly included conformal and topological field theory, integrable models, quantum groups and noncommutative geometry. The reader will find all of them presented in the contributions to this volume. Along with some pedagogical exposition of well-known facts each paper also includes original, new results in the field. We hope that this volume will be useful both for the experts as well as the beginners in the field.

Finally, we would like to thank the Academy of Finland for its generous support for the Mathematical Physics Research Project which also made this Student Seminar possible. We also acknowledge the generous support provided by the Research Institute for Theoretical Physics (TFT) and Research Institute for High Energy Physics (SEFT) for their support in organizing this meeting.

Uppsala
June 3rd 1994

Anton Alekseev
Antti Niemi

Contents

Contents

The New Results on Lattice Deformation of Current Algebra

L. D. Faddeev[1,*] and A. Yu. Volkov[2]

[1] Steklov Mathematical Institute,
Fontanka 27, St.Petersburg, Russia
and
Research Institute for Theoretical Physics
P.O.Box 9, FIN-00014 University of Helsinki, Finland

[2] Physique - Mathématique, Université de Bourgogne,
B.P.138 Dijon Cédex, France

The topic "Quantum Integrable Models" was reviewed in the literature and presented to the conferences and schools many times. Only the reports of our own have been done on quite a few occasions (see, e.g., [1], [2]). So here we shall try to present a fresh approach to the description of the ingredients of construction of integrable models. It has gradually evolved in the process of our joint work. Whereas our goal was the Sugawara construction for the lattice affine algebra (known now as the St.Petersburg algebra), (see, e.g., [1]), some technical developments happen to be new and useful for the already developed subjects. Here we shall underline this development.

1   Dynamical Variables

We shall work on the lattice, using discrete variables for both space and time. The space will be supposed to be one-dimensional (as usual in this business), so we shall use the set of integers n for space coordinates. The dynamical variables will be labelled by n. The first choice is the set of operators w_n with commutation relations

$$w_n w_{n+1} = q^2 w_{n+1} w_n$$

$$w_n w_m = w_m w_n \qquad |n - m| \geq 2.$$

---

*Supported by the Russian Academy of Sciences and Academy of Finland

Here
$$q = e^{i\gamma h},$$

where the parameter γ will play the role of coupling constant. In case of "finite volume" $n = 1, \ldots, N$ and periodic boundary conditions

$$w_n = w_{n+N}$$

we are to use these relations for $n = 1, \ldots n, N - 1$ and adjoin the last relation

$$w_N w_1 = q^2 w_1 w_N.$$

For even N we have two central elements

$$C_1 = \prod w_{odd}$$

$$C_2 = \prod w_{even},$$

(no attention to order as the factors commute), and for odd N there is one

$$C = w_1 \ldots w_N.$$

Thus we have for fixed C the system with $\frac{N}{2} - 1$ (even N) or $\frac{N-1}{2}$ (odd N) degrees of freedom.

Variables w_n allow for several interpretations in terms of a suitable continuous limit. Putting

$$w_n = e^{2\Delta p(x)},$$

where Δ is the lattice spacing, $x = n\Delta$, and assuming for safety the classical limit $\hbar \to 0$ we have in the limit $\Delta \to 0$ the classical variables $p(x)$ with the Poisson relation

$$\{p(x), p(y)\} = \gamma \delta'(x - y) \tag{1}$$

characteristic of the $U(1)$ current algebra.

This allows us to call w_n the lattice current variables. The typical dynamics for $p(x)$ is of chiral (massless) type – moving left or right – as in the free case

$$\dot{p} = p', \tag{2}$$

or in MKdV case

$$\dot{p} = p^2 p' + p'''. \tag{3}$$

Here space-time will be represented as a rectangular lattice.

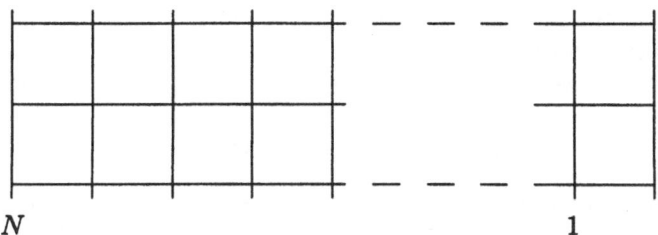

$$N \qquad\qquad\qquad\qquad 1$$

However, a different assignment, i.e.

$$w_{2n} = e^{2\varphi_n(x)}$$

$$w_{2n+1} = e^{\Delta(p_n - p_{n-1})},$$

where φ_n, p_n are the canonical pairs with Poisson brackets

$$\{\varphi_n, p_m\} = \gamma\delta_{nm}$$

allowing to use w_n to treat the massive models.

Here the space-time lattice is to be treated as lightlike, with w_n attached to the vertices of the "initial saw".

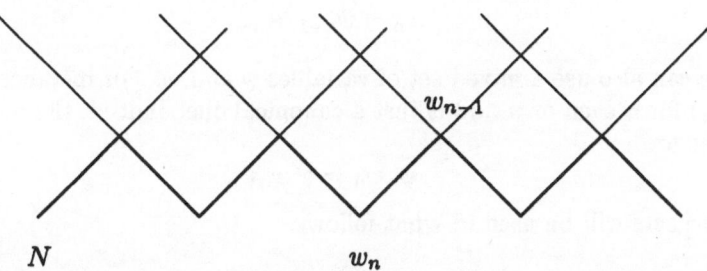

$$N \qquad\qquad\qquad w_n$$

The current variables w_n are local, but not ultralocal – the near neibourghs do not commute. It is the use of such variables in the framework of integrable models that is the novelty we want to advertize.

In the CFT one uses also the nonlocal variables – the so-called vertex variables besides the current ones. The typical classical Poisson relations are of the form

$$\{\psi(x), \psi(y)\} = \frac{1}{2}\gamma\text{sign}(x - y) \tag{4}$$

Clearly, that the variables

$$p = \psi'$$

satisfy then the relations (1). The quantum lattice version of (4), which we shall use in what follows, can be introduced for odd N, $N = 2M$, what we shall suppose to be the case from now on. The relations look as follows

$$\psi_m \psi_n = q \psi_n \psi_m,$$

if $n - m$ is odd and $0 < n - m < 2M$. There are also quasiperiodicity conditions

$$\psi_{2n+2M} = C_1 \psi_{2n}$$

$$\psi_{2n+1+2M} = C_1 \psi_{2n+1}$$

where C_1 and C_2 are now dynamical variables with commutation relations

$$\psi_{odd} C_1 = q^2 C_1 \psi_{odd}$$

$$\psi_{even} C_2 = q^2 C_2 \psi_{even}$$

and, of course,

$$[\psi_{even}, \psi_{even}] = [\psi_{odd}, \psi_{odd}] = 0; [C_1, C_2] = 0.$$

Now the number of degrees of freedom is $M + 1$. The constraints $C_1 = c_1$, $C_2 = c_2$, where c_1 and c_2 are constants, reduce the number o degrees of freedom to $M - 1$.

In this reduced system we can return to the w-variables via

$$w_n = \psi_{n+1} / \psi_{n-1}.$$

One can also use a mixed set of variables ψ and w. For instance, set of pairs (w_n, ψ_n) for n even or n odd is just a canonical one. Indeed, the only nontrivial relations are

$$\psi_n w_n = q^2 w_n \psi_n$$

All these sets will be used in what follows.

2   Massless case: Virasoro algebra on the lattice

Here we follow [3].

Classical continuous current variables $p(x)$ define Poisson action of the Virasoro algebra through the generators

$$S(x) = p^2 + p'(x). \tag{5}$$

Indeed, it follows from (1) that $S(x)$ have Poisson brackets

$$\{S(x), S(y)\} = 2\gamma(S(x) + S(y))\delta'(x - y) - \gamma\delta'''(x - y),$$

characteristic of the Virasoro algebra. The Hamiltonian

$$H_1 = \frac{1}{\gamma} \int S(x) dx$$

leads to the free equation of motion (2). The next one

$$H_2 = \frac{1}{\gamma} \int S^2(x) dx$$

gives the MKdV equation (3). We shall define the lattice quantum analogue of the first dynamical system. It is natural to try to realize the following lattice analogue of the equation of motion

$$w_{n+1}(t + \Delta) = w_n(t).$$

The evolution operator U such that

$$w_n(t + \Delta) = U w_n(t) U^{-1}$$

is thus the shift

$$U w_n U^{-1} = w_{n-1}.$$

We want to express U as a function of variables w . Here is the answer. Let $r(w)$ be a function, defined through the functional equaiton

$$\frac{r(qw)}{r(q^{-1}w)} = w. \tag{6}$$

Then it is almost trivial to check that the ordered product of $2N - 1$ factors

$$U = r(w_{2N}) \ldots r(w_2) \tag{7}$$

realise the shift, if a necessary condition

$$C_1 = C_2$$

is satisfied.

The solution of (6) is given by the θ-series

$$r(w) = \sum_{n=-\alpha}^{\infty} q^{n^2} w^n, \tag{8}$$

which rapidly converges for $|q| < 1$. More interesting case of $|q| = 1$ needs more care. If q is an odd root of unity

$$q^{2Q+1} = 1,$$

we can take a finite sum $-Q < n < Q$ in (8).

The form (7) of shift does not lead to the form of Hamiltonian H_1 in the classical limit. This, however, can be achieved in the following way. Let $s(w)$ be a solution of the equation

$$\frac{s(qw)}{s(q^{-1}w)} = 1 + w.$$

It is clear, that $r(w)$ and $s(w)$ are connected by

$$r(w) = s(w)s(\frac{1}{w}).$$

Now the function $s(w)$ being essentially the q-exponent, it has the following important property: let u and v be a Weyl pair

$$uv = q^2 vu.$$

Then there exists a normalization of s such that

$$s(v)s(u) = s(u + v + q^{-1}uv). \tag{9}$$

This allows to express the U operator in the form

$$U = s(w_{2N+1}^{-1})s(w_{2N})s(w_{2N}^{-1})\ldots s(w_{n+1}^{-1})s(w_n)\ldots s(w_2)$$

(we make use of the shift property of U), or

$$U = s(s_{2N})\ldots s(s_2), \tag{10}$$

(using property (9)) , where

$$s_n = w_{n+1}^{-1} + w_n + q^{-1}w_{n+1}^{-1}w_n.$$

The last expression was introduced some time ago in [4] [5] as quantum lattice analogue of the formula (5). Indeed, using the connection of w_n and $p(x)$, we get

$$\frac{1}{4}(1 + s_n) = (1 + \Delta^2 S(x)).$$

Formula (10) is the mail result of this section. We were able to show quantum lattice analogue of all classical continuous formalism pertinent to the $U(1)$ current algebra. Note, that instead of locality of classical expression for H_1 we have multiplicative locality for U.

3 Massive case: SG model on the lattice

Here following [6] [7] and a pioneer paper by Hirota [8], we present the lattice analogue of the Sine-Gordon model equation of motion

$$\partial_t^2 \varphi - \partial_x^2 \varphi + m^2 \sin \varphi = 0$$

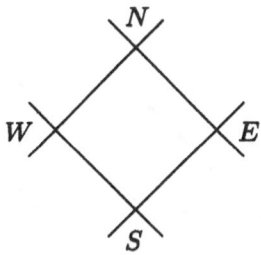

generated by the Hamiltonian

$$H = \frac{1}{2\gamma} \int [\pi^2 + \varphi_x^2 + m^2(1 - \cos\varphi)]dx$$

and Poisson brackets

$$\{\pi(x), \varphi(y)\} = \gamma\delta(x - y).$$

We shall use the variables w and ψ in their light-cone interpretation. They will enter into the family of Lax operators, generating the equation of motion via the zero-curvature condition. We shall associate Lax operator with any elementary wedge of our lattice; let A and B be its initial and final vertices; then

$$L_{AB}(\lambda) = \begin{pmatrix} v_A v_B^{-1} & \lambda v_A^{-1} v_B^{-1} \\ \lambda v_A v_B & v_A^{-1} v_B \end{pmatrix},$$

where

$$v = \psi^{1/2}.$$

For the elementary square with vertices denoted geographically as W, N, E, S the zero curvature condition

$$L_{WS}(\lambda\kappa)L_{SE}(\frac{\lambda}{\kappa}) = L_{WN}(\frac{\lambda}{\kappa})L_{NE}(\lambda\kappa)$$

leads to the equation of motion

$$\psi_N = \psi_S f(q^{-1}\psi_W \psi_E^{-1}), \tag{11}$$

where

$$f(w) = \frac{1 + \kappa^2 w}{\kappa^2 + w}.$$

By a suitable change of variables equation (11) can be transformed into

$$\chi_N = \frac{1}{\chi_S} f(q^{-1}\chi_W \chi_E). \tag{12}$$

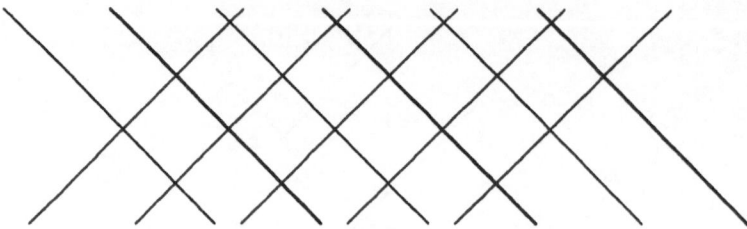

To achieve this we are to put alternatively $\chi = \psi$ or $\chi = \psi^{-1}$ on the even and odd SW lines of our lattice. It can be done periodically only if M is even.

Equation (12) was introduced first by Hirota [8] and leads to the classical continuous SG equation if we put

$$\psi = e^{i\varphi}$$

so that

$$\varphi_N + \varphi_S - \varphi_E - \varphi_W = \Delta(\partial_t^2 \varphi - \partial_x^2 \varphi),$$

and

$$m^2 = \frac{\kappa^2}{\Delta^2}.$$

To get the evolution operator, generating quantum equations (11, ref12), we are, following [9], to construct the "fundamental" Lax operator, entering the fundamental commutation relation in one of the alternative forms, for instance

$$l(\frac{\lambda}{\mu}, w)L_{AB}(\lambda)L_{BC}(\mu) = L_{AB}(\mu)L_{BC}(\lambda)l(\frac{\lambda}{\mu}, w), \qquad (13)$$

where

$$w = \psi_A \psi_C^{-1}.$$

One can check that the FCR () is equivalent to the functional equation

$$\frac{l(qw)}{l(q^{-1}w)} = \frac{1 + \lambda w}{\lambda + w}$$

generalizing those of section 2.

Then the general scheme of construction of evolution operator through the fundamental Lax operator, introduced in [10], developed in [11] and applied to Sine-Gordon model in [6], leads to the following expressions:

$$U = \prod r(w_{even}) \prod r(w_{odd}) \qquad (14)$$

for equation in the form () and

$$U = U_0 \prod r(\hat{w}_{even}) \prod r(\hat{w}_{odd}) \qquad (15)$$

for equation in the form (12).

The values w in (14) are expressed through ψ as $w_n = \psi_{n+1}\psi_{n-1}^{-1}$. The variables $\hat{w}$ are related to χ by

$$\hat{w}_n = \chi_{n+1}\chi_{n-1}.$$

The product in (14) and (15) is over the initial saw. Finally, the U_0 is the inversion operator

$$U_0^{-1}\chi_n U_0 = \chi_n^{-1}.$$

The integrability of equations (11) and (12) follows from the fact, that the trace of the monodromy (the product of Lax operators $L_{AB}(\lambda)$) along the initial saw) generates the commuting conservation laws. More on this is in [7].

Conclusion

We have shown how the nonultralocal variables of current or vertex type can be used in the construction of integrable models on the space-time lattice. There exist other examples of analogous construction, in particular for the lattice WZNW model [12]. So this construction seems to be an interesting counterpart of a more traditional approach, based on ultralocal canonical pairs, reviewed, e.g., in [2].

References

[1] Faddeev L.D., in *"Integrable Systems, Quantum Groups and Quantum Field Theories"*, ed. by L.A.Ibort, M.A.Rodrigues, Dordrecht - Boston - London, (NATO ASI Series, Sec.C, vol.409), 1993.

[2] Faddeev L.D., The Bethe Ansatz.-*Andrejewsky Lectures at Humboldt University in Berlin*, Jan. 1993, Preprint SFB 288.

[3] Faddeev L.D., Volkov A.Yu., *Phys.Lett.B*.

[4] Faddeev L.D., Takhtajan L.A., in *Lecture Notes in Physics*, vol.246, (Springer, Berlin, 1986), p.66.

[5] Volkov A.Yu., *Phys. Lett.* A167, (1992), 345.

[6] Faddeev L.D., Volkov A.Yu., *THeor. Math. Phys.* 92, (1992), 207.

[7] Faddeev L.D., Volkov A.Yu., *Lett. Math. Phys.* 1994, to be published.

[8] Hirota R., *J.Phys. Soc. Jap.* , 43, (1977), 207q.

[9] Tarasov V.O., Takhtajan L.A., Faddeev L.D., *Theor. Math. Phys.* 57, (1983), 163.

[10] Faddeev L.D., Reshetikhin N.Yu., *Ann. Phys.*, 167, (1986), 227.

[11] Destri C., De Vega H.J. *Nucl. Phys.* B290, (1987), 363.

[12] Faddeev L.D. Embedding of the WZNW model into the Sine-Gordon chein. *- Talks at the Schrödinger Institute Workshop*, March 1993 and LHEPT-ENS seminar, April 1993.

Baxterization, Dynamical Systems, and the Symmetries of Integrability

C.-M. Viallet*

Laboratoire de Physique Théorique et des Hautes Energies,
Centre National de la Recherche Scientifique,
Université de Paris 6- Paris 7, Tour 16, 1er étage, boîte 126.
4 Place Jussieu, F–75252 PARIS Cedex 05, FRANCE

Abstract. We resolve the 'baxterization' problem with the help of the automorphism group of the Yang-Baxter (resp. star-triangle, tetrahedron, ...) equations. This infinite group of symmetries is realized as a non-linear (birational) Coxeter group acting on matrices, and exists as such, *beyond the narrow context of strict integrability*. It yields among other things an unexpected elliptic parametrization of the non-integrable sixteen-vertex model. It provides us with a class of discrete dynamical systems, and we address some related problems, such as characterizing the growth of the complexity of iterations.

1 Introduction

The results presented here originate in a long-standing collaboration of M. Bellon, J.M. Maillard and the author [1, 2, 3, 4, 5, 6, 7], augmented by further elaborations (as [8, 9]).

The Yang-Baxter equations and their variations (star-triangle equations, tetrahedron equations,...) are nowadays considered as a characteristic feature of integrability of mechanical systems (classical hamiltonian systems, as well as quantum and statistical systems) [10, 11, 12, 13, 14, 15, 16, 17, 18, 19, 20]....
This field has expanded very much for the last twenty years. We shall not dwell here on its innumerable developments, and relegate general considerations to the end.

We shall start from the Yang-Baxter equations and view them as an algebraic system of equations on matrix entries.

We first recall (section 2) what the two most common forms of the equations for vertex models are.

---

*work supported by CNRS

In section 3 we introduce certain groups generated by involutions (Coxeter groups) [21], together with some realizations in terms of birational transformations of projective spaces. These realizations are generically denoted Γ in the sequel.

In section 4 we build up, from the groups Γ, an infinite group of automorphisms (denoted $\mathcal{A}$) of the Yang-Baxter equations. One should keep in mind that the equations we analyze form an *overdetermined* system, making the existence of a large group of symmetry quite remarkable.

We next show, for the paradigmatic example of the Baxter symmetric eight-vertex model and with both a picture and algebraic results, how our automorphism group reconstructs the well known elliptic curves of solutions. This is a first solution to the 'baxterization' problem [22].

In section 6 we describe more groups of matrix transformations. These are very similar to the ones of section 3, but are adapted to the star-triangle relations, tetrahedron equations, and higher dimensional generalizations.

In section 7 we complete the solution to the baxterization problem, emphasizing the notions of Γ-covariant versus Γ-invariant quantities.

We next produce (section 8) an algorithm to calculate the covariant polynomials as well as the algebraic invariants.

One of the features of Γ is that it is defined outside of the space of solutions of the Yang-Baxter equations, and we may look at its action *beyond integrability*. It becomes an ordinary discrete dynamical system on the space of parameters, possibly non-conservative. In section 9, we describe this action for the general 16-vertex model, i.e. the most general two-state vertex model on a two-dimensional square lattice, which is known *not to be integrable*. The outcome is quite surprising: the orbits of Γ stay within 1-dimensional subvarieties of the 15-dimensional space of parameters. This reveals the actuality of an amazingly large number of algebraically independent invariants of Γ, and yields an elliptic parametrization of the non-integrable model, without reference to the Yang-Baxter equations.

In section 11 we turn to the notion of complexity of the iterations and show how it fits with the existence of algebraic invariants.

In the last section (12), we exemplify some of our results on a definite system, related to a three dimensional vertex model.

2   The Yang-Baxter equations (vertex models)

Many presentations of the Yang-Baxter equations appear in the literature, dividing into two classes:

The first class contains a parameter, also called spectral parameter [18], and reads, for vertex models:

$$\sum_{\alpha_1,\alpha_2,\alpha_3} R^{i_1 i_2}_{\alpha_1 \alpha_2}(\lambda_1, \lambda_2) R^{\alpha_1 i_3}_{j_1 \alpha_3}(\lambda_1, \lambda_3) R^{\alpha_2 \alpha_3}_{j_2 j_3}(\lambda_2, \lambda_3) \qquad (1)$$
$$= \sum_{\beta_1,\beta_2,\beta_3} R^{i_2 i_3}_{\beta_2 \beta_3}(\lambda_2, \lambda_3) R^{i_1 \beta_3}_{\beta_1 j_3}(\lambda_1, \lambda_3) R^{\beta_1 \beta_2}_{j_1 j_2}(\lambda_1, \lambda_2).$$

In this system of equations, R is a matrix of size $q^2 \times q^2$, whose entries are functions of the parameters λ_i, usually through the difference $\lambda_i - \lambda_j$ (at least when λ labels a point on an elliptic curve).

The parameter first appeared as parametrizing the solution. It was later understood as a "spectral parameter" in the context of the quantum inverse scattering method. The origin of the parameter is one of the issues we wish to address.

A given family of models is obtained when one specifies the size q and the relations between the entries of R. Typical relations are equalities between different entries or vanishing of some others (see for example section 5).

Since the entries of R are allowed to be different for the three copies of R entering (1), one may rewrite (1) with three explicitly different matrices A, B, and C of the same size $q^2 \times q^2$.

$$\sum_{\alpha_1,\alpha_2,\alpha_3} A^{i_1 i_2}_{\alpha_1 \alpha_2} B^{\alpha_1 i_3}_{j_1 \alpha_3} C^{\alpha_2 \alpha_3}_{j_2 j_3} = \sum_{\beta_1,\beta_2,\beta_3} C^{i_2 i_3}_{\beta_2 \beta_3} B^{i_1 \beta_3}_{\beta_1 j_3} A^{\beta_1 \beta_2}_{j_1 j_2} \tag{2}$$

or shortly

$$A_{12} B_{13} C_{23} = C_{23} B_{13} A_{12} \tag{3}$$

with now usual notations. The latter form has the interest of not referring to any explicit parametrization.

The second class contains no parameter, and is sometimes called "constant" Yang-Baxter equation. It reads:

$$R_{12} R_{13} R_{23} = R_{23} R_{13} R_{12} \tag{4}$$

that is to say like (3) but with $A = B = C = R$.

To go from equation (1) to equation (4) requires essentially an adequate choice of $\lambda_1, \lambda_2, \lambda_3$, namely $\lambda_1 = \lambda_2 = \lambda_3$. The reverse move, i.e. to recover the parameter dependence from a constant solution was given the name of 'baxterization' [22].

The *use* of the spectral parameter is clear in the context of the Bethe ansatz, since it serves building up the generating functional for commuting conserved quantities, but we shall not be concerned with this aspect. Its *origin* will be explained in the next sections.

3   Some operations on matrices: the groups Γ

We describe here some elementary operations on matrices of various sizes. The matrices we consider are defined up to an overall multiplicative factor. The space of parameters is thus some projective space $\mathbf{P}_n$ with the entries of the matrix as homogeneous coordinates.

Let us start with the matrix R of a two-dimensional q-state vertex model on a square lattice. R has the structure of a tensor product and the indices of R are pairs of indices:

$$R^{ij}_{kl} \qquad i,j,k,l = 1..q \tag{5}$$

We may define the inverse I up to a factor (well defined in the projective space)

$$\sum_{\alpha\beta}(IR)^{ij}_{\alpha\beta}R^{\alpha\beta}_{kl} = \mu\,\delta^i_k\delta^j_l \qquad i,j,k,l = 1..q \tag{6}$$

with μ an arbitrary multiplicative factor, the transposition t:

$$(tR)^{ij}_{kl} = R^{kl}_{ij} \qquad i,j,k,l = 1..q \tag{7}$$

as well as two *partial transpositions* t_l (index l for left) and t_r respectively by:

$$(t_l R)^{ij}_{kl} = R^{kj}_{il} \qquad (t_r R)^{ij}_{kl} = R^{il}_{kj} \qquad i,j,k,l = 1..q \tag{8}$$

Of course

$$t = t_l\,t_r = t_r\,t_l, \qquad I^2 = t^2 = t_l^2 = t_r^2 = 1, \quad \text{and} \quad I\,t = t\,I \tag{9}$$

However

$$t_l\,I \neq I\,t_l \quad \text{and}\, t_r\,I \neq I\,t_r \tag{10}$$

The two partial transpositions *do not commute with the inversion while their product t does.* The transformation $t_l I$ and $t_r I$ *are of infinite order.* Notice of course that they may have finite orbits when acting on certain non-generic matrices.

The group Γ generated by t_l, t_r and I essentially consists of the iterates of $t_l I$ and its inverse It_l (or $t_r I$ and It_r) up to multiplication by an element of its center. It is generated by involutions realized as rational (possibly linear) transformations of some projective space $\mathbf{P}_n$.

We shall describe in section 6 more instances of groups Γ acting as rational transformations of matrices, and generated by involutions.

4 The automorphisms of the Yang-Baxter equations

Suppose we have a solution (A, B, C) of (3):

$$A_{12}B_{13}C_{23} = C_{23}B_{13}A_{12} \tag{11}$$

We may take the partial transpose t_1 of (11) along space 1 and get:

$$(t_1 B_{13})\,(t_1 A_{12})\,C_{23} = C_{23}\,(t_1 A_{12})\,(t_1 B_{13}) \tag{12}$$

Taking the partial transpose t_2 of (12) yields

$$(t_1 B_{13})\,(t_2 C_{23})\,(t_1 t_2 A_{12}) = (t_1 t_2 A_{12})\,(t_2 C_{23})\,(t_1 B_{13}) \tag{13}$$

that is to say:

$$(t_l B)_{13}\,(t_l C)_{23}\,(tA)_{12} = (tA)_{12}\,(t_l C)_{23}\,(t_l B)_{13} \tag{14}$$

Multiplying both sides of (14) to the left and the right by the inverse ItA of tA gives:

$$(tIA)_{12} \, (t_lB)_{13} \, (t_lC)_{23} = (t_lC)_{23} \, (t_lB)_{13} \, (tIA)_{12} . \tag{15}$$

which shows that the transformation

$$\mathcal{K}_A : (A, B, C) \longrightarrow (tIA, t_lB, t_lC) \tag{16}$$

takes a solution of (3) into a solution of (3).

We could similarly construct $\mathcal{K}_B$ (resp. $\mathcal{K}_C$) which inverts and transposes B (resp. C) and acts by partial transpositions on A and C (resp. A and B).

The automorphisms $\mathcal{K}_A, \mathcal{K}_B, \mathcal{K}_C$ are involutive and generate an infinite group of automorphisms $\mathcal{A}$ of the Yang-Baxter equations. It is possible to show that $\mathcal{A}$ is isomorphic to the Weyl group of an affine Lie algebra of type $A_2^{(1)}$ [3, 4].

The group Γ appears when we look at the action of $\mathcal{A}$ on one of the individual copies of R entering equation (1). However, the action of Γ is defined without reference to (1), and we will now concentrate on it, and somehow forget about (1).

5 Baxterization of the Baxter model

The exemplary two-dimensional integrable vertex model is the Baxter model [23, 24, 12], with matrix of Boltzmann weights:

$$R = \begin{pmatrix} a & 0 & 0 & d \\ 0 & b & c & 0 \\ 0 & c & b & 0 \\ d & 0 & 0 & a \end{pmatrix} \tag{17}$$

where one sees the vanishing conditions and equality relations between the entries. It is straightforward to see that the form (17) is stable by all the generators of Γ. We say that the pattern (17), i.e. the collection of vanishing conditions and equalities between entries is admissible (see [1, 2]).

The transformations t_l and t_r read:

$$a \to a, \quad b \to b, \quad c \to d, \quad d \to c,$$

and I reads

$$a \to \frac{a}{a^2 - d^2}, \quad b \to \frac{b}{b^2 - c^2}, \quad c \to \frac{-c}{b^2 - c^2}, \quad d \to \frac{-d}{a^2 - d^2}.$$

One may look at the action of Γ on R. Take one starting point in the space $\mathbf{P}_3$ of parameters, iterate It_l on it and just look at the orbit. One gets the following picture, showing a three-dimensional perspective of the orbit, in inhomogeneous coordinates $u = b/a$, $v = c/a$, $w = d/a$, with starting point (∗):

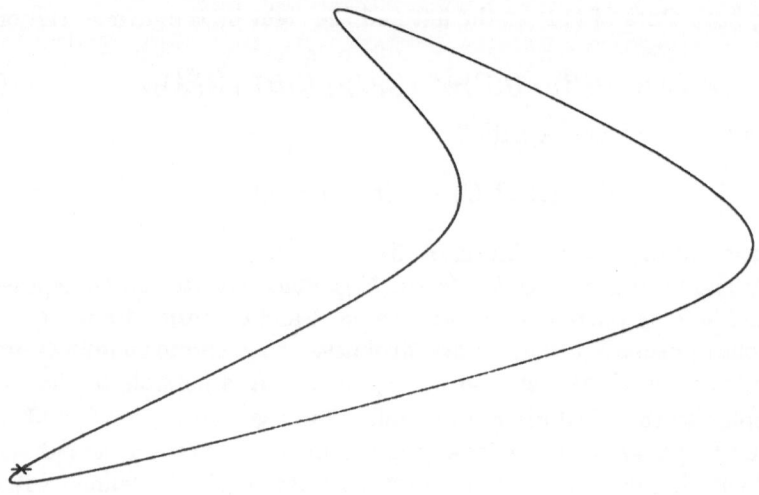

Figure 1: 'Baxterization' of the Baxter model: a perspective view

One sees from this figure that the orbit lies inside a curve. The action of Γ must have at least two algebraically independent invariants. Indeed there are two such invariants:

$$\Delta_1 = \frac{a^2 + b^2 - c^2 - d^2}{ab + cd}, \qquad \Delta_2 = \frac{ab - cd}{ab + cd}. \tag{18}$$

The curves $\Delta_1 = constant$, $\Delta_2 = constant$, are precisely the elliptic curves which appear in the solution of (1), *but we have obtained them with the form (17) as only input, and in particular without having to solve (1)*. The resolution of (1) becomes straightforward once the adequate parametrization of the above curves is used. The orbit we get, although discrete, tends to be dense inside one dimensional curves. The phenomenon is similar to the iteration of a rotation (obtained as the composition of two axial symmetries) with irrational rotation number.

Warning: It is in no way the existence of these curves which ensures the compatibility of the Yang-Baxter equations!

6 Other transformation groups

There are many other equations like (1). They also have automorphism groups generated by (bi)rational transformations. This means we may construct many different realizations Γ. Here are a few more.

6.1   From spin model with interactions along the edges

As was shown in [3], the star-triangle equation also has an infinite group of automorphisms. The matrices under consideration are ordinary $q \times q$ matrices m, of entries m_{ij}. The generators of Γ are the matrix inverse I as previously

$$\sum_\alpha (Im)_{i\alpha}\, m_{\alpha j} = \mu\, \delta_{ij} \tag{19}$$

with μ an arbitrary multiplicative constant, and the element-by-element inverse (Hadamard inverse):

$$J: \qquad m_{ij} \longrightarrow 1/m_{ij} \tag{20}$$

I and J are two non-commuting rational involutions. The product $\varphi = IJ$ is a birational transformation of infinite order.

6.2   From vertex models in three (and more) dimensions

The matrices under consideration in this case again have multi-indices. The matrices are of size $q^3 \times q^3$, with entries of the form:

$$R^{ijk}_{lmn} \tag{21}$$

By examining the symmetries [4, 5] of the tetrahedron equations [25, 26, 27]:

$$R_{123} R_{145} R_{246} R_{356} = R_{356} R_{246} R_{145} R_{123}. \tag{22}$$

with the usual notations, one is lead to the group Γ generated by the inverse I and the *three* partial transpositions t_g, t_m and t_d with

$$(t_g R)^{i_g i_m i_d}_{j_g j_m j_d} = R^{j_g i_m i_d}_{i_g j_m j_d}, \tag{23}$$

and similar definitions for t_m and t_d.

One could as well consider multi-index matrices of size $q^d \times q^d$ written in the form $M^{i_1 i_2 \ldots i_d}_{j_1 j_2 \ldots j_d}$. Section 4 deals with $d = 2$. The tetrahedron equation corresponds to $d = 3$, and so on. There exist d different partial transpositions $t_1, t_2, \ldots, t_d$ with obvious definitions. Combined with the inverse I, they generate an *infinite* group Γ. We will return to the question of the size of these groups in sections 8 and 11.

6.3   More groups

One may further enrich the representations by imposing constraints on the entries of the matrices, provided the transformations are compatible with these constraints (see [1] for the notion of admissible patterns). This yields realizations on projective spaces of lower dimensions.

If the dimension n of the projective space is 1, the only birational transformations are homographies. If $n = 2$, the group, called Cremona group [28], is already much larger. Its elements may however be written as products of linear transformations and the basic blow-up J. The interesting subgroup made out of polynomial and polynomially invertible transformations of the plane has a more elementary structure [29, 30]. When n is larger than 2, the structure of the group of birational transformations is much more complex.

Let us stress that, at the level of the realization, there may exist additional relations between the generators, possibly making it finite.

7 Baxterization

It appears clearly in sections 4 and 5 that, for the Baxter model, the existence of the parameter comes from the existence of sufficiently many algebraic invariants of Γ to confine its orbits, and therefore the ones of $\mathcal{A}$, to curves. The general situation is more subtle, but it always amounts to the following: find the Γ-covariant varieties passing through a constant solution, and/or look for solutions only on covariant varieties.

Different situations may appear, and we illustrate them by specific examples:
– the simplest one was encountered in section 5, recalling that the only input information was the form of the matrix (17). The parameter is just the uniformizing parameter for the elliptic curve which pops out of the action of Γ on any of its generic points. In this case, the action of Γ *has INVARIANTS to imprison the orbit*, and the entire space of parameters is foliated by invariant subvarieties (curves for the Baxter model). Each generic spectral curve has an infinite group of automorphisms, and they consequently all have genus 0 or 1. The solution of (3) is obtained in the following way: choose any point A in the space of parameters. Through A passes one curve of the foliation. Choose B anywhere on this curve. The last matrix C is completely determined then. What is special about the Baxter model is that there is no restriction on the first choice (A), within the space $\mathbf{P}_3$ of parameters. This is far from being the general occurrence.

– a more subtle situation is encountered for the chiral Potts model (as well as for the free fermion condition [31, 32, 33]). Integrability appears on a algebraic variety of which all points have a periodic orbit under Γ. This variety is automatically globally invariant by Γ, but there may very well be NO INVARIANT of the action of Γ. Besides, the resolution of the Yang-Baxter equations may lead to consider further algebraic conditions. Solutions are to be found on a possibly isolated subvariety of the space of parameters. In this situation, the spectral curve may have genus higher than 0 or 1, since it has only a finite number of automorphisms [34, 35, 36].

– a similar situation is encountered in the case of the Jaeger-Higman-Sims model [37, 38]. The parameter space of the model is $\mathbf{P}_2$. The matrix of Boltzmann weights belongs to a 3-dimensional abelian algebra of matrices of size 100×100, with three generators $1, A, \mathcal{J} - 1 - A$ where 1 is the unit matrix, A the adjacency matrix of the Higman-Sims graph, and $\mathcal{J}$ the matrix with all entries equal to 1. Two products exist in this algebra, respectively the usual matrix product and the element by element product. The symmetry group is generated by the inverses I and J, corresponding to these two products, as in section (6.1), equations (19, 20). The two inverses are related by a collineation C (linear map of $\mathbf{P}_2$):

$$I = C^{-1}JC$$

In terms of homogeneous coordinates $[x_0, x_1, x_2]$:

$$J : [x_0, x_1, x_2] \rightarrow [x_1 x_2, x_0 x_2, x_1 x_2]$$

$$C = \begin{bmatrix} 1 & 22 & 77 \\ 1 & -8 & 7 \\ 1 & 2 & -3 \end{bmatrix}$$

We can find three Γ-invariant subvarieties of $\mathbf{P}_2$: one line D, and two hyperbolae H1, H2:

$$
\begin{aligned}
D : &\quad x_2 - x_1 = 0 \\
H1 : &\quad x_2{}^2 + 3\,x_1\,x_2 - 3\,x_0\,x_2 - x_0\,x_1 = 0 \\
H2 : &\quad x_1\,x_2 + 2\,x_1{}^2 - 2\,x_0\,x_2 - x_0\,x_1 = 0
\end{aligned}
$$

Hyperbola $H1$ is the one of reference [39]. The two hyperbolae contain an infinite number of singular points, and we know this prevents the existence of an invariant [9], but not of covariant polynomials, and a fortiori not of covariant ideals.

8 Covariants and invariants

It is fortunately possible to go further in the analysis of the invariant subvarieties [9].

We start from a group G generated by by ν involutions $I_1, I_2, \ldots, I_k$, ($k = 1 \ldots \nu$), *verifying no relations other than the involution property*. The group G is infinite and there are two essentially different situations.

If $\nu = 2$, the group is the infinite dihedral group $\mathbf{Z}_2 \times \mathbf{Z}$, and all elements may uniquely be written $I_1^\alpha (I_1 I_2)^q$, with $\alpha = 0, 1$ and $q \in \mathbf{Z}$. The number of elements of given length l is 2.

If $\nu \geq 3$, the number of elements of length l grows exponentially with l, and the group is in a sense bigger (still countable).

As an example, for the groups described in section 6.2, the number ν of generators depends on the dimension d of the lattice: it is just 2^{d-1} so that if $d = 2$, G is generated by two involutions and if $d \geq 3$, G is generated by more than three involutions.

We then construct various realizations Γ of G by explicit transformations of some projective space. They are obtained by specifying the realization of the generators. We use the same notation I_k for the generators of G and their representatives in Γ. The realizations Γ of G may be written as polynomial transformations in terms of the homogeneous coordinates.

Each involution I_k defines a characteristic polynomial ϕ_k of degree $d_k^2 - 1$ in the following manner. The I_k being involutions, I_k^2 appears as the multiplication by a degree $d_k^2 - 1$ polynomial $\phi_k(x_0, \ldots, x_n)$.

A Γ-covariant polynomial P verifies:

$$P(\gamma(x)) = a(\gamma, x)P(x) \qquad \forall \gamma \in \Gamma, \forall x \in \mathbf{P}_n \qquad (24)$$

The coefficient $a(\gamma, x)$ has to fulfill the cocycle condition:

$$a(\gamma_1 \gamma_2, x) = a(\gamma_1, \gamma_2\, x)a(\gamma_2, x) \qquad (25)$$

Indeed relation (24) demands

$$\begin{aligned} P(\gamma_1 \gamma_2 x) &= a(\gamma_1 \gamma_2, x)P(x) \\ &= a(\gamma_1, \gamma_2\, x)P(\gamma_2 x) \\ &= a(\gamma_1, \gamma_2\, x)a(\gamma_2, x)P(x) \end{aligned}$$

The cocycle a will be completely determined by the values of $a(I_k, x), k = 1 \ldots \nu$. These values may be found easily: when applied to $\gamma_1 = \gamma_2 = I_k$, condition (25) shows that $a(I_k, x)$ has to divide a suitable power of ϕ_k.

Finding an invariant $\Delta = P/Q$ implies finding two polynomials P and Q of the same degree, which transform the same way (are covariant with the same cocycle) under all the generators, i.e:

$$P(I_k(x)) = a(I_k, x) \cdot P(x) \qquad \text{and} \qquad Q(I_k(x)) = a(I_k, x) \cdot Q(x) \qquad (26)$$

Once the cocycle a is chosen, solving (26) becomes a handable linear problem, of which the compatibility can be further studied [9]. We have proved that the proliferation of singularities impeaches the existence of any invariants, but the converse is not true. One of the outcomes is that, in this general setting, the existence of invariants is exceptional.

Finding all invariant subvarieties is more tricky. What we have obtained so far is invariant subvarieties *of codimension 1*, or subvarieties determined by invariants of Γ. Smaller subvarieties have more than one equation, and (24) has to be replaced by a matrix relation:

$$\Pi(\gamma(x)) = A(\gamma, x)\Pi(x) \qquad \forall \gamma \in \Gamma, \forall x \in \mathbf{P}_n \qquad (27)$$

where Π is a vector constructed from the (non canonical) list of equations, and A a matrix. In other words we demand that the ideal defining the subvariety be invariant by Γ, keeping in mind that the subvariety we look for does not have to be a complete intersection.

Let us describe an example, coming from the hard hexagon model [40, 41]. Consider the two involutions of $\mathbf{P}_4$ given in terms of homogeneous coordinates $[x_0, x_1, x_2, x_3, x_4]$:

$$i_1 \longrightarrow [x_5\, x_1\, x_4, x_4\left(x_0\, x_5 - x_2{}^2\right), -x_2\, x_1\, x_4, x_1\left(x_0\, x_5 - x_2{}^2\right), x_0\, x_1\, x_4]$$

$$i_2 \longrightarrow [x_4\, x_2\, x_5, -x_1\, x_2\, x_5, x_5\left(x_0\, x_4 - x_1{}^2\right), x_0\, x_2\, x_5, x_2\left(x_0\, x_4 - x_1{}^2\right)]$$

The two inverses are related by a linear transformation, i.e $i_2 = \tau\, i_1 \tau$ with

$$\tau : [x_0, x_1, x_2, x_4, x_5] \longrightarrow [x_0, x_2, x_1, x_5, x_4] \tag{28}$$

and $\phi_1 = i_1^2 = x_1^2 x_4^2 \left(x_0\, x_5 - x_2^2\right)^2$. If

$$\Pi_1 = x_0^2 x_2^2 x_4 + x_0^2 x_1^2 x_5 - x_0^3 x_4\, x_5 - x_4^2 x_5\, x_2^2 - x_4\, x_5^2 x_1^2 + x_0\, x_4^2 x_5^2 - $$
$$- x_1^2 x_2^2 x_0$$
$$\Pi_2 = x_0^2 - x_4\, x_5 - \lambda\, x_1\, x_2$$

with λ a free parameter, then we do have relation (27) with a matrix

$$A_1 = \left[\begin{array}{cc} x_1^3 x_4^3 \left(x_0\, x_5 - x_2^2\right)^2 & 0 \\ -x_4 & -x_4^2 \left(x_0\, x_5 - x_2^2\right) \end{array} \right]$$

and

$$A_1(i_1(x)) A_1(x) = \begin{pmatrix} \phi_1^5(x) & 0 \\ 0 & \phi_1^2(x) \end{pmatrix}$$

Both Π_1 and Π_2 are unchanged by τ, Π_1 is covariant, and consequently the variety $\Pi_1 = 0$ is invariant by Γ. In contrast, Π_2 is not covariant and only the intersection of $\Pi_2 = 0$ with $\Pi_1 = 0$ is invariant by Γ.

9 The sixteen-vertex model

We abandon here the strict context of integrability, and turn to problems of *discrete dynamical systems, defined in the space of parameters of the models.*

The field of discrete dynamical system is a well developed one. The notion of integrable map goes back, in the context of hamiltonian dynamics, to Poincaré, who founded the subject ([42, 43], see also [44, 45]). The last 30 years have seen the subject expand a lot, specifically with the advent of computer calculations. In order to gain simplicity, one was lead long ago to studying iterations of polynomial and polynomially invertible transformations of the plane [46]), eventually renouncing to hamiltonian structures. Analytic maps have also been fruitfully analyzed, especially in one complex dimension [47, 48]. More recently, some remarkable multi-parameter families of maps, eventually integrable, have been constructed from soliton equations [49, 50, 51, 52, 53, 54].

We want to analyze the behaviour of iterations of a typical infinite order element of our groups Γ such as the product of two generating involutions. The first example we will examine is of great interest for statistical mechanics on the lattice. It comes from the general 2-state vertex model on the square lattice in dimension 2, i.e. the sixteen-vertex model. Take R to be the general 2×2 matrix. This matrix up to a multiplicative constant represents an element of the 15-dimensional projective space $\mathbf{P}_{15}$. A typical orbit of the adequate Γ, when projected to a 2-plane of coordinates, looks like:

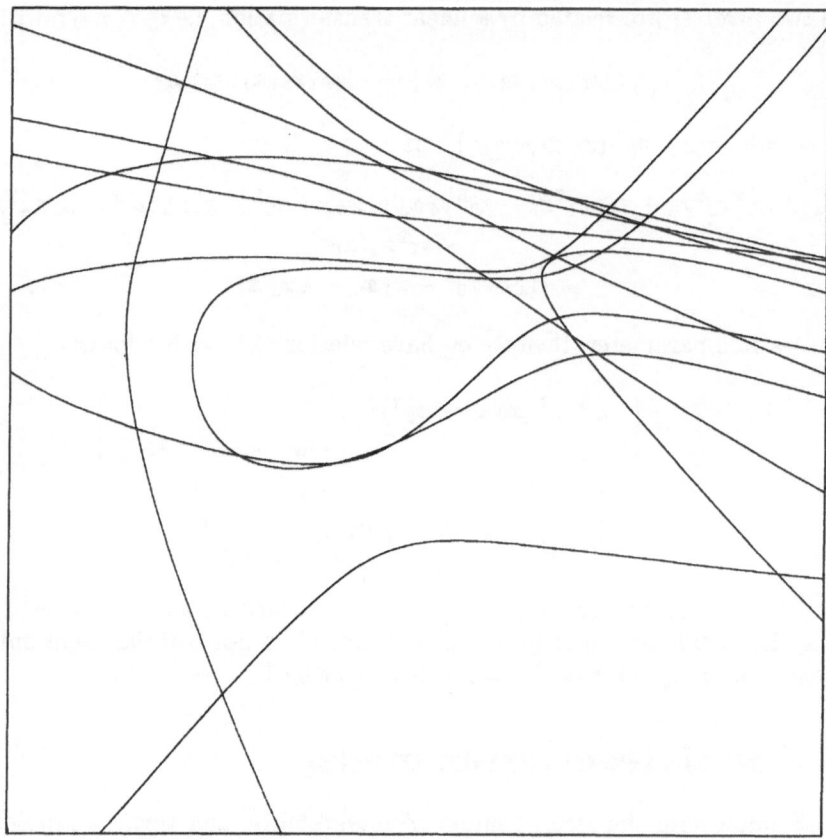

Figure 2: One generic orbit of Γ *for the 16-vertex model*

This figure shows how efficient the graphical method can be. The situation is very favorable since the orbit is confined to a low dimensional variety (a curve). What can be proved [6], following the ideas of section (8), is the existence of a collection of invariants, of algebraic rank 14. More precisely, there are 17 invariants with 3 relations. As in Figure 1, the discrete orbits are generically dense in a curve, and this helps making the 'graphical detector of invariants' so useful.

10 More pictures

One should not believe that Γ-orbits are always curves. Let us introduce a class of transformations in $\mathbf{P_2}$ which contains in particular the transformations (24). Consider Γ generated by I and J, given in homogeneous coordinates $[x_0, x_1, x_2]$ by:

$$J : [x_0, x_1, x_2] \rightarrow [x_1 x_2, x_0 x_2, x_1 x_2] \qquad (29)$$
$$I = C^{-1} J C \qquad (30)$$

with C a projective linear transformation of $\mathbf{P_2}$. This class of realizations, parametrized by the collineation matrix C already contains all kinds of behaviours. Moreover, many nice examples obtained from matrix inversions do implement relation (30).

10.1   Non-symmetric Z_7

Suppose m is the following cyclic 7×7 matrix:

$$m = \begin{pmatrix} x & y & y & z & y & z & z \\ z & x & y & y & z & y & z \\ z & z & x & y & y & z & y \\ y & z & z & x & y & y & z \\ z & y & z & z & x & y & y \\ y & z & y & z & z & x & y \\ y & y & z & y & z & z & x \end{pmatrix} \tag{31}$$

The matrix inverse verifies (30) with[1]

$$C_{Z_7} = \begin{bmatrix} 2 & 6 & 6 \\ 2 & -1 - i\sqrt{7} & -1 + i\sqrt{7} \\ 2 & -1 + i\sqrt{7} & -1 - i\sqrt{7} \end{bmatrix} \tag{32}$$

The iteration of $\varphi = IJ$, with I the matrix inversion and J the element by element inversion, yields the following picture, in the inhomogeneous coordinates $u = y/x$, $v = z/x$, if we represent a number of orbits at the same time:

---

[1] This form is a particular value of an eigenmatrix of conference digraph [55]

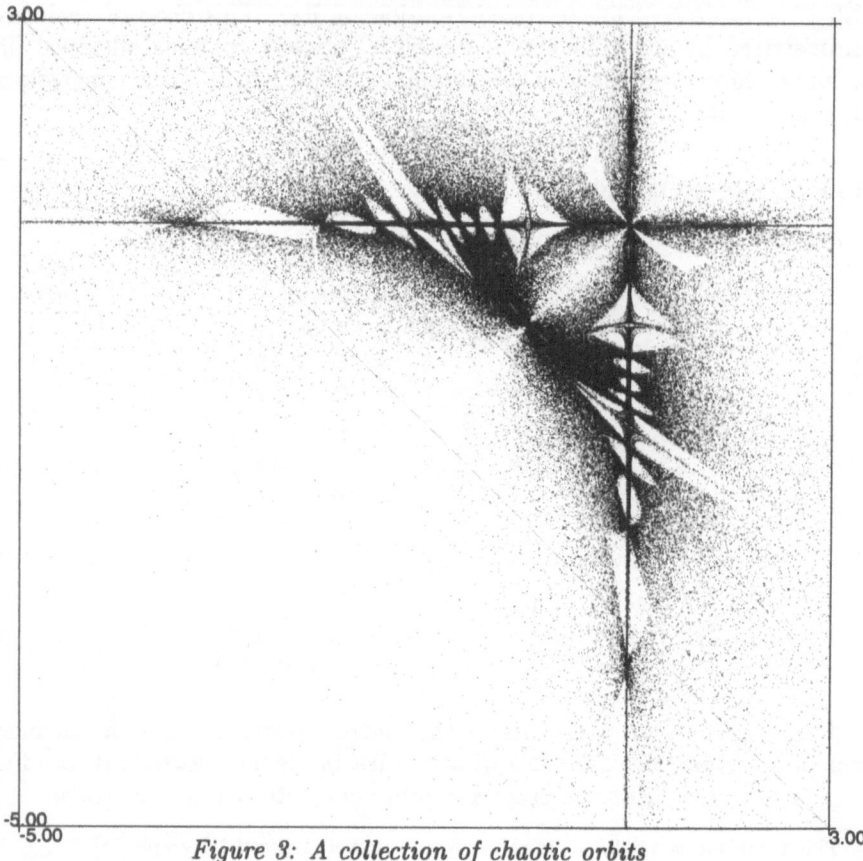

Figure 3: A collection of chaotic orbits

Figure 3 shows there should not be any invariant. Indeed, as proved in [9], there cannot be any because Γ has an infinite number of singular points. The first singular points are visible on the figure. Their location is very easy to compute.

10.2 A finite diagram model

Take

$$C = \begin{bmatrix} 2 & 0 & 2 \\ 1 & 1 & -1 \\ -1 & 1 & 1 \end{bmatrix} \tag{33}$$

This has been introduced in [9] and the denomination 'finite diagram model' refers to the finite number of singular points of Γ. Although such a transformation was not directly obtained from a matrix inverse, it (or similar ones) appears in properly defined reductions of the inversion of matrices of larger sizes [56]. We have found no invariant for this model. The graphical method is quite ineffectual at generic points. It proves nevertheless useful in the vicinity of a fixed point.

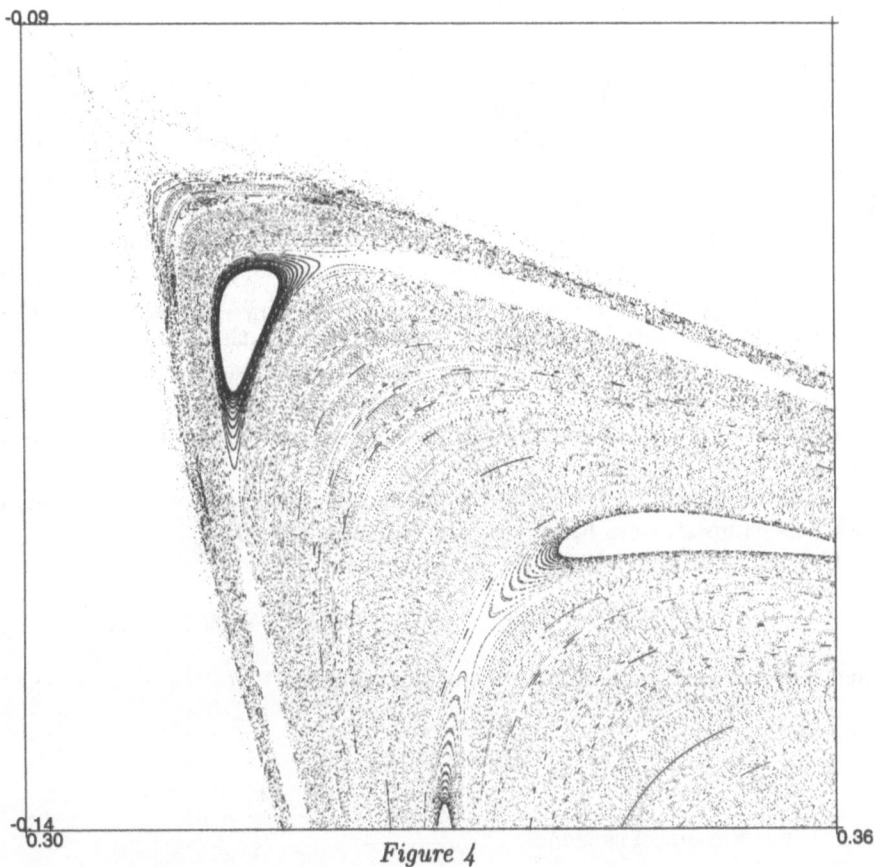

Figure 4

Figure 4 represents a collection of orbits not far away from a fixed point of $(IJ)^2$. The fixed point $[\alpha^2, \alpha, -1], \alpha = (1 + \sqrt{5})/2$, $(x_1/x_0 \simeq .618, x_2/x_0 \simeq -.382)$ is an elliptic fixed point at which the eigenvalues of the linear differential are the roots of $s^2 - 17/16\,s + 1$. We see that the orbits explode when one gets away from the fixed point. This is exactly the situation encountered in [57], see [58].

11   The complexity of iterations

More insight into the behaviour of the iterations can be gained by looking at their complexity, in a way we will define.

The first grasp at the complexity of Γ is through its size (section 8), but we shall not comment on this here.

We may satisfy ourselves with examining the simple case $\nu = 2$, where the group is essentially two copies of $\mathbf{Z}$, and describe in the next section a specific example (section 12) with $\nu = 3$.

The simplest situation is the one where Γ acts by transformations of the 2-dimensional projective space $\mathbf{P}_2$. We have given in section 10 a class of such a realization, and the study of 'the dynamics of singularities' as explored in [9] proves extremely useful.

It is nevertheless something else we have in mind here: suppose $\varphi = IJ$ is of degree w. Its iterates φ^k do not have to be of degree w^k. Indeed there is a simple mechanism for the lowering of the degree $d(k)$ of φ^k; for one has to factorize out common factors from the expressions of the homogeneous coordinates of $\varphi^{(k)}$. We may analyze the sequence $d(k)$, as a function of the order of the iteration k. This measure of the complexity of the iterations actually coincides with the one introduced by Arnold [59]; see [60] for the particular case of bi-polynomial transformations of the plane. It brings a variety of behaviours, lying between the generic exponential growth and periodicity, with the particular instance of polynomial (or polynomially bounded) growth. In a given system of coordinates the notion of degree is well defined. Of course, changes of coordinates may change the degree, but allowed birational changes will preserve the nature of the growth.

We have calculated the degree of the successive iterates of φ in a number of cases, the simplest ones being given by the class (29,30).

If $C = C_{Z_7}$ given by (32), the first terms are:

$$1, \quad 4, \quad 12, \quad 33, \quad 88, \quad 232, \quad 609, \quad \ldots \tag{34}$$

This may be seen as the first terms of the sequence:

$$d(k) = \sum_{i=0}^{k} f_{2i}$$

with f_i the Fibonacci sequence:

$$1, \quad 2, \quad 3, \quad 5, \quad 8, \quad 13, \quad 21, \quad 34, \quad \ldots$$

This behaviour is found for any *generic* elements of the four parameter family [61] of collineations,

$$\begin{bmatrix} 1 & p & q-p-1 \\ 1 & s & -s-1 \\ 1 & r & -r-1 \end{bmatrix} \tag{35}$$

Notice that the Jaeger-Higman-Sims model belongs to this family (compare collineations (24) and (35)), and has the same sequence (34).

For the 'finite diagram model' with C given by (33) the degree reaches its maximal value:

$$d(k) = 4^k$$

For

$$C(w) = \begin{bmatrix} 1 & w-1 & w \\ 1 & -1 & 0 \\ 1 & 0 & -1 \end{bmatrix}, \tag{36}$$

Γ is known to have an invariant, and the orbits lie within elliptic curves of equation

$$P_w^2(x_0, x_1, x_2)Q_w(x_0, x_1, x_2) - \lambda(x_1 + x_2)^4(x_0 - x_2)^2(x_0 - x_1) = 0$$

where

$$P_w = (1-w)(x_2^2 - x_0 x_1) + (w-3)x_2(x_0 - x_1),$$
$$Q_w = (1-w^2)(x_1^3 - x_0 x_2^2) + (w^2 - 4w - 1)x_1^2(x_0 - x_2) + 2(w-1)^2 x_1 x_2(x_0 - x_1)$$

and the sequence of degrees starts with:

1, 4, 10, 20, 34, 53, 75, 102, 132, 167, 206, 249, 295, 347, 402, 461, ...

that is to say the second derivative $d(k+1) - d(k-1) - 2d(k)$ is periodic of period 12. Notice that (36) has the form (35). It is just not generic.

As a last example, in the Baxter model, and also for the 16-vertex model, the iteration of $\varphi = t_g I$ yields the sequence of degrees:

$$1, \quad 3, \quad 9, \quad 19, \quad 33, \quad 51, \quad 73, \quad 99, \quad 129, \quad 163, \quad 201, \quad \ldots \qquad (37)$$

from which one infers

$$d(k) = 2k^2 + 1$$

We have examined many more examples, and they all lead to the conjecture [9]: *chaotic behaviour implies exponential growth, as regular behaviour implies polynomial bounds*.

The proof of this conjecture should follow from considerations on addition on elliptic curves [62] at least for realizations in $\mathbf{P}_2$, the main idea being that φ is a constant shift on some curve.

Another approach goes through the study of the factorization properties: the reduction of the degree $d(k)$ from exponential to polynomially bounded comes from a sufficiently regular factorization process. An analysis of this process was successfully undertaken in [63, 56] for various realizations Γ and confirms up to now our conjecture. We shall give one more example in the next section.

12 An example

In [7] we introduced a restriction of the general 2-state model on the cubic lattice in three dimensions, by imposing the following relations on the entries of the R-matrix:

$$R_{j_1 j_2 j_3}^{i_1 i_2 i_3} = R_{-j_1, -j_2, -j_3}^{-i_1, -i_2, -i_3} \qquad (38)$$

$$R_{j_1 j_2 j_3}^{i_1 i_2 i_3} = 0 \quad \text{if} \quad i_1 i_2 i_3 j_1 j_2 j_3 = -1 \qquad (39)$$

R is the direct product of two times the same 4×4 submatrix [5]. It is further possible to impose that this 4×4 matrix is symmetric, since the product $t = t_l t_m t_r$ acts as its transposition.

$$R_{j_1 j_2 j_3}^{i_1 i_2 i_3} = R_{i_1 i_2 i_3}^{j_1 j_2 j_3} \qquad (40)$$

We use the following notations for the 10 homogeneous entries of this 4×4 submatrix:

$$
\begin{pmatrix}
a & d_1 & d_2 & d_3 \\
d_1 & b_1 & c_3 & c_2 \\
d_2 & c_3 & b_2 & c_1 \\
d_3 & c_2 & c_1 & b_3
\end{pmatrix}. \tag{41}
$$

The generators of Γ are the inversion of R, which acts as a matrix inverse on (41), and the permutations of entries t_l, t_m, and t_r being respectively:

$$
\begin{aligned}
t_l : &\quad c_3 \leftrightarrow d_3, &\quad c_2 \leftrightarrow d_2 \\
t_m : &\quad c_1 \leftrightarrow d_1, &\quad c_3 \leftrightarrow d_3 \\
t_r : &\quad c_2 \leftrightarrow d_2, &\quad c_1 \leftrightarrow d_1
\end{aligned}
$$

A numerical iteration gives the following picture of an orbit, projected on a 2-plane of coordinates.

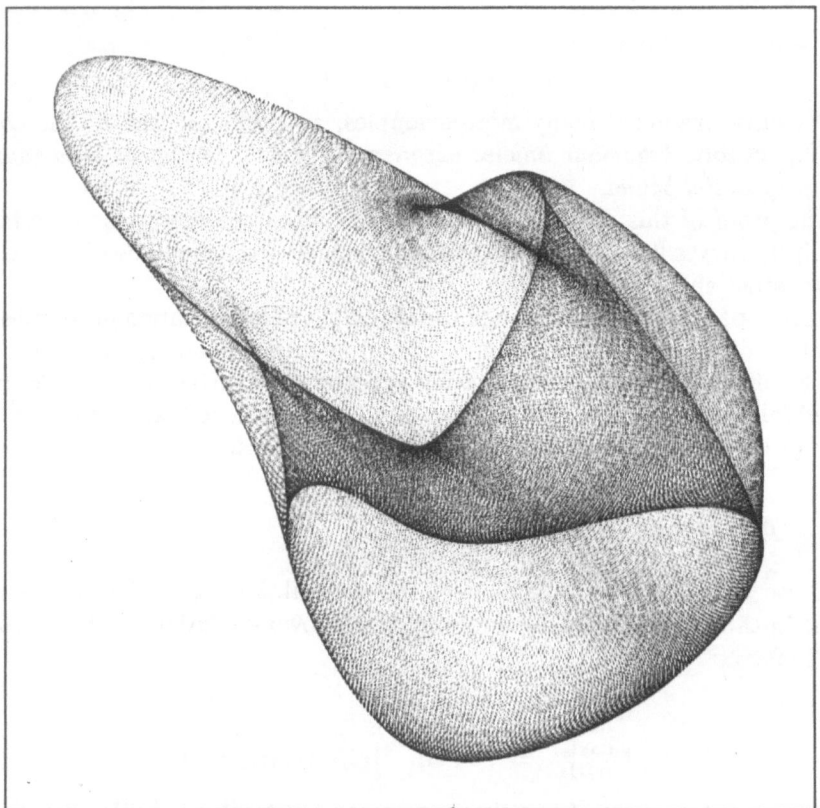

Figure 5: A 'two-dimensional' orbit

This is another example of the efficiency of the graphical detector: we see that the orbit is confined to a 2-dimensional surface. This may be easily proven.

If one defines

$$p_3 = ab_3 + b_1b_2 - c_3^2 - d_3^2, \qquad q_3 = c_1d_1 - c_2d_2, \qquad (42)$$

together with the polynomial p_2 and p_1 obtained by permutation of the indices 1,2, and 3, and (with a corrected misprint from [7]):

$$\begin{aligned}
r_3 &= ab_3 - b_1b_2 + c_3^2 - d_3^2, \\
s_3 &= (a + b_3)c_3 - d_1d_2 - c_1c_2, \\
t_3 &= (b_1 + b_2)d_3 - d_1c_2 - c_1d_2.
\end{aligned}$$

then

- The polynomials $p_i, q_i, i = 1, 2, 3$ form a five dimensional space, and any ratio of these polynomials is invariant by the *whole group* Γ. This means that the orbit of Γ is confined to a 5-dimensional subspace of the original 9-dimensional parameter space.

- If we consider the subgroup generated by I and t_r, then r_3, s_3, t_3 are three more covariant polynomials, and they furnish three more invariants, proving that the orbit is confined to 2 dimensions.

As far as the complexity is concerned, the non-chaotic image fits with the sequence of degrees, which is the same as (37). The analysis of the degrees can be performed easily: Let M be a generic matrix of the form (41), and let $M_0 = t_r M$ be the starting point of the iteration. We iterate $t_r I$ defined polynomially in terms of the entries, and keep track of the common factors (factorized out from the result of the action of K). Let $M_{k+1} = t_r I(M_k)$, f_k the extracted factor, and $\Delta_k = det(M_k)$. One gets $d_0 = det(t_r M)$, $f_0 = 1$, $d_1 = det(t_r I t_r M)$, $f_2 = 1$. Then a factorization of the determinants appears regularly, defining in the course an additional sequence δ_k such that:

$$\begin{aligned}
&\Delta_2 = \Delta_0^3 \delta_2 \quad \Delta_3 = \Delta_1^3 \delta_3 \quad \Delta_4 = \delta_2^3 \delta_4 \quad \Delta_5 = \delta_3^2 \delta_5 \quad \cdots \quad \Delta_n = \delta_{n-2}^3 \delta_n \\
&\cdots \qquad\qquad f_3 = \Delta_0^2 \quad\quad f_4 = \Delta_1^2 \quad f_5 = \Delta_2^2/\Delta_0^6 = \delta_2^2 \quad \cdots \quad f_n = \delta_{n-3}^2
\end{aligned}$$

Denoting by u_k, v_k, g_k, x_k the degrees of respectively Δ_k, δ_k, f_k and of the matrix elements of M_k in terms of the entries of M_0, this implies

$$\begin{aligned}
u_k &= 3\, v_{k-2} + v_k \\
g_k &= 2\, v_{k-3} \\
u_k &= 4\, x_k \\
x_k &= 3\, x_{k-1} - g_k
\end{aligned}$$

resolved by

$$x_{k+3} - 3\, x_{k+2} + 3\, x_{k+1} - x_k = 0 \quad \text{and} \quad d(k) = 2k^2 + 1$$

This is an instance of the general result obtained in [56] (formula (3.19), for $q = 4$), giving a good flavour of the method.

13 Conclusion and perspectives

Inversion relations have proved to be a powerful tool in the study of statistical mechanical models, leading in particular to exact relations on the partition functions, even for non-integrable or higher dimensional models [64, 65, 66].

The description often appeals to a diagrammatic representation [67], which we have not used here.

The Yang-Baxter equations in their various forms have brought an enormous amount of exact results in the field. They have unified branches of physics and mathematics such as ice models [68, 69] and solitonic wave equations [70, 71], gave new insights into knot polynomials [72, 73, 74, 75], and produced successful offsprings like quantum groups [76, 77, 78].

What [3, 4] contain is a concrete link between discrete dynamical systems and quantum integrability, producing a large number of interesting dynamical systems. We have been able to use a little part of the paraphernalia available from the field of dynamical systems –the simplest being numerical calculations–. One outcome is that the rational nature of the transformations captures a good proportion of the algebro-geometric content of the Yang-Baxter equations [33, 6].

One issue for future work is the resolution of the tetrahedron equations, and the finding of truly three-dimensional integrability. Indeed there is a competition then between the higher overdetermination of the systems of equations to solve, and the increasing size of the corresponding group Γ. This antagonism might be resolved either by a relative triviality of the solutions, which are disguised two-dimensional solutions, either by a low complexity of the actual realization of Γ, and this leaves room for interesting solutions. The most natural to conjecture is that Γ has a *finite realizations*, or that its trajectories lie on abelian varieties [79].

The remarkable elliptic parametrization [6] of the 16-vertex model should lead to interesting properties of the corresponding transfer matrices, through controlled functional relations on the exact partition function and/or via a construction of physically relevant spaces of states. This applies as well to three-dimensional models [7], and prompts us to a return to basics, i.e: Bethe Ansatz [80].

Acknowledgment. *I would like to thank F. Jaeger for a very useful discussion, and J. Avan, M. Bellon, J.-M. Maillard, G. Rollet, M. Talon for continued stimulating exchanges.*

References

[1] M.P. Bellon, J-M. Maillard, and C-M. Viallet, *Integrable Coxeter Groups.* Physics Letters **A 159** (1991), pp. 221–232.

[2] M.P. Bellon, J-M. Maillard, and C-M. Viallet, *Higher dimensional mappings.* Physics Letters **A 159** (1991), pp. 233–244.

[3] M.P. Bellon, J-M. Maillard, and C-M. Viallet, *Infinite Discrete Symmetry Group for the Yang-Baxter Equations: Spin models*. Physics Letters **A 157** (1991), pp. 343–353.

[4] M.P. Bellon, J-M. Maillard, and C-M. Viallet, *Infinite Discrete Symmetry Group for the Yang-Baxter Equations: Vertex Models*. Phys. Lett. **B 260** (1991), pp. 87–100.

[5] M.P. Bellon, J-M. Maillard, and C-M. Viallet, *Rational Mappings, Arborescent Iterations, and the Symmetries of Integrability*. Physical Review Letters **67** (1991), pp. 1373–1376.

[6] M.P. Bellon, J-M. Maillard, and C-M. Viallet, *Quasi integrability of the sixteen-vertex model*. Phys. Lett. **B 281** (1992), pp. 315–319.

[7] M.P. Bellon, S. Boukraa, J-M. Maillard, and C-M. Viallet, *Towards three-dimensional Bethe Ansatz*. Phys. Lett. **B 314** (1993), pp. 79–88.

[8] M.P. Bellon, J-M. Maillard, G. Rollet, and C-M. Viallet, *Deformations of dynamics associated to the chiral Potts model*. Int. J. Mod. Phys. **B 6** (1992), pp. 3575–3584.

[9] G. Falqui and C.-M. Viallet, *Singularity, complexity, and quasi–integrability of rational mappings*. Comm. Math. Phys. **154** (1993), pp. 111–125.

[10] L. Onsager, *Crystal statistics: I. A two-dimensional model with an order-disorder transition*. Phys. Rev **65** (1944), pp. 117–149.

[11] J.B. McGuire, *Studies of exactly solvable one-dimensional N-body problems*. J. Math. Phys. **5** (1964), pp. 622–636.

[12] R.J. Baxter. *Exactly solved models in statistical mechanics*. London Acad. Press, (1981).

[13] C.N. Yang, *Some exact results for the many-body problem in one dimension with repulsive delta-function interaction*. Phys. Rev. Lett. **19**(23) (1967), pp. 1312–1315.

[14] L.A. Takhtajan and L.D. Faddeev, *The quantum inverse problem and the XYZ Heisenberg model*. Russian Math. Surveys **34**(5) (1979), pp. 11–68.

[15] L.D. Faddeev, E.K. Sklyanin, and L.A. Takhtajan, *The quantum inverse problem method*. Theor. Math. Phys **40** (1980), p. 688.

[16] E.K. Sklyanin, *Quantum version of the method of inverse scattering*. transl. of Zap. Nauch. Sem. LOMI Steklov **95** (1980), pp. 55–128.

[17] P.P. Kulish and E.K. Sklyanin. volume 151 of *Lecture Notes in Physics*, pages 61–119, (1982).

[18] L.D. Faddeev. Integrable models in $1+1$ dimensional quantum field theory. In *Les Houches Lectures (1982)*, Amsterdam, (1984). Elsevier.

[19] M. Gaudin. *La fonction d'onde de Bethe*. Collection du C.E.A. Série Scientifique. Masson, Paris, (1983).

[20] E.K. Sklyanin. Quantum inverse scattering method. selected topics. In *Quantum Groups and Quantum Integrable Systems*, Singapore, (1992). World Scientific. Proceedings of the Nankai Lectures 1991 (and preprint hep-th-9211111).

[21] H.S.M. Coxeter and W.O.J. Moser. *Generators and relations for discrete groups*. Springer Verlag, second edition, (1965).

[22] V.F.R. Jones, *Baxterization*. Int. J. Mod. Phys. **B 4** (1990), pp. 701–713. proc. of 'Yang-Baxter equations, conformal invariance and integrability in statistical mechanics and field theory', Canberra, 1989.

[23] R.J. Baxter, *Eight-vertex model in lattice statistics*. Phys. Rev. Lett. **26**(14) (1971), pp. 832–833.

[24] R.J. Baxter, *One dimensional anisotropic Heisenberg chain*. Phys. Rev. Lett. **26**(14) (1971), p. 834.

[25] A.B. Zamolodchikov, *Tetrahedron equations and the relativistic S-matrix of straight-strings in 2+1 dimensions*. Comm. Math. Phys. **79** (1981), p. 489.

[26] M. T. Jaekel and J. M. Maillard, *Symmetry relations in exactly soluble models*. J. Phys. **A15** (1982), pp. 1309–1325.

[27] J.M. Maillet and F. Nijhoff, *Integrability for multidimensional lattice models*. Phys. Lett. **B 224** (1989), pp. 389–396.

[28] M. Demazure, *Sous-groupes algébriques de rang maximum du groupe de Cremona*. Ann. Scient. Ec. Norm. Sup. 4e série **t.3** (1971), pp. 507–588.

[29] H. Jung, *Über ganze birationale Transformationen der Ebene*. J. Reine Angew. Math. **184** (1942), pp. 161–172.

[30] S. Friedland and J. Milnor, *Dynamical properties of plane polynomial automorphisms*. Ergod. Theory Dyn. Systems **9** (1989), pp. 67–99.

[31] C. Fan and F.Y. Wu, *General lattice statistical model of phase transition*. Phys. Rev. **B 2** (1970), pp. 723–733.

[32] B.U. Felderhof, *Diagonalization of the transfer matrix of the free-fermion model. II*. Physica **66** (1973), pp. 279–297.

[33] I.M. Krichever, *Baxter's equations and algebraic geometry*. Funct. Anal. and its Appl. **15** (1981), pp. 92–103.

[34] H. Au-Yang, B.M. Mc Coy, J.H.H. Perk, S. Tang, and M.L. Yan, *Commuting transfer matrices in the chiral Potts models: solutions of the star–triangle equations with genus* ≥ 1. Phys. Lett. **A123** (1987), p. 219.

[35] R.J. Baxter, J.H.H. Perk, and H. Au-Yang, *New solutions of the star–triangle relations for the chiral Potts model.* Phys. Lett. **A128** (1988), p. 138.

[36] D. Hansel and J. M. Maillard, *Symmetries of models with genus* > 1. Phys. Lett. A **133** (1988), p. 11.

[37] F. Jaeger, *Strongly regular graphs and spin models for the Kauffman polynomial.* Geometriæ Ded. **44** (1992), pp. 23–52.

[38] P. de la Harpe, *Spin models for link polynomials, strongly regular graphs and Jaeger's Higman-Sims model.* Pacific J. Math. **162**(1) (1994), pp. 57–96.

[39] V.F.R Jones, *On a certain value of the Kauffman polynomial.* Comm. Math. Phys. **125** (1989), pp. 459–467.

[40] R.J. Baxter and P.A. Pearce, *Hard hexagons: interfacial tension and correlation length.* J. Phys. **A**(15) (1982), pp. 897–910.

[41] J. Avan, M. Talon, J.M. Maillard, and C.M. Viallet, *New local relations for lattice models.* Int. J. Mod. Phys. **B** 4 (1990), pp. 1895–1912.

[42] H. Poincaré. *Oeuvres. Tomes I–XI.* Gauthier–Villars, Paris, (1952).

[43] H. Poincaré. *Les méthodes nouvelles de la mécanique céleste.* Gauthier–Villars, Paris, (1892).

[44] G. D. Birkhoff, *Dynamical systems with two degrees of freedom.* Trans. Amer. Math. Soc. **18** (1917), pp. 199–300.

[45] A.S. Wightman, *The mechanics of stochasticity in classical dynamical systems.* Perspectives in Statistical Physics (1981), pp. 343–363. Reprinted in *"Hamiltonian dynamical systems"*, R.S. MacKay and J.D. Meiss editors, Adam Hilger (1987).

[46] M. Henon, *A two-dimensional mapping with a strange attractor.* Comm. Math. Phys. **50** (1976), pp. 69–77.

[47] C.L. Siegel, *Iteration of analytic functions.* Ann. Math. **43** (1942), pp. 807–812.

[48] J.-C. Yoccoz, *Conjugaison différentiable des difféomorphismes du cercle dont le nombre de rotation vérifie une condition diophantienne.* Ann. Sc. E.N.S. 4eme série t. **17** (1984), pp. 333–359.

[49] G.R.W. Quispel, J.A.G. Roberts, and C.J. Thompson, *Integrable Mappings and Soliton Equations.* Phys. Lett. A **126** (1988), p. 419.

[50] G.R.W. Quispel, J.A.G. Roberts, and C.J. Thompson, *Integrable Mappings and Soliton Equations II*. Physica **D34** (1989), pp. 183–192.

[51] J. Moser and A.P. Veselov, *Discrete versions of some classical integrable systems and factorization of matrix polynomials*. Comm. Math. Phys. **139** (1991), pp. 217–243.

[52] B. Grammaticos, A. Ramani, and V. Papageorgiou, *Do integrable mappings have the Painlevé property?* Phys. Rev. Lett. **67** (1991), pp. 1825–1827.

[53] V.G. Papageorgiou, F.W. Nijhoff, and H.W. Capel, *Integrable mappings and nonlinear integrable lattice equations*. Phys. Lett. **A147** (1990), pp. 106–114.

[54] O. Ragnisco. Restricted flows of toda hierarchy as integrable maps. In *Proceedings of the XIX I.C.G.T.M.P*, (1992).

[55] Takuya Ikuta. *Non-existence of spin models corresponding to non symmetric association schemes of class 2 on $4m + 2$ vertices with $m \geq 1$*. preprint.

[56] S. Boukraa, J-M. Maillard, and G. Rollet. *Integrable mappings and polynomial growth*. LPTHE preprint 93-26, to appear in Physica A.

[57] M. Henon, *Numerical study of quadratic area preserving maps*. Q. J. Appl. Math. **27** (1969), pp. 291–312.

[58] M.C. Gutzwiller. *Chaos in Classical and Quantum Mechanics*. Spinger Verlag, New-York, Berlin, Heidelberg, (1991). Interdisciplinary Applied Mathematics.

[59] V.I. Arnold, *Dynamics of complexity of intersections*. Bol. Soc. Bras. Mat. **21** (1990), pp. 1–10.

[60] A.P. Veselov, *Growth and Integrability in the Dynamics of Mappings*. Comm. Math. Phys. **145** (1992), pp. 181–193.

[61] A. Neumaier, *Duality in coherent configurations*. Combinatorica **9**(1) (1989), pp. 59–67.

[62] M. Bellon. *Addition on curves and complexity of quasi-integrable rational mappings*. in preparation.

[63] S. Boukraa, J-M. Maillard, and G. Rollet. *Determinantal identities on integrable mappings*. LPTHE preprint 93-25.

[64] R.J. Baxter, *The Inversion Relation Method for Some Two-dimensional Exactly Solved Models in Lattice Statistics*. J. Stat. Phys. **28** (1982), pp. 1–41.

[65] M. T. Jaekel and J. M. Maillard, *Inverse functional relations on the Potts model*. J. Phys. **A15** (1982), pp. 2241–2257.

[66] D. Hansel, J.M. Maillard, J. Oitmaa, and M.J. Velgakis, *Analytical properties of the anisotropic cubic Ising model.* J. Stat. Phys. **48** (1987), pp. 69–80.

[67] R.J. Baxter, *Solvable eight-vertex model on an arbitrary planar lattice.* Phil. Trans. R. Soc. London **289** (1978), p. 315.

[68] E.H. Lieb, *Exact solution of the F model of an antiferroelectric.* Phys. Rev. Lett. **18**(24) (1967), pp. 1046–1048.

[69] E.H. Lieb, *Exact solution of the problem of the entropy of two-dimensional ice.* Phys. Rev. Lett. **18**(17) (1967), pp. 692–694.

[70] V.E. Zakharov and L.D. Faddeev, *The Korteweg-de Vries equation, a completely integrable hamiltonian system.* Funct. Anal. and its Appl. **5** (1971), pp. 280–287.

[71] L.D. Faddeev and L.A. Takhtajan. *Hamiltonian methods in the theory of solitons.* Springer Verlag, Heidelberg, (1986).

[72] L.H. Kauffman, *State models and the Jones polynomial.* Topology **26** (1987), pp. 395–407.

[73] V.G. Turaev, *The Yang-Baxter equation and invariants of links.* Invent. Math. **92** (1988), pp. 527–553.

[74] V.F.R. Jones, *On knots invariants related to some statistical mechanical models.* Pacific J. Math. **137** (1989), pp. 311–334.

[75] F.Y. Wu, *Knot theory and statistical mechanics.* Rev. Mod. Phys. **64**(4) (1992), pp. 1099–1131.

[76] M. Jimbo, *A q–difference analogue of $U(\mathcal{G})$ and the Yang–Baxter equation.* Lett. Math. Phys. **10** (1985), p. 63.

[77] V.G. Drinfel'd. Quantum groups. In *Proceedings of the International Congress of Mathematicians.* Berkeley, (1986).

[78] S.L. Woronowicz, *Twisted SU(2) Group. An Example of Noncommutative Differential Calculus.* Publ. Res. Inst. Math. Sci. **23** (1987), pp. 117–181.

[79] I.G. Korepanov. *Vacuum Curves, Classical Integrable Systems in Discrete Space-Time and Statistical Physics.* preprint hep-th 9312197, (1993).

[80] H. Bethe, *Zur Theorie der Metalle. I. Eigenwerte und Eigenfunktionen der linearen Atomkette.* Z. Phys. **71** (1931), pp. 205–226.

A Lecture on the Calogero-Sutherland Models

V. Pasquier

Service de Physique Théorique de Saclay *,
F-91191 Gif sur Yvette Cedex, France

Abstract. In these lectures, I review some recent results on the Calogero-Sutherland model and the Haldane Shatry-chain. The list of topics I cover are the following: 1) The Calogero-Sutherland Hamiltonien and fractional stastistics. The form factor of the density operator. 2) The Dunkl operators and their relations with monodromy matrices, Yangians and affine-Hecke algebras. 3) The Haldane-Shastry chain in connection with the Calogero-Sutherland Hamiltonian at a specific coupling constant.

1 Introduction

The Calogero-Sutherland model has recently attracted some attention mainly because, in spite of its simplicity, it yields nontrivial results which contribute to shape our understanding of fractional statistics [1]. Its most remarkable property is that its ground state is given by a Jastrow wave-function which is the one dimensional analogue of the Laughlin wave-function. The excitations are described by quasiparticles which carry a fraction of the quantum numbers of the fundamental particles. The wave functions are simple enough to allow the computation of physical quantities such as the dynamical correlation functions. Recently, it has become clear that the spin version of this model is closely related to the known spin chains such as the XXX chain. The wave functions of the spin models can be obtained as simply as those of the Calogero-Sutherland model by diagonalising simple differential operators known as the Dunkl operators [2].

In these lectures, I restrict to the simplest models and I introduce some techniques useful in their study. These lectures are based on work done in collaboration with D.Bernard, M.Gaudin and D.Haldane [3] and with F.Lesage and D.Serban [4].

In the first part, I review Sutherlands method to diagonalise the Calogero-Sutherland Hamiltonian [5]. I then give the expression of the form factor of the density operator. On this example, I describe the fractional character of the quasiparticles which propagate in the intermediate states.

---

* Laboratoire de la Direction des Sciences de la Matière du Commissariat à l'Energie Atomique.

In the second part, I review the spin generalisation of the Calogero-Sutherland model and I diagonalise their Hamiltonian using the Dunkl operators. I then exhibit a representation of the Yangian which commutes with the Hamiltonian. This is achieved by "quantizing" the spectral parameters of a monodromy matrix obeying the Yang-Baxter equation. The consistancy of this procedure requires that the spectral parameters (the Dunkl operators) obey the defining relations of a affine-Hecke-algebra. Finally, I obtain a representation of this algebra which degenerates to the Dunkl operators using some operators defined by Yang in his study of the δ-interacting gas [6].

In the last part, I review the long range interacting spin model known as the Haldane-Shastry chain [7][8]. Although closely related to the Calogero-Sutherland model, this chain is more difficult to study because the Dunkl operators cannot be used to obtain the wave-functions in a straightforward way. I exhibit a representation of the affine Hecke algebra in terms of parameters. This representation becomes reducible at the special point where it commutes with the Haldane-Shastry Hamiltonian. I then use a correspondance with the Calogero-Sutherland models to obtain the eigenvalues and some of the eigenvectors of this Hamiltonian.

The most unphysical feature of the models discussed here is the long range character of the particle particle interaction which behaves as $1/x^2$ in the thermo-dynamical limit. More realistic models with short range potential $(1/\sinh(x)^2)$ can be defined [9] but are more difficult to study (see [10] for some recent progress). Some aspects of the conformal limit of the Haldane-Shastry chain are also discussed in [11].

2   The Calogero-Sutherland Hamiltonian

This model describes particles on a circle interacting with a long range potential [3,4]. The positions of the particles are denoted by x_i, $1 \le i \le N$, $0 \le x_i \le L$ and the total momentum and Hamiltonian which give their dynamics are respectively given by:

$$P = \sum_{j=1}^{N} \frac{1}{i} \frac{\mathrm{d}}{\mathrm{d}x_j}$$

$$H = -\sum_{j=1}^{N} \frac{1}{2} \frac{\mathrm{d}^2}{\mathrm{d}x_j^2} + \beta(\beta - 1)\frac{\pi^2}{L^2} \sum_{i<j} \frac{1}{\sin^2\left(\frac{\pi}{L}(x_i - x_j)\right)}$$

(2.1)

From now on, we shall work on the unit circle and set $\theta_j = 2\pi\frac{x_j}{L}$ and $z_j = e^{i\theta_j}$. The wave functions solution of the equation $H\psi = E\psi$ can be given the following structure:

$$\psi(\theta) = \phi(\theta)\Delta^{\beta}(\theta)$$

(2.2)

with

$$\Delta(\theta) = \prod_{i<j} \sin\left(\frac{\theta_i - \theta_j}{2}\right)$$

(2.3)

Φ is a symmetric polynomial in the variables $z_j = e^{i\theta_j}$ and z_j^{-1}.

To understand the simplicity of the spectrum, one must write the effective Hamiltonian $\tilde{H} = \Delta^{-\beta} H \Delta^{\beta}$ acting on ϕ.
Following Sutherland [5], we obtain:

$$\tilde{H} = \sum_{j=1}^{N} \left(z_j \frac{\partial}{\partial z_j} \right)^2 + \beta \sum_{i \neq j} \frac{z_i + z_j}{z_i - z_j} \left(z_i \frac{\partial}{\partial z_i} - z_j \frac{\partial}{\partial z_j} \right) \qquad (2.4)$$

The remarkable property of this Hamiltonian is that acting upon symmetric polynomials of a given homogeneity in the variables z_j it is realised as a triangular matrix. So, it's eigenvalues can be read on the diagonal of the matrix and there are simple algorithms to find the eigenvectors.

More precisely, let us define the following basis of symmetric polynomials indexed by a partition: $\lambda_1 \geq \lambda_2 ... \geq \lambda_N \geq 0$.

$$m_{\{\lambda\}} = \sum z^{\lambda} \qquad (2.5)$$

which is the sum over distinct permutations of the monomial $z_1^{\lambda_1} z_2^{\lambda_2} ... z_N^{\lambda_N}$.

First, it is easy to see that the subspace of polynomials of a given homogeneity ($| \lambda | = \sum_{i=1}^{N} \lambda_i$) is preserved. Then, inside this subspace, we can define an order on the partition by saying that $\lambda \geq \mu$ if $\lambda_1 \geq \mu_1$, $\lambda_1 + \lambda_2 \geq \mu_1 + \mu_2, ..., \lambda_1 + ... + \lambda_N \geq \mu_1 + \mu_2 + ... + \mu_N$. It follows from Sutherland's argument that:

$$\tilde{H} m_{\lambda} = \sum_{\mu \leq \lambda} C_{\lambda\mu} m_{\mu} \qquad (2.6)$$

that is to say $\tilde{H}$ is a triangular matrix. The eigenvalues are given by the diagonal elements:

$$E_{\lambda} = C_{\lambda\lambda} = \sum_{i=1}^{N} \left(\lambda_i + (\frac{N+1}{2} - i)\beta \right)^2 - ((\frac{N+1}{2} - i)\beta)^2 \qquad (2.7)$$

The momentum is given by:

$$P_{\lambda} = \sum_{i=1}^{N} \left(\lambda_i + (\frac{N+1}{2} - i)\beta \right) - (\frac{N+1}{2} - i)\beta \qquad (2.8)$$

This set of states is not complete, but the complete set can be obtained by multiplying these wave functions by $(z_1 z_2 ... z_N)^p$ where p is a positive or negative integer. Taking this into account, a complete basis is given in terms of N momenta $k_i = \lambda_i + (\frac{N+1}{2} - i)\beta + p$ and the energy and momentum are given by:

$$P |k_1 ... k_N\rangle = \left(\sum_{i=1}^{N} k_i \right) |k_1 ... k_N\rangle$$

$$H |k_1 ... k_N\rangle = \left(\sum_{i=1}^{N} k_i^2 \right) |k_1 ... k_N\rangle \qquad (2.9)$$

This looks like a free particle spectrum, in particular for $\beta = 0, 1$, the Hamiltonian is a pure kinetic energy term and it reduces to the usual boson and fermion description of the states. To understand the difference, let us consider the case of integer values of β. One sees from the expression of the k_i that they are integers and obey the constraint:

$$k_i - k_{i+1} \geq \beta \tag{2.10}$$

If $\beta = 1$, this is just the Pauli-exclusion principle, but if $\beta = 2, 3, ...$ the Pauli principle is replaced by a stronger exclusion principle.

2.1 Particle Hole Excitations

We shall now describe the excitations of the ground state of such a system and show that they can naturally be described in terms of quasiparticles. To illustrate this, we shall consider the matrix elements (form factor):

$$\langle \alpha \mid \rho(x) \mid 0 \rangle \tag{2.11}$$

of the density operator

$$\rho(x) = \sum_{i=1}^{N} \delta\left(\theta_i - x\right) \tag{2.12}$$

where the state $\mid 0 >$ stands for the ground state and $\mid \alpha >$ for some excited state. The results which we present here were motivated by the evaluation of the density density correlation function obtained by Simons, Lee and Altshuler in the matrix models [12] (see also [1]).

The question we address is for which states $\mid \alpha >$ is the matrix element non zero?

In the case $\beta = 1$ which corresponds to free fermions, it is easy to answer using second-quantization arguments. The operator $\rho(x)$ can be represented as $\psi^+(x)\psi(x)$ where $\psi(x)$ is a fermion field. Acting on the vacuum, the operator $\psi(x)$ anihilates a particle from the Fermi sea ($\mid k \mid \leq k_F$ where k_F is the Fermi momentum) and the operator $\psi^+(x)$ creates a particle out of the Fermi sea. The states $\mid \alpha >$ are therefore described by $(N - 1)$ momenta k_i less than the Fermi momentum k_F and 1 momentum k_N larger than k_F or equivalently in terms of a hole of momentum less than k_F and a particle of momentum larger than k_F.

For β arbitrary integer, the calculation is based on the properties of the wave functions of $\tilde{H}$ (the Jack polynomials) established by Macdonald and Stanley [13][14]. The final result is however simple enough to be described here. Let us view the ground state as a Fermi-sea filled with N particles subject to the constraint $|k_i - k_j| \geq \beta$:

$$k_i = \beta\left(i - \frac{N+1}{2}\right), \quad 1 \leq i \leq N \tag{2.13}$$

The Fermi momentum is given by the largest value of k_i :

$$k_F = \beta\frac{N-1}{2}. \tag{2.14}$$

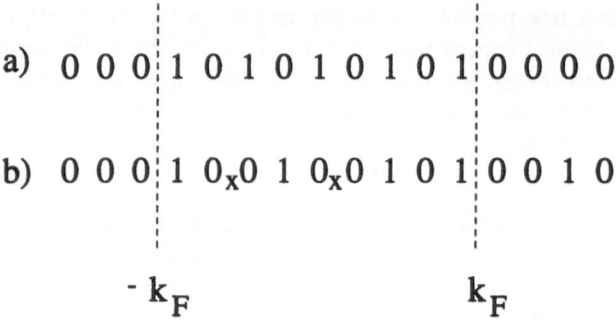

a) $0\ 0\ 0\ 1\ 0\ 1\ 0\ 1\ 0\ 1\ 0\ 1\ 0\ 1\ 0\ 0\ 0\ 0$

b) $0\ 0\ 0\ 1\ 0\ 0_x 0\ 1\ 0\ 0_x 0\ 1\ 0\ 1\ 0\ 0\ 1\ 0$

$-\mathbf{k}_F$ $\mathbf{k}_F$

Fig. 1: The Fermi sea

This is represented on Figure 1 a) for $\beta = 2$, $N = 5$. The ones stand for the integers k_i occupied by a particle and the 0 for the remaining integers.

It turns out that the states $\mid \alpha >$ which propagate are those for which $(N-1)$ momenta are inside the Fermi-sea $|k_i| \le k_F$ and one momentum is out $k_N > k_F$ (see Figure 1 b)). Such states can be completely described by β holes of momentum less than k_F and a particle of momentum larger than k_F. A hole corresponds to a sequence of β consecutive integers not occupied by a momentum (represented by a cross on the Figure).

The outcome of the computation [4] (These results have also been conjectured in [15] and [16]) is that these holes are eigenstates of a Calogero-Sutherland Hamiltonian with coupling constant β^{-1} and have a mass equal to $\frac{1}{\beta}$ the mass of the particles. In the thermodynamic limit the form factor is equal to :

$$\langle \alpha \mid \rho(0) \mid 0 \rangle = (w^2-1)^{\frac{\beta-1}{2}} \prod_{i=1}^{\beta}(1-v_i^2)^{\frac{\beta-1}{2}-1} \prod_{i<j}|v_i-v_j|^{1/\beta} \frac{(\Sigma_i v_i - \beta w)}{\prod_{i=1}^{\beta}(v_i - w)} \quad (2.15)$$

where v_i and w are respectively the rapidities of the holes and of the particle describing the intermediate state. Other form factors such that $\langle \lambda \mid \psi(x) \mid 0 \rangle$ confirm this description of the excitations. Although the description given here is reminiscent of a second quantized formalism, I am not aware that such a formalism exists.

3   The Spin Calogero-Sutherland Models

A step towards the unification of the Calogero-Sutherland models and the integrable spin chains consists in introducing spin generalizations of the model studied in the last section.

The Hamiltonian we consider is

$$H = \sum_{j=1}^{N}\left(z_j \frac{\partial}{\partial z_j}\right)^2 - \sum_{i \ne j}\beta\,(\beta + P_{ij})\frac{z_i z_j}{(z_i - z_j)^2} \quad (3.1)$$

The particles have an internal spin degree of freedom and the permutation P_{ij} exchanges the spins of the particles i and j.

The eigenstates have the following structure:

$$\psi\left(z_i, \sigma_i\right) = \Phi\left(z_i, \sigma_i\right) \prod_{i<j} \left(z_i - z_j\right)^{\beta} \tag{3.2}$$

where the wave function $\phi\left(z_i, \sigma_i\right)$ is completely antisymmetric under the simultaneous permutations of the spins and coordinates.

For our purpose, it is more convenient to look at the effective Hamiltonian acting on $\Phi\left(z_i, \sigma_i\right)$:

$$\tilde{H} = \sum_{j=1}^{N} \left(z_i \frac{\partial}{\partial z_j}\right)^2 + \beta \sum_{i \neq j} \frac{z_i + z_j}{z_i - z_j} \left(z_i \frac{\partial}{\partial z_i} - z_j \frac{\partial}{\partial z_j}\right) - \beta \sum_{i \neq j} (P_{ij} + 1) \frac{z_i z_j}{(z_i - z_j)^2} \tag{3.3}$$

It turns out that this Hamiltonian can be diagonalized in a similar way as the one introduced in the last section. It has the same additive spectrum but the selection rules for the momenta differ from (2.10).

To do this, let us define the following Dunkl operators:

$$d_i = z_i \frac{\partial}{\partial z_i} - \beta i - \beta \sum_{j>i} \frac{z_i}{z_i - z_j} (K_{ij} - 1) + \beta \sum_{j<i} \frac{z_j}{z_j - z_i} (K_{ij} - 1) \tag{3.4}$$

where K_{ij} permutes the coordinates z_i and z_j :

$$\begin{aligned}[K_{ij}, z_k] &= 0 \quad \text{for} \quad k \neq i, j \\ K_{ij} z_j &= z_i K_{ij}\end{aligned} \tag{3.5}$$

Some motivation for the form of these operators and the algebra which they obey will come later in the text (see also [17][18]), but first, let us see how they can be used to solve the model. We shall show that the diagonalization of H can be reduced to the simultaneous diagonalization of the Dunkl operators d_i. Moreover in a convenient basis the d_i are represented by triangular matrices.

The d_i obey the defining relations of a degenerate affine-Hecke algebra:

$$\begin{aligned}a) &\quad [d_i, d_j] = 0 \\ b) &\quad [K_{ii+1}, d_k] = 0 \quad \text{if} \quad k \neq i, i+1 \\ c) &\quad K_{i,i+1} d_i - d_{i+1} K_{i,i+1} + \beta = 0\end{aligned} \tag{3.6}$$

From these relations, one deduce that the quantity:

$$D(u) = \prod_{i=1}^{N} (u - d_i) = u^N + \sum_{k=1}^{N} c_k u^{N-k} \tag{3.7}$$

commutes with the permutations K_{ij}.

By computing explicitly $\sum d_i^2 = c_1^2 - 2c_2$, one sees that it is equal to the Hamiltonian $\tilde{H}$ except that the P_{ij} are replaced by $-K_{ij}$. Since $D(u)$ commutes with the permutations, we can restrict to the space of antisymmetric wave functions $\phi(z_i, \sigma_i)$. In this subspace $-K_{ij}$ and P_{ij} coincide and $\sum d_i^2$ is equal to $\tilde{H}$. Thus, instead of $\tilde{H}$, it is easier to simultanesouly diagonalize the d_i 's.

If we consider the explicit expression (3.4) of the d_i's, one sees that they preserve the space of homogeneous polynomials of a given degree in the variables z_i. It then follows from simple arguments that these operators are realized by triangular matrices in the monomial basis $z^\lambda = z_1^{\lambda_1}...z_n^{\lambda_n}$ [3]. Their eigenvalues are equal (up to a permutation) to the components of the multiplet:

$$\delta_i = \lambda_i + \beta(N - i) \tag{3.8}$$

The eigenvectors of $\sum d_i^2$ corresponding to a given eigenvalue form a representation of the permutation algebra. The wave functions $\phi(z_i, \sigma_i)$ are obtained by antisymmetrizing these eigenvectors with an arbitrary spin function $X(\sigma_i)$. In particular, if we symmetrize the wave functions, we recover the Jack polynomials of the last section . If we antisymmetrize them, we recover the Jack polynomials at coupling $\beta + 1$ multiplied by the Vandermonde determinant $\prod_{i<j}(z_i - z_j)$. The mathematical properties of the Jack polynomials can probably be extanded to these wave-functions.

3.1 Dunkl Operators and Monodromy Matrices

It is clear from the explicit construction of the wave functions of H that it's spectrum is highly degenerate and we shall now interpret these degeneracies from a symmetry principle. For this, we need to introduce the so-called monodromy matrix $T(u)$ solution of the equation:

$$R^{ab}T^aT^b = T^bT^aR^{ab} \tag{3.9}$$

where T^a in the operator valued matrix $T(u_a) \otimes 1$, T^b is $1 \otimes T(u_b)$ and R^{ab} is given by

$$R^{ab} = u_a - u_b + \beta P_{ab} \tag{3.10}$$

where P^{ab} permutes the two components of the tensor product $V^a \otimes V^b$.

Consistency of the above relations requires that R^{ab} is a solution of the Yang-Baxter equation in $V^a \otimes V^b \otimes V^c$:

$$R^{ab}R^{ac}R^{bc} = R^{bc}R^{ac}R^{ab} \tag{3.11}$$

Assume that $T(u)$ has a series expansion in $1/u$, the coefficients of this expansion define an algebra which is called the Yangian [19]. In this section, we exhibit a representation of the Yangian which commutes with $\tilde{H}$. For this, let us define the so-called "L operators "[20] defined as :

$$L_i = u - d_i - \beta P_{ai} \tag{3.12}$$

where P^{ai} permutes the components of the tensor product $V^a \otimes V^i$ and d_i is a coefficient which we identify with the Dunkl operator (3.4). This is consistent since the d_i's commute among themselves and with the permutations P_{ij}. A well known representation of the monodromy matrix $T(u)$ is then given by [20]:

$$T^a(u) = L_1(u)L_2(u)...L_N(u) \tag{3.13}$$

Now, let us act with this monodromy matrix on the space of wave functions $\phi(z_i, \sigma_i)$. The coordinates z_i are permuted by the K_{ij} and the spins σ_i are permuted by the P_{ij}. We shall see that the antisymmetric wave functions $\phi(z_i, \sigma_i)$ are preserved by the action of the monodromy matrix $T(u)$. To show this, we define a projection Π which substitutes $-P_{ij}$ for K_{ij} to the right of an expression.

The subspace of antisymmetric wave functions is preserved if the following condition is satisfied:

$$\Pi\left((K_{ij} + P_{ij})\, T(u)\right) = 0 \tag{3.14}$$

which is equivalent to the condition:

$$\Pi\left((K_{ii+1} + P_{ii+1})\, L_i(u) L_{i+1}(u)\right) = 0 \tag{3.15}$$

when we expand this relation in powers of u, the coefficients of u and u^2 give relations which are trivially satisfied. Let us here consider the coefficient of 1. We omit the Π in front of the expression:

$$0 = (K_{12} + P_{12})\left(d_1 d_2 - \beta d_1 P_{a2} - \beta d_2 P_{a1} + \beta^2 P_{a1} P_{a2}\right) \tag{3.16}$$

using (3.6) and the fact that $[K_{12}, d_1 d_2] = 0$ we obtain:

$$-\beta P_{a1} P_{a2} P_{12} + \beta P_{12} P_{a1} P_{a2}$$
$$-d_2 P_{12} P_{a1} - d_1 P_{12} P_{a2} \tag{3.17}$$
$$+P_{a1}(d_1 P_{12} - \beta) + P_{a2}(d_2 P_{12} + \beta) = 0$$

which is easily shown. Notice that the unexpected term equal to β in (3.6)c) is necessary for the proof of this relation.

It follows from this analysis that the monodromy matrix $T(u)$ generates a representation of the Yangian algebra in the subspace of antisymmetric wave functions $\phi(z_i, \sigma_i)$. It is also easy to see that this representation commutes with the Hamiltonian $\tilde{H}$ and therefore, the Yangian is the symmetry algebra which explains the degeneracies of $\tilde{H}$.

3.2 Yang's Representation of Affine-Hecke Algebras

In this section, we establish a relation between the Dunkl operators (3.4) and some operators defined by Yang in his study of the δ interacting gas (see also [21] for a discussion in the elliptic case).

To make this relation more transparent, we shall consider the case of the affine-Hecke algebra defined by :

a) $[y_i, y_j] = 0$

b) $[g_i, y_j] = 0$ if $j \neq i, i+1$ (3.18)

c) $g_i y_i - y_{i+1} g_i^{-1} = 0$

and the g_i obey the relations of a Hecke algebra :

$$[g_i, g_j] = 0 \quad \text{if} \quad |i - j| \geq 2$$

$$g_i g_{i+1} g_i = g_{i+1} g_i g_{i+1}$$ (3.19)

$$(g_i - q)(g_i + q^{-1}) = 0$$

This algebra replaces the permutations and the d_i's if one repeats the arguments of the last section replacing the rational solution (3.10) of the Yang-Baxter equation (3.11) with a trigonometric solution. This is considered in [3]. This algebra is also studied in the works of I.Cherednick [17]. Here we shall content ourselves to obtain a natural representation of the affine-Hecke algebra (3.18) which in some limit degenerates to the Dunkl operators (3.4).

For this, let us consider some realization of the Hecke-algebra (3.19) acting in the tensor product $V \otimes^N$ where the operator g_i acts in $V_i \otimes V_{i+1}$. To ease the forthcoming discussion, we rename $g_i : g_{i,i+1}$. Following Yang's argument [22], we shall exhibit a set of operators which commute together. Then we shall see that in some limit these operators can be identified with the y_j.

The starting point is the trigonometric form of the L operator [20] given by:

$$L_i^a(v) = \left(\frac{v}{v_i} g_{ia} - g_{ia}^{-1}\right) K_{ai}$$ (3.20)

where g_{ai} acts in $V_a \otimes V_i$ and K_{ai} permutes the two components of $V_a \otimes V_i$. Using this form we can construct a monodromy matrix solution of (3.9) given by :

$$T^a(u) = L_1(u) L_2(u) ... L_N(u) S^a$$ (3.21)

where S^a is a matrix acting in V^a and such that $[R^{ab}, S^a \otimes S^b] = 0$. It then follows from the Yang-Baxter equation (3.9) that the trace of this monodromy matrix $\bar{T}(v)$ defines a commuting set of transfer matrices:

$$[\bar{T}(v), \bar{T}(w)] = 0 \qquad \forall u, \forall v$$ (3.22)

To define the trace of the monodromy matrix, we consider T^a as a matrix acting in V^a and take its trace in the usual sense.

Hence, the operators $\tilde{y}_i = \bar{T}(v_i)$ form a commuting set . These operators play an important role in the study of the δ interacting gas [22]. They can be explicitly computed using the fact that:

$$L_i^a(v_i) = (q - q^{-1}) K_{ai}$$ (3.23)

If one then writes $\bar{T}(v_i)$ in the form $\bar{T}(v_i) = \text{tr } K_{ia}\theta$ where θ is an operator that contains only the permutations K_{ij} with $i, j \neq a$ and uses the following property of the trace:

$$\text{tr } K_{ai}\theta = \theta \tag{3.24}$$

one finally obtains :

$$\tilde{y}_i = X_{ii+1}X_{ii+2}...X_{iN}S^iX_{i1}...X_{ii-1} \tag{3.25}$$

where the operators X_{ij} are given by:

$$X_{ij} = \left(\frac{v_i}{v_j}g_{ij} - g_{ij}^{-1}\right)K_{ij} \tag{3.26}$$

The operators $\tilde{y}_i$ constructed in this way obey the relation (3.18)a) but not (3.18)b) c). It is then not difficult to see that these relations are satisfied in the limit $0 \ll v_1 \ll v_2... \ll v_N$. In this limit the operators y_j are given by [3][23]:

$$y_i = g_{ii+1}^{-1}K_{ii+1}...g_{iN}^{-1}K_{iN}S^iK_{1i}g_{1i}...K_{i-1i}g_{i-1i} \tag{3.27}$$

To make contact with the Dunkl operators d_i, we must use the following representation of the Hecke algebra acting in the space of polynomials in N variables z_i :

$$g_{ii+1} = qK_{ii+1} - (q - q^{-1})\frac{z_i}{z_i - z_{i+1}}(K_{ii+1} - 1) \tag{3.28}$$

and S^i is defined by:

$$(S^if)(z_1, ..., z_N) = f(z_1, ..., tz_i, ..., z_N) \tag{3.29}$$

If one sets $q = t^\beta$ and let $t \to 1$, the operators y_i tend to the d_i defined in (3.4).

4   The Haldane-Shastry Chain

Among this class of models, the Haldane Shastry chain [7][8] is the closest to the known integrable spin chains such as the XXX chain [22]. It is also the less well understood. Here, we shall present these chains from the point of view of the Dunkl operators.

The Hamiltonian is the $\beta = \infty$ limit of (3.1), it is equal to :

$$H = \sum_{i<j} \frac{P_{ij}}{\sin^2(\theta_i - \theta_j)} \tag{4.1}$$

where the variables θ_j take the values $\frac{\pi j}{N}$, $0 \leq j < N$. The main difficulty comes from the fact that this Hamiltonian, unlike (3.1), cannot be obtained from the static limit of the Dunkl operators as $\sum d_i^2$ because in this limit this quantity is a number. In this section, we define a representation of the degenerate affine Hecke algebra (3.6) labelled by N complex numbers ω_k. When the ω_k are arbitrary, the representation is irreducible, when they are equal to $\exp(i\frac{2\pi k}{N})$, it is reducible and one can find the Haldane Shastry Hamiltonian in its center. We shall then use a correspondence with the $\beta = 1$ representation (3.4) to diagonalise the Hamiltonian (4.1).

Consider the representation of the degenerate affine Hecke algebra (3.6) given by:

$$\tilde{d}_i = \sum_{j>i} \frac{z_i}{z_i - z_j} K_{ij} - \sum_{j<i} \frac{z_j}{z_j - z_i} K_{ij} \tag{4.2}$$

It acts on the Hilbert space defined by the permutations :

$$\mathcal{H}_N = \{|P_1...P_N > /P_i \in Z_N,\ P_i \neq P_j \quad \text{for all i,j}\} \tag{4.3}$$

in the following way :

$$\begin{aligned}
K_{ij}|...P_i...P_j... > &= |...P_j...P_i... > \\
z_j|P_1...P_N > &= \omega_{P_j}|P_1...P_N >
\end{aligned} \tag{4.4}$$

In the case where the ω_k's are arbitrary, there is no reason for this representation to be reducible. On the other hand, if $\omega_k = e^{i\frac{2\pi k}{N}}$, the operator C defined by:

$$C|P_j > = |(P_j + 1) > \tag{4.5}$$

obviously commutes with the d_j's and the K_{ij}. So, in this case the representation is reducible and one can verify that the Hamiltonian given by :

$$\tilde{H} = \sum_{i<j} \frac{z_i z_j}{(z_i - z_j)^2} K_{ij} \tag{4.6}$$

is also in the center. No systematic way to obtain the operators which commute with the K_{ij}'s and the d_i's is known. In what follows we restrict to the case where $\omega_k = \exp(i\frac{2\pi k}{N})$.

The Hamiltonian (4.6) can easilly be related to the Haldane Shastry Hamiltonian and we shall now indicate how to obtain its eigenvalues and some of its eigenvectors. We shall construct a class of states on which the two representations of the d_i (3.4) (for $\beta = 1$) and (4.2) act in the same way. On these states, any conserved quantity of the spin chain has its analogue in the dynamical model and can thus be diagonalised .

For $M < N$ let us define the Hilbert space $\mathcal{H}_M$

$$\mathcal{H}_M = \{|P_1...P_M > /P_i \in Z_N,\ P_i \neq P_j \quad \text{for all i,j}\} \tag{4.7}$$

which is uniquely embedded in $\mathcal{H}_N$ if one requires that :

$$K_{ij}|P_1...P_M > = |P_1...P_M > \quad \text{for i,j} > M \tag{4.8}$$

In $\mathcal{H}_M$ we consider the following states :

$$|\alpha> = \Psi(\omega^{P_1}, ..., \omega^{P_M})|P_1, ..., P_M > \qquad (4.9)$$

where :

$$\Psi(z_i) = \Phi(z_i)\prod_{i<j}(z_i - z_j) \qquad (4.10)$$

and $\Psi(z_i)$ is a polynomial of degree between 1 and $N - 1$ in the variables z_i. On these states, $\tilde{d}_i = d_i - \frac{N+1}{2}$ for $1 \le i \le M$. It results from the fact that a polynomial $P(z)$ of degree between 1 and $N - 1$ satisfies the relation :

$$(z\partial_z - \frac{N+1}{2})P(z)|_{z=\omega_k} = \sum_{l \ne k} \frac{\omega_k}{\omega_k - \omega_l}P(\omega_l) \qquad (4.11)$$

The same relation enables to show that in this subspace $\tilde{H}$ acts as $\sum_{k=1}^{M}(d_k + 1/2 + N/2)(d_k + 1/2 - N/2)$ and to find its spectrum:

$$E = \sum_{k=1}^{M} \epsilon(f_k) \quad \text{with} \quad \epsilon(f_k) = (f_k + 1/2)^2 - N^2/4 \qquad (4.12)$$

$f_k + \frac{N+1}{2}$ are the degrees of the highest monomial of $\Psi(z_i)$. Therefore, the f_k satisfy the inequalities:

$$-\frac{N-1}{2} \le f_1 \le f_2 ... \le \frac{N-1}{2} - 1 \qquad (4.13)$$

References

[1] F.D.M. Haldane, To appear in "Proceedings of the 16th Taniguchi Symposium", 1993.
[2] C.F.Dunkl, Trans.Amer.Math.Soc, 311 (1989) 167.
[3] D.Bernard, M.Gaudin, F.D.M.Haldane and V.Pasquier, J.Phys.A 26 (1993) 5219.
[4] F.Lesage, V.Pasquier and D.Serban saclay preprint april 94.
[5] B. Sutherland,Phys.Rev A 5 (1972), 1372.
[6] C.N.Yang Phys.rev.Letters, 19 (1967) 1312.
[7] F.D.M Haldane, Phys.Rev.Lett. 60 (1988) 635.
[8] B.S.Shastry, Phys.Rev.Lett. 60 (1988) 639.
[9] V.I.Inotzemtsev, J.Stat.Phys.59 (1990) 1143.
[10] I.V.Cherednick, preprint march 94.

[11] D.Bernard, V.Pasquier and D.Serban saclay preprint april 94.

[12] B.D. Simons, P.A. Lee, and B.L. Altshuler, Phys. Rev. Lett. V.70, No26, (1993) 4122.

[13] I.G. Macdonald, Séminaire Lotharingien, Publ. I.R.M.A. Strasbourg, 1988
      .

[14] R. P. Stanley, Adv. in Math., 77, (1989) 76-115.

[15] F.D.M. Haldane, To appear in proceedings of the International Colloquium in Modern Field Theory, Tata institute, 1994.

[16] J. Minanan and A.P. Polychronackos CERN preprint april 94.

[17] I.V.Cherednick, Invent.Math.106 (1991) 411.

[18] A.P.Polychronakos,Phys.Rev.Letters, 69 (1992) 703.

[19] V.G.Drinfeld, Funct.Anal.Appl.20 (1988) 56.

[20] L.D.Faddeev Les Houches lectures, Elsevier Science Publishers (1984).

[21] V.M Buchstaber, G.Felder and A.P. Veselov preprint march 94.

[22] M.Gaudin, "la fonction d'onde de Bethe" Masson (1981).

[23] I.V.Cherednick, Commun.Math.Phys (1992) 150.

Quantum Group and Magnetic Translations.
Bethe-Ansatz Solution for Bloch Electrons in a Magnetic Field

P.B Wiegmann[1]* and A.V.Zabrodin[2]

[1] James Frank Institute and Enrico Fermi Institute of the
University of Chicago,
5640 S.Ellis Ave.,Chicago Il 60637
and
Landau Institute for Theoretical Physics

[2] Enrico Fermi Institute and Mathematical Disciplines Center of
the University of Chicago,
5640 S.Ellis Ave.,Chicago Il 60637
and
the Institute of Chemical Physics, Kosygina St. 4, SU-117334,
Moscow, Russia

Abstract. We present a new approach to the problem of Bloch electrons in magnetic field, by making explicit a natural relation between magnetic translations and the quantum group $U_q(sl_2)$. The approach allows to express the spectrum and the Bloch function as solutions of the Bethe-Ansatz equations typical for completely integrable quantum systems

1 Introduction

Several times a peculiar problem of Bloch electrons in magnetic field emerged with a new face to describe another physical application [1],[2],[3],[4],[5]

i) it resembles some properties of the integer Hall effect [5],

ii) its spectrum has an extremely rich structure of Cantor set, and exibits a *multifractal* behaviour [4],[6], [7] (see also [8] for a review)

---

*e-mail: wiegmann@control.uchicago.edu

iii) it describes the localization phenomenon in *quasiperiodic* potential (see e.g.[8] and references therein).

iv) it has been recently conjectured that the symmetry of magnetic group may appear dynamically in strongly correlated electronic systems[9],[10].

In this paper we show that this problem (some times called Hofstadter problem) and a class of quasiperiodic equations are solvable by the *Bethe- Ansatz*. We made explicit a long time anticipated connection of the group of magnetic translations with the *Quantum Group $U_q(sl_2)$* and with Quantum Integrable Systems. The result of the paper is the algebraic *Bethe- Ansatz equations* for the spectrum. Although we do not solve the Bethe-Ansatz equation here, we are confident that they provide a basis for analytical study of the multyfractal properties of the spectrum.

2   Model and Result

The Hamiltonian of a particle on a two dimensional square lattice in magnetic field is

$$H = \sum_{<n,m>} e^{iA_{\vec{n},\vec{m}}} c_{\vec{n}}^{\dagger} c_{\vec{m}} \tag{1}$$

$$\prod_{plaquette} e^{iA_{\vec{n},\vec{m}}} = e^{i\Phi} \tag{2}$$

where $\Phi = 2\pi\frac{P}{Q}$ is a flux per plaquette, P and Q are mutualy prime integers. In the most conventional Landau gauge $A_x = A_{\vec{n},\vec{n}+\vec{1}_x} = 0, A_y = \Phi n_x$ the Bloch wave function is

$$\psi(\vec{n}) = e^{i\vec{k}\vec{n}}\psi_{n_x}(\vec{k}), \quad \psi_n = \psi_{n+Q} \tag{3}$$

where $n_x \equiv n = 1...Q$ is a coordinate of the magnetic cell. With these substitution the Schrodinger equation turns into a famous one-dimensional quasiperiodic difference equation ("Harper's" equation):

$$e^{ik_x}\psi_{n+1} + e^{-ik_x}\psi_{n-1} + 2\cos(k_y + n\Phi)\psi_n = E\psi_n \tag{4}$$

The spectrum of this equation has Q bands and feels the difference between rational and irrational numbers - if the flux is irrational, the spectrum is singular continuum - uncountable but measure zero set of points (Cantor set). If the flux is rational, then the spectrum has Q bands.

To ease the reference we state the main result of the paper:

It is known that due to gauge invariance, the energy depends on a single parameter $\lambda = \cos(Qk_x) + \cos(Qk_y)$. We find that the spectrum at $\lambda = 0$ ("mid" band spectrum) is given by the sum of roots z_l

$$E = iq^Q(q - q^{-1}) \sum_{l=1}^{Q-1} z_l, \tag{5}$$

of the *Bethe-Ansatz* equations for the Quantum Group $U_q(sl_2)$

$$\frac{z_l^2 + q}{qz_l^2 + 1} = q^Q \prod_{m=1, m\neq l}^{Q-1} \frac{qz_l - z_m}{z_l - qz_m}, \quad l = 1...Q-1. \tag{6}$$

with

$$q = e^{\frac{i}{2}\Phi}, \tag{7}$$

Another version of the Bethe-Ansatz equations is presented in the end of the paper. Solution of the model with anisotropic hopping (when the coefficient in front of cos in Eq.(4) differs from 2 is also available. It will be published elsewhere. The Quantum Group symmetry is more transparent in another gauge $A_x = -\frac{\Phi}{2}(n_x + n_y)$, $A_y = \frac{\Phi}{2}(n_x + n_y + 1)$. In this gauge a discrete coordinate of the Bloch function $\psi(\vec{n}) = \exp{(i\vec{p}\vec{n})}\psi_n(\vec{p})$, turns to $n = n_x + n_y$.

It is defined in two magnetic cells : $n = 1...2Q$, $p_{\pm} = (p_x \pm p_y)/2 \in [0, \Phi/2]$. An equivalent form of the Harper Eq. (4) for ψ_n is

$$2e^{\frac{i}{4}\Phi + ip_+}\cos(\frac{1}{2}\Phi n + \frac{1}{4}\Phi - p_-)\psi_{n+1} +$$
$$2e^{-\frac{i}{4}\Phi - ip_+}\cos(\frac{1}{2}\Phi n - \frac{1}{4}\Phi - p_-)\psi_{n-1} = E\psi_n \tag{8}$$

In the new gauge the "midband" $\lambda = 0$ corresponds to the point $\vec{p} = (\frac{1}{2}\pi, \frac{1}{2}\pi)$ [1]. The advantage of this gauge is that the wave function turns into the polynomial with roots z_m

$$\Psi(z) = \prod_{m=1}^{Q-1} (z - z_m) \tag{9}$$

at the points $z = q^{2l}$

$$\psi_n = \Psi(q^n) \tag{10}$$

3 Group of Magnetic Translations

The wave function of a particle in a magnetic field forms a represantation of the *group of magnetic translations* [1]: let generators of translations be

$$T_{\vec{\mu}}(\vec{i}) = e^{iA_{\vec{i},\vec{i}+\vec{\mu}}} \mid \vec{i} >< \vec{i} + \vec{\mu} \mid \tag{11}$$

They form the algebra

$$T_{\vec{\mu}} = T_{-\vec{\mu}}^{-1}, \quad T_{\vec{n}}T_{\vec{m}} = q^{-\vec{n}\times\vec{m}}T_{\vec{n}+\vec{m}},$$
$$T_y T_x = q^2 T_x T_y, \quad T_y T_{-x} = q^{-2}T_{-x}T_y... \tag{12}$$

---

[1] The doubling of period in comparison with original Harper's equation is artificial: using a simple transformation (multiplying ψ_n by $e^{-i\Phi n^2/4}$) one comes to an equation with coefficients of period Q. We are indebted to Alexandre Abanov for clarifying connection between two gauges

with q given by the Eq.(7). The Hamiltonian (1) can be expressed

$$H = T_x + T_{-x} + T_y + T_{-y} \tag{13}$$

This group is also equivalent to the Heisenberg-Weyl group: $[\hat{p}, \hat{q}] = i\Phi$, $T_x = \exp \hat{q}$, $T_y = \exp \hat{p}$.

4 Quantum Group

The algebra $U_q(sl_2)$ (a q-deformation of the universalenveloping of the sl_2) is generated by the elements A, B, C, D, with the commutation relations [11],[12], [13],[14],[15]

$$\begin{aligned}
AB &= qBA, \; BD = qDB, \\
DC &= qCD, \; CA = qAC, \\
AD &= 1, \; [B, C] = \frac{A^2 - D^2}{q - q^{-1}}
\end{aligned} \tag{14}$$

The center of this algebra is a $q-$analog of the Casimir operator

$$c = \left(\frac{q^{-\frac{1}{2}}A - q^{\frac{1}{2}}D}{q - q^{-1}} \right)^2 + BC \tag{15}$$

In the classical limit $q \to 1 + \frac{i}{2}\Phi$, the quantum group turns to the sl_2 algebra: $(A - D)/(q - q^{-1}) \to S_3$, $B \to S_+$, $C \to S_-$, $c \to \vec{S}^2 + 1/4$.

The commutation relations (14) are simply another way to write the Yang-Baxter equation

$$R_{a_1 a_2}^{b_1 b_2}(u/v) L_{b_1 c_1}(u) L_{b_2 c_2}(v) = L_{a_1 b_1}(u) L_{a_2 b_2}(v) R_{b_1 b_2}^{c_1 c_2}(u/v) \tag{16}$$

where generators A, B, C, D are matrix elements of the L- operator

$$L(u) = \begin{bmatrix} \frac{uA - u^{-1}D}{q - q^{-1}} & u^{-1}C \\ uB & \frac{uD - u^{-1}A}{q - q^{-1}} \end{bmatrix} \tag{17}$$

Here u is a spectral parameter and R- matrix is the L- operator in the spin 1/2-representation. It is given by the same matrix (17) with elements: $A = q^{\frac{1}{2}\sigma_3}$, $D = q^{-\frac{1}{2}\sigma_3}$, $B = \sigma_+$, $C = \sigma_-$, where $\vec{\sigma}$ are the Pauli matrices.

Finite dimensional representations (except some representations of dimension Q) of the $U_q(sl_2)$ can be expressed in the weight basis, where A and D are diagonal matrices: $A = diag\,(q^j, ..., q^{-j})$. An integer or halfinteger j is the spin of the representation, and $2j + 1$ is its dimension. The value of the Casimir operator (15) in this representation is given by the q- analog of $(j + 1/2)^2$

$$c = \left(\frac{q^{j+1/2} - q^{-j-1/2}}{q - q^{-1}} \right)^2 = [j + 1/2]_q^2 \tag{18}$$

Representations can be realized by polynomials $\Psi(z)$ of the degree $2j$:

$$A\Psi(z) = q^{-j}\Psi(qz), \ D\Psi(z) = q^j\Psi(q^{-1}z),$$
$$B\Psi(z) = z(q - q^{-1})^{-1}\left(q^{2j}\Psi(q^{-1}z) - q^{-2j}\Psi(qz)\right)$$
$$C\Psi(z) = -z^{-1}(q - q^{-1})^{-1}\left(\Psi(q^{-1}z) - \Psi(qz)\right) \tag{19}$$

This again the q-analog of the representation of the sl_2 algebra by a differential operator:

$$S_3 = z\frac{d}{dz} - j, \ S_+ = z(2j - z\frac{d}{dz}), \ S_- = \frac{d}{dz} \tag{20}$$

5 Magnetic Translations as a Special Representation of the Quantum Group

Dimension of our physical space of states is $2j + 1 = Q$. This is a very special dimension when $q^{2j+1} = \mp 1$ for P - odd (even). The Casimir operator (18) in this case is

$$c = -4(q - q^{-1})^{-2}, \ for \ P - odd$$
$$c = 0, \ for \ P - even \tag{21}$$

In this special case [16],[17] representation of the quantum group can be naturally (but not unambiguously [18]) expressed in terms of Magnetic Translations. Say one may choose

$$T_{-x} + T_{-y} = \pm i(q - q^{-1})B,$$
$$T_x + T_y = i(q - q^{-1})C,$$
$$T_{-y}T_x = \pm q^{-1}A^2, \ T_{-x}T_y = \pm qD^2 \tag{22}$$

where the upper sign corresponds to an odd P and the lower to an even P (representation of $U_q(sl_2)$ in terms of different but related Weyl basis can be found in Ref.[16],[17],[18],[21]). It is straightforward to check that this representation obeys commutation relations (12) and (14) and gives correct values (21) of the Casimir operator (15).

The Hamiltonian (1,13) now can be expressed in terms of the quantum group generators

$$H = i(q - q^{-1})(C \pm B) \tag{23}$$

whereas the Schrodinger equation becomes a difference functional equation

$$i(z^{-1} + qz)\Psi(qz) - i(z^{-1} + q^{-1}z)\Psi(q^{-1}z) = E\Psi(z) \tag{24}$$

The original Harper's discrete equation in the form (8) at the point $\vec{p} = (\pi/2, \pi/2)$ can be obtained from the functional equation (24), by setting $z = q^l$ and $\psi_l = \Psi(q^l)$. The advantage to use the extention of ψ_l to a complex plane z is that the representation theory of the quantum group garantees that in a proper gauge the extended wave function would be polynomial (9).

In addition to representation (19) having the the highest and the lowest weight , in the special dimension $q^{2j+1} = \pm 1$ there is a parametric family of representations having in general no highest or lowest weights [17],[21]. The parameter describes the anisotropy of the hopping amplitude in the Hamiltonian (1) or the strength of the potential (i.e.the coefficient in front of cos) in the Harper Equation (4). In this case the wave function is not polynomial in any gauge. Nevertheless the Bethe Ansatz solution of the anisotropic problem is also possible but acqueres much heavier mathematical techique. We postpone it for a more extended paper.

6   Functional Bethe-Ansatz

Among various methods of the theory of quantum integrable systems the *functional Bethe-Ansatz* [19] seems to be the most direct way to diagonalize the Hamiltonian (23).

We know that the solution of Eq.(24) is polynomial

$$\Psi(z) = \prod_{m=1}^{Q-1} (z - z_m) \tag{25}$$

at the points $z = q^{2l}$

$$\psi_n = \Psi(q^n) \tag{26}$$

Let us substitute it in the Eq.(24) and divide both sides by $\Psi(z)$. We obtain

$$i(z^{-1} + qz) \prod_{m=1, m\neq l}^{Q-1} \frac{qz - z_m}{z - z_m}$$
$$-i(z^{-1} + q^{-1}z) \prod_{m=1, m\neq l}^{Q-1} \frac{q^{-1}z - z_m}{z - z_m} = E \tag{27}$$

The l.h.s. of this equation is a meromorphic function, whereas the r.h.s. is a constant. To make them equal we must null all residues of the l.h.s.. They appear at $z = 0$, at $z = \infty$ and at $z = z_m$. The residue at $z = 0$ vanishes automatically.

The residue at $z = \infty$ is $-iq^Q + iq^{-Q}$. Its null determines the deqree of the polynom.

Comparing the coefficients of z^{Q-1} in the both sides of Eq.(24),we obtainthe energy given by Eq.(5).

Finally, annihilation of poles at $z = z_m$ gives the Bethe-Ansatz equations (6) for roots of the polynomial (9). We write them here in a more conventional form by setting $z_l = \exp(2\varphi_l)$

$$\frac{\cosh(2\varphi_l - i\frac{\Phi}{4})}{\cosh(2\varphi_l + i\frac{\Phi}{4})} = \mp \prod_{m=1, m\neq l}^{Q-1} \frac{\sinh(\varphi_l - \varphi_m + i\frac{\Phi}{4})}{\sinh(\varphi_l - \varphi_m - i\frac{\Phi}{4})} \tag{28}$$

Another form of the Bethe-Ansatz equations is given in the next section.

7 Miscellaneous Results

7.1 Another Form of the Bethe-Ansatz Equations

As we already mentioned the representation of the quantum group by magnetic translations is not unique. This means that there is another gauge where the wave function is a polynom. Consider the gauge $A_x = -A_y = -\Phi n_x$. Then instead of Harper's equation(4) we obtain an equivalent equation for the gauge transformed wave function $\psi_n \to \phi_n = \exp(i\frac{\Phi}{2}n(n-1))\psi_n$ (for an odd Q it respects the periodic conditions (3)). At $\lambda = 0$, it has the form

$$e^{-i\Phi n}\phi_{n+1} + e^{i\Phi(n-1)}\phi_{n-1} - 2\cos(n\Phi)\phi_n = E\phi_n \qquad (29)$$

This choice of the gauge corresponds to another representation of the quantum group by magnetic translations. Say for an odd P we have

$$T_{-x} + T_{-y} = -i(q - q^{-1})q^{-\frac{1}{2}}BD,$$
$$T_x + T_y = -i(q - q^{-1})q^{-\frac{1}{2}}CA,$$
$$T_{-y}T_x = q^{-1}A^2, \; T_{-x}T_y = qD^2 \qquad (30)$$

Then the Hamiltonian (13) turns into quadratic form in the $U_q(sl_2)$ generators

$$H = -i(q - q^{-1})q^{-\frac{1}{2}}(CA + BD) \qquad (31)$$

The representation (19) (for $q^{2j+1} = -1$) now gives another functional equation

$$z^{-1}\Psi(q^2 z) + q^{-2}z\Psi(q^{-2}z) - (z + z^{-1})\Psi(z) = E\Psi(z) \qquad (32)$$

which is identical to (29) on the set of points $z = q^{2l}$, where $l = 0...Q - 1$. The Bethe-Ansatz may be obtained in the similar way:

$$z_l^2 = q^Q \prod_{m=1, m \neq l}^{Q-1} \frac{q^2 z_l - z_m}{z_l - q^2 z_m}, \; l = 1...Q - 1. \qquad (33)$$

The energy is given again by the sum of roots

$$E = -q(q - q^{-1})\sum_{l=1}^{Q-1} z_l, \qquad (34)$$

Inspite of the difference Eqs.(33,34) must be equivalent to the Eqs.(6,5).

7.2 Quadratic Form of Quantum Group Generators

A limited number of other interesting solvable discrete equations may be obtained from a general quadratic form of quantum group generators. Their "classical" version ($q \to 1$) would be differential equations generated by quadratic forms of sl_2 -generators (20) These differential equations are known in the literature as so-called "quasi exactly soluable " problems of quantum mechanics [20],[22]. A quadratic form can be considerd as trace of a monodromy matrix of an integrable model with nonperiodic boundary conditions

$$\tau = tr K_+(u)L(su)K_-(u)\sigma_2 L^t(su^{-1})\sigma_2 \qquad (35)$$

where $K_\pm$ are c-number matrices - solutions of "reflective " Yang-Baxter equations (RYB) [23],[24]. These matrices describe all boundary conditions consistent with integrability There is a 3-parametric family of boundary matrices $K_\pm$ which generates a general quadratic form of A, B, C, D. Becides, there is a parameter s in (35) which one can introduce in the L-operator (17) preserving integrability. For a particular

choice of boundary K-matrices and parameter s τ is proportional to the hamiltonian (31). Therefore, one can say, that the Hofstadter problem is equivalent to an integrable magnet of spin $(Q-1)/2$ on one site with a proper boundary condition.

7.3   q-Analog of Orthogonal Polynomials

There is an intriguing connection between a wave function (25) and q-generalization of orthogonal polynomials [25]. They satisfy the difference equation (q-analog of differential hypergeometrical equation)

$$A(z)P_n(q^2 z) + A(z^{-1})P_n(q^{-2}z) - (A(z) + A(z^{-1}))P_n(z) = \qquad (36)$$
$$= (q^{-2n} - 1)(1 - abcdq^{2n-2})P_n(z)$$

where $A(z) = (1-az)(1-bz)(1-cz)(1-dz)/((1-z^2)(1-q^2z^2))$ and a, b, c, d are parameters and n is a degree of the polynomial. Choosing $c = -d = q, a = -b = 0$ we arrive at the equation for the q-Hermite polynomials $H_n^{(q)}$

$$H_n^{(q)}(q^2 z) - z^2 H_n^{(q)}(q^{-2}z) = q^{-2n}(1 - z^2)H_n^{(q)}(z) \qquad (37)$$

(these are polynomials in $z + z^{-1}$ of degree n). For an odd Q at $n = (Q-1)/2$ this yields a zero energy solution to eq.(32) $\Psi^{(E=0)}(iz) = z^{(Q-1)/2}H_{(Q-1)/2}^{(q)}(z)$.

Another choice $c = -d = q$, $a = -b = q$ and then the replacement q by $q^{1/2}$ gives the q-Legendre equation

$$\frac{1-qz^2}{1-z^2}P_n^{(q)}(qz) + \frac{q-z^2}{1-z^2}P_n^{(q)}(q^{-1}z) = (q^{-n} + q^{n+1})P_n^{(q)}(z) \qquad (38)$$

Then, comparing with the Eq.(24) we conclude that the zero mode solution is given by the q-Legendre polynomial $\Psi^{(E=0)}(iz) = z^{(Q-1)/2}P_{(Q-1)/2}^{(q)}(z)$.

The almost immidiate and the most interesting task now is to solve the Bethe-Ansatz equations in the limit $P, Q \to \infty$ when the flux $\Phi/2\pi$ is irrational. In all previous examples of integrable systems it was always possible to derive an integral equation for a distribution function of roots z_l. We hope that it would be also possible for this problem and after all allows to obtain fractal properties of the spectrum analytically.

Acknowledgements
We would like to thank P.G.O.Freund, A.Abanov, A.Gorsky, A.Kirillov, E.Floratos, J.-L.Gervais and J.Schnittger for interesting discussions. A.Z is grateful to the Mathematical Disciplines Center of the University of Chicago for the hospitality and support. P.W. acknowledges the hospitality of Weizmann Institute of Science and Laboratoire de Physique Theorique de l'Ecole Normale Supérieure where this work was completed. This workwas supported in part by NSF under the Research Grant 27STC-9120000.

References

[1] J.Zak, *Phys. Rev.* **134** 1602 (1964)

[2] M.Ya.Azbel, *Sov. Phys.JETP* **19** 634 (1964)

[3] G.H.Wannier,*Phys.Status Solidi* **88**, 757 (1978)

[4] D.R.Hofstadter,*Phys. Rev. B* **14** 2239 (1976)

[5] D.J.Thouless,   M.Kohmoto,   P.Nightingale,   M.den   Nijs, *Phys.Rev.Lett* **49**,405 (1982)

[6] S.Aubry and G.Andre *Ann.Israel Phys.Soc.* **3**, 131 (1980)

[7] D.J.Thouless, *Phys. Rev.* **28** 4272 (1983)

[8] H.Hiramoto, M.Kohmoto, *Int.J.Mod.Phys. B* **6**, 281 (1992)

[9] P.W.Anderson in Proc. of Nobel Symp. 73 *Physica Scripta* **T27**, (1988)

[10] P.B.Wiegmann in Proc. of Nobel Symp. 73 *Physica Scripta* **T27**, (1988)

[11] P.P.Kulish, N.Yu.Reshetikhin *Zap.nauch.semin.LOMI* **101**, 112 (1980)

[12] E.K.Sklyanin, *Uspekhi Mat.Nauk* **40** 214 (1985)

[13] V.G.Drinfeld , *Dokl.Acad.Nauk* **283** 1060 (1985)

[14] M.Jimbo *Lett.Math. Phys.* **10** 63 (1985)

[15] N.Yu.Reshetikhin, L.A.Takhtadjan, L.D.Faddeev,*Algebra i Analiz* **1**, 178 (1989)

[16] P.Roche, D.Arnaudon *Lett.Math. Phys.* **17**, 295 (1989).

[17] E.K.Sklyanin *Func.Anal.Appl.* **17**, 273 (1983)

[18] E.G.Floratos *Phys.Lett* . **B233**, 235 (1989).

[19] E.K.Sklyanin *Zap.nauch.semin.LOMI* **134**, 112 (1983)

[20] A.V.Turbiner *Comm.Math. Phys.* **118** 467 (1988)

[21] V.V.Bazhanov and Yu.G.Stroganov *J.Stat.Phys.* **59**. 799 (1990)

[22] A.Ushveridze *Sov.Journal Part.Nucl.* **20**, 185 (1989), *ibid* **23**, 25 (1992)

[23] E.K.Sklyanin *J.Phys. A* **21**, 2375 (1988)

[24] I.Cherednik *Theor.Mat. Fys.* **61** 35 (1984)

[25] see e.g.G.Gasper and M.Rahman,*Basic Hypergeometric Seriaes* (1990), Cambridge Univ.Press and N.Vilenkin, A.Klimyk, *Representation of Lie Groups and Special Functions*,vol.3 (1992) Kluwer Acad.Publ.

Symplectic Geometry of the Chern-Simons Theory

A.Yu. Alekseev *†
A. Z. Malkin ‡§

Institute of Theoretical Physics, Uppsala University,
Box 803 S-75108, Uppsala, Sweden.

Abstract. This article is a review of two original papers [1], [2]. We begin with a description of Kirillov symplectic form and quantum mechanics on a coadjoint orbits of a simple Lie group. This theory may be generalized for the case of a Poisson-Lie group. Both these theories are important for understanding of the Chern-Simons model which may be treated as a 3D gauge theory interacting with coadjoint orbits sitting on Wilson lines. Due to topological nature of the Chern-Simons theory one can get rid of the gauge fields in exchange of modification of coadjoint theories. We discover that this modification is exactly the same as we find in the Poisson-Lie case.

1   Introduction

The nonabelian Chern-Simons theory in 3 dimensions has been solved in [3] using its relation to the 2-dimensional Wess-Zumino-Novikov-Witten model. Recently it has been proved that the Chern-Simons theory on the cylinder (Cartesian product of a Riemann surface and a real axis) may be efficiently reduced to the 2 dimensional topological gauged WZNW model [4], [5]. We learn from these examples that the topological 3-dimensional theory can be related to some solvable two-dimensional theory either conformal or topological.

Here we advocate another approach to 3D-topological theories and demonstrate it on the example of the Chern-Simons model. Namely, instead of dealing with some 2-dimensional model we reduce the problem to a solvable quantum mechanics. The natural question in the Chern-Simons theory is to evaluate a

---

*On leave of absence from Steklov Mathematical Institute, Fontanka 27, St.Petersburg, Russia

†Supported by Swedish Natural Science Research Council (NFR) under the contract F-FU 06821-304

‡On leave of absence from St.-Petersburg University.

§Supported in part by a Soros Foundation Grant awarded by the American Physical Society.

correlation function on some 3D manifold with several Wilson lines inserted. It was suggested in [3] and advocated in [6] that the Chern-Simons theory may be represented as a 3D gauge theory interacting with some quantum mechanical systems living on the Wilson lines. These systems give a physical interpretation of the representation theory of Lie algebras [7]. The models of these family are designed in such a way that their Hilbert spaces coincide with particular irreducible representations of a given Lie algebra. Our aim in this paper is to get rid of the gauge fields in the model and end up with somewhat modified quantum mechanics on the Wilson lines. We restrict ourselves to the geometry of the cylinder and fulfil the described program.

When the gauge field disappears from the system the quantum mechanics on the Wilson lines changes. Fortunately, this particular way to modify the orbit quantum mechanics has been studied previously [8], [1]. It corresponds to the generalization of the notion of the Lie group to Poisson-Lie group when the group manifold carries a nontrivial Poisson bracket. After quantization this idea leads to a definition of quantum groups.

We always stay here on the classical level of consideration as quantum effects in the Chern-Simons model lead only to a finite renormalization of the coupling constant.

The paper is organized as follows. In Section 2 we remind the construction of Kirillov symplectic form and then define quantum mechanical systems appropriate for description of Wilson lines. Section 3 is devoted to machinery of Poisson-Lie groups. There we introduce the modified symplectic structures which will replace naive Kirillov form in the Chern-Simons model. In Section 4 we turn to the main point of the paper and first represent the Chern-Simons theory on a cylinder as an interacting theory of 3D gauge fields and Wilson line quantum mechanics following [6]. As it was pointed out in [3] the problem reduces to analysis of the moduli space of flat connections on a Riemann surface with marked points. We reexamine the symplectic structure of this space and discover that it splits into the direct sum of several terms. Some of these terms may be naturally assigned to the Wilson lines. They coincide with certain symplectic forms related to Poisson-Lie groups and described in Section 3. The other terms have a similar structure and take into account topology of a 3D-manifold. More exactly, each handle of the Riemann surface is roughly speaking equivalent to two marked points.

2 Geometric quantization and Wilson lines

For the purpose of selfconsistency we collect in this section some well-known results concerning Poisson and symplectic structures associated to Lie groups. The most important part of our brief survey is a theory of coadjoint orbits. We concentrate on Kirillov symplectic form and the corresponding action for the dynamical system on the orbit. It appears that a Wilson line observable may be represented as a quantum partition function for such system.

2.1 Kirillov form

Let us fix notations. The main object of our interest is a simple Lie group G. We denote the corresponding Lie algebra by $\mathcal{g}$. The linear space $\mathcal{g}$ is supplied with Lie commutator $[,]$. If $\{\varepsilon^a\}$ is a basis in $\mathcal{g}$, we can define structure constants f_c^{ab} in the following way:

$$[\varepsilon^a, \varepsilon^b] = \sum_c f_c^{ab} \varepsilon^c \ . \tag{2.1}$$

The Lie group G has a representation which acts in $\mathcal{g}$. It is called adjoint representation:

$$\varepsilon^g \equiv Ad(g^{-1})\varepsilon \ . \tag{2.2}$$

For a matrix realization of the group G the adjoint action is represented by conjugation:

$$\varepsilon^g \equiv g^{-1}\varepsilon g \ . \tag{2.3}$$

The corresponding representation of the algebra $\mathcal{g}$ is realized by the commutator:

$$ad(\varepsilon)\eta = [\varepsilon, \eta] \ . \tag{2.4}$$

We denote elements of the algebra $\mathcal{g}$ by small Greek letters.

Let us introduce a space $\mathcal{g}^*$ dual to the Lie algebra $\mathcal{g}$. There is a canonical pairing $<,>$ between $\mathcal{g}^*$ and $\mathcal{g}$ and we may construct a basis $\{l_a\}$ in $\mathcal{g}^*$ dual to the basis $\{\varepsilon^a\}$ so that

$$< l_a, \varepsilon^b >= \delta_a^b \ . \tag{2.5}$$

We use small Latin letters for elements of $\mathcal{g}^*$. Each vector ε from $\mathcal{g}$ defines a linear function on $\mathcal{g}^*$:

$$H_\varepsilon(l) =< l, \varepsilon > \ . \tag{2.6}$$

In particular, a linear function H^a corresponds to an element ε^a of the basis in $\mathcal{g}$. By duality the group G and its Lie algebra $\mathcal{g}$ act in the space $\mathcal{g}^*$ via the coadjoint representation:

$$< Ad^*(g)l, \varepsilon >=< l, Ad(g^{-1})\varepsilon > \ , \tag{2.7}$$

$$< ad^*(\varepsilon)l, \eta >= - < l, [\varepsilon, \eta] > \ . \tag{2.8}$$

The space $\mathcal{g}$ can be considered as a space of left-invariant or right-invariant vector fields on the group G. Let us define the universal right-invariant one-form θ_g on G which takes values in $\mathcal{g}$:

$$\theta_g(\varepsilon) = -\varepsilon \ . \tag{2.9}$$

We treat ε in the l.h.s. of formula (2.9) as a right-invariant vector field whereas in the r.h.s. as an element of $\mathcal{g}$. Since the one-form θ_g and the vector field ε are right-invariant the result does not depend on the point g of the group. θ_g is known as Maurer-Cartan form.

Similarly, the universal left-invariant one-form μ_g can be introduced:

$$\mu_g(\varepsilon) = \varepsilon \quad , \quad \mu_g = Ad(g^{-1})\theta_g \quad , \tag{2.10}$$

where ε is a left-invariant vector field, Ad acts on values of θ_g.

In the case of matrix group G the invariant forms θ_g and μ_g look like follows:

$$\theta_g = \delta g \, g^{-1} \quad , \tag{2.11}$$

$$\mu_g = g^{-1}\delta g \quad . \tag{2.12}$$

For any group G there exist two covariant differential operators ∇_L and ∇_R taking values in the space $\mathcal{g}^*$. These are left and right derivatives:

$$< \nabla_L f, \varepsilon > (g) = -\frac{\delta}{\delta t} f(exp(t\varepsilon)g) \quad , \tag{2.13}$$

$$< \nabla_R f, \varepsilon > (g) = \frac{\delta}{\delta t} f(g \, exp(t\varepsilon)) \quad , \tag{2.14}$$

where exp is the exponential map from a Lie algebra to a Lie group. The simple relation for left and right derivatives of the same function f holds:

$$\nabla_R f = -Ad^*(g^{-1})\nabla_L f \quad . \tag{2.15}$$

The space $\mathcal{g}^*$ carries a natural Poisson structure invariant with respect to the coadjoint action of G on $\mathcal{g}^*$. Let us remark that the differential of any function on $\mathcal{g}^*$ is an element of the dual space , i.e. of the Lie algebra $\mathcal{g}$. It gives us a possibility to define the following Kirillov-Kostant Poisson bracket:

$$\{f, h\}(l) = < l, [\delta f(l), \delta h(l)] > \quad . \tag{2.16}$$

In particular, for linear functions H_ε the r.h.s. of (2.16) simplifies:

$$\{H_\varepsilon, H_\eta\} = H_{[\varepsilon, \eta]} \quad , \tag{2.17}$$

$$\{H^a, H^b\} = \sum_c f_c^{ab} H^c \quad . \tag{2.18}$$

The last formula simulates the commutation relations (2.1).

In general situation the space $\mathcal{g}^*$ supplied with Poisson bracket (2.16) is not a symplectic manifold. The Kirillov-Kostant bracket is degenerate. For example, in the simplest case of $\mathcal{g} = su(2)$ the space $\mathcal{g}^*$ is 3-dimensional. The matrix of Poisson bracket is antisymmetric and degenerates as any antisymmetric matrix in an odd-dimensional space.

The relation between symplectic and Poisson theories is the following. Any Poisson manifold with degenerate Poisson bracket splits into a set of symplectic leaves. A symplectic leaf is defined so that its tangent space at any point consists of the values of all hamiltonian vector fields at this point:

$$v_h(f) = \{h, f\} \quad . \tag{2.19}$$

Each symplectic leaf inherits the Poisson bracket from the manifold. However, being restricted onto the symplectic leaf the Poisson bracket becomes nondegenerate and we can define the symplectic two-form Ω so that:

$$\Omega(v_f, v_h) = \{f, h\} \ . \tag{2.20}$$

The relation (2.20) defines Ω completely because any tangent vector to the symplectic leaf may be represented as a value of some hamiltonian vector field.

If we choose dual bases $\{e_a\}$ and $\{e^a\}$ in tangent and cotangent spaces to the symplectic leaf we can rewrite the bracket and the symplectic form as follows:

$$\{f, h\} = - \sum_{ab} P^{ab} < \delta f, e_a > < \delta h, e_b > \ , \tag{2.21}$$

$$\Omega = \sum_{ab} \Omega_{ab} \, e^a \otimes e^b = \frac{1}{2} \sum_{ab} \Omega_{ab} \, e^a \wedge e^b \ . \tag{2.22}$$

Using definition (2.20) of the form Ω and formulae (2.21), (2.22) one can check that the matrix Ω_{ab} is inverse to the matrix P^{ab}:

$$\sum_c \Omega_{ac} P^{cb} = \delta_a^b \ . \tag{2.23}$$

For the particular case of the space g^* with Poisson structure (2.16), there exists a nice description of the symplectic leaves. They coincide with the orbits of coadjoint action (2.7) of the group G. Starting from any point l_0, we can construct an orbit

$$\mathcal{O}_{l_0} = \{l = Ad^*(g)l_0 \ , \ g \in G\} \ . \tag{2.24}$$

Any point of g^* belongs to some coadjoint orbit. The orbit $\mathcal{O}_{l_0}$ can be regarded as a quotient space of the group G over its subgroup S_{l_0}:

$$\mathcal{O}_{l_0} \approx G/S_{l_0} \ , \tag{2.25}$$

where $S_{l_0'}$ is defined as follows:

$$S_{l_0} = \{g \in G \ , \ Ad^*(g)l_0 = l_0\} \ . \tag{2.26}$$

In the case of $G = SU(2)$ the coadjoint action is represented by rotations in the 3-dimensional space g^*. The orbits are spheres and there is one exceptional zero radius orbit which is just the origin. The group S_{l_0} is isomorphic to $U(1)$ and corresponds to rotations around the axis parallel to l_0. For the exceptional orbit $S_{l_0} = G$ and the quotient space G/G is a point.

Let us denote by p_{l_0} the projection from G to $\mathcal{O}_{l_0}$:

$$p_{l_0} : \quad g \longrightarrow l_g = Ad^*(g)l_0 \ . \tag{2.27}$$

We may investigate the symplectic form Ω on the orbit directly. However, for technical reasons it is more convenient to consider its pull-back $\Omega_{l_0}^G = p_{l_0}^* \Omega$ defined on the group G itself. We reformulate the famous Kirillov's result in the following form. Let $\mathcal{O}_{l_0}$ be a coadjoint orbit of the group G and p_{l_0} be the projection (2.27). The Poisson structure (2.16) defines a symplectic form Ω on $\mathcal{O}_{l_0}$.

Theorem 1 *The pull-back of Ω along the projection p_{l_0} is the following:*

$$\Omega_{l_0}^G = \frac{1}{2} < \delta l_g \stackrel{\wedge}{,} \theta_g > \quad . \tag{2.28}$$

We do not prove formula (2.28) but the proof of its Poisson-Lie counterpart in subsection 3.4 will fill this gap. Let us make only few remarks. First of all, the form $\Omega_{l_0}^G$ actually is a pull-back of some two-form on the orbit $\mathcal{O}_{l_0}$. Then, $\Omega_{l_0}^G$ is a closed form:

$$\delta \Omega_{l_0}^G = 0 \quad . \tag{2.29}$$

This is a direct consequence of the Jacobi identity for the Poisson bracket (2.16). The form $\Omega_{l_0}^G$ is exact, while the original form Ω belongs to a nontrivial cohomology class. The left-invariant one-form

$$\alpha = < l_g, \theta_g > = < l_0, \mu_g > \tag{2.30}$$

satisfies the equation

$$\delta \alpha = \Omega_{l_0}^G \quad . \tag{2.31}$$

In physical applications the form α defines an action for a hamiltonian system on the orbit:

$$S = \int \alpha \quad . \tag{2.32}$$

This action plays a crucial role in the representation of a Wilson line via functional integral. (see section 2.2).

The rest of this subsection is devoted to the cotangent bundle T^*G of the group G. Actually, the bundle T^*G is trivial. The group G acts on itself by means of right and left multiplications. Both these actions may be used to trivialize T^*G. So we have two parametrizations of

$$T^*G = G \times \mathcal{G}^* \tag{2.33}$$

by pairs (g, l) and (g, m) where l and m are elements of $\mathcal{G}^*$. In the left parametrization G acts on T^*G as follows:

$$\text{L} \qquad h : (g, m) \longrightarrow (hg, m) \quad , \tag{2.34}$$

$$\text{R} \qquad h : (g, m) \longrightarrow (gh^{-1}, Ad^*(h)m) \quad . \tag{2.35}$$

In the right parametrization left and right multiplications change roles:

$$\text{L} \qquad h : (g, l) \longrightarrow (hg, Ad^*(h)l) \quad , \tag{2.36}$$

$$\text{R} \qquad h : (g, l) \longrightarrow (gh^{-1}, l) \quad . \tag{2.37}$$

The two coordinates l and m are related:

$$l = Ad^*(g)m \quad . \tag{2.38}$$

The cotangent bundle T^*G carries the canonical symplectic structure Ω^{T^*G} [9]. Using coordinates (g, l, m), we write a formula for Ω^{T^*G} without proof:

$$\Omega^{T^*G} = \frac{1}{2}(< \delta m \overset{\wedge}{,} \mu_g > + < \delta l \overset{\wedge}{,} \theta_g >) \ . \tag{2.39}$$

The symplectic structure on T^*G is a sort of universal one. We can recover the Kirillov two-form (2.28) for any orbit starting from (2.39). More exactly, let us impose in (2.39) the condition:

$$m = m_0 = const \ . \tag{2.40}$$

It means that instead of T^*G we consider a reduced symplectic manifold with the symplectic structure (for justification see subsection 3.3):

$$\Omega_r = \frac{1}{2} < \delta l, \theta_g > \ , \tag{2.41}$$

where l is subject to constraint

$$l = Ad^*(g)m_0 \ . \tag{2.42}$$

Formulae (2.41), (2.42) reproduce formulae (2.27), (2.28) and we can conclude that the reduction leads to the orbit O_{m_0} of the point m_0 in $\mathcal{g}^*$.

2.2 Functional integral for a coadjoint orbit

Our main motivation to consider geometric quantization and Kirillov symplectic form is the application of this theory to the Chern-Simons model. More exactly, we rewrite the expression for a Wilson line observable as a certain functional integral over a coadjoint orbit of the group G. It is convenient to restrict ourselves to the case of G being a simple Lie group as it is the main example which we are interested in in the framework of the Chern-Simons theory.

First, let us remind that the quantization of Kirillov-Kostant bracket (2.16) reproduces the Lie algebraic commutator (2.1). So, we expect that after quantization the Lie algebra $\mathcal{g}$ acts in the Hilbert space of the corresponding quantum system. If we start with an orbit we expect that the corresponding representation is irreducible. This guess is based on the observation that before quantization the group action can move any given point on the orbit to any other point. Procedure of geometric quantization [10] provides a mathematical proof of this conjecture. However, in this paper we use a physical language and treat the quantization procedure in the framework of path integral formulation.

To begin with we need an action which describes our physical system. As we live on the orbit, our nearest concern is to introduce some efficient coordinates. Actually, it has been done in the previous subsection where we parametrized a point on the orbit by the group element (2.27):

$$T = v^{-1}Dv. \tag{2.43}$$

Here we introduced special notations for the case of the simple group G. We denote a point on the orbit represented by matrix from g by T. The fixed point D is a diagonal matrix which defines the orbit. The group G acts by conjugations:

$$T^g = g^{-1}Tg. \tag{2.44}$$

We remind T gives a momentum mapping corresponding to this action. In terms of D and $v \in G$ Kirillov form (2.28) looks as:

$$\varpi = TrD(\delta vv^{-1})^2. \tag{2.45}$$

So, the action for such a system may be written as

$$S_D(v) = \int TrD(\delta vv^{-1}) - \int H dt. \tag{2.46}$$

Here Hamiltonian H is an arbitrary function on the orbit. For our purposes it is convenient to choose it to be a linear function:

$$H = iTr(AT), \tag{2.47}$$

where $A = A(t)$ is a time-dependent source. The main problem of this theory is to evaluate the partition function:

$$Z_D(A) = \int \mathcal{D}v e^{iS_D(v)}. \tag{2.48}$$

We shall consider this integral with periodic boundary conditions. Strictly speaking, it is not well-defined because of the gauge symmetry with respect to the left action of the diagonal subgroup of G: $u \to hu$. However, this symmetry may be taken into account by the standard renormalization of the integration measure. As for any functional integral, we can rewrite the partition function using an ordered exponent of the Hamiltonian:

$$Z_D(A) = Tr_{\mathcal{H}} P exp(\int \sum_a A_a(t)\hat{T}^a dt). \tag{2.49}$$

Here $\hat{T}^a$ is an operator corresponding to T^a after quantization, $\mathcal{H}$ is a Hilbert space of the resulting theory. As we discussed, this Hilbert space is expected to be an irreducible representation of the Lie algebra g. The problem is how to find out which representation we get starting from the action $S_D(u)$. The answer looks like follows. Let us represent the highest weight $w(D)$ of the corresponding representation as a diagonal matrix. Then

$$w(D) = D - \rho, \tag{2.50}$$

for ρ being a half sum of positive roots of g.

Let us conclude that we obtained a nice representation for a Wilson line observable in the Chern-Simons theory. Namely, such an observable may be always represented as a partition function in the auxiliary theory on the certain coadjoint orbit:

$$W_{w(D)}(\Gamma) = Z_D(A(t)), \tag{2.51}$$

where $A(t)$ is the restriction of the gauge field A on the curve Γ.

For further information on the orbit functional integral we send the reader to original papers [11],[7],[12]. The representation (2.51) has been applied to the Chern-Simons theory in [6].

3 Symplectic structures associated to Poisson-Lie groups

In this Section we develop machinery of Poisson-Lie groups and find out how Kirillov form modifies when we introduce a nontrivial Poisson bracket on a group manifold. We follow the approach of [1].

3.1 Heisenberg double of Lie bialgebra

One of the ways to introduce deformation leading to Poisson-Lie groups is to consider the bialgebra structure on $\mathcal{G}$. Following [13], we consider a pair $(\mathcal{G}, \mathcal{G}^*)$, where we treat $\mathcal{G}^*$ as another Lie algebra with the commutator $[,]^*$. For a given commutator $[,]$ in $\mathcal{G}$ we can not choose an arbitrary commutator $[,]^*$ in $\mathcal{G}^*$. The axioms of bialgebra can be reformulated as follows. The linear space

$$\mathcal{D} = \mathcal{G} + \mathcal{G}^* \tag{3.1}$$

with the commutator $[,]_{\mathcal{D}}$:

$$[\varepsilon, \eta]_{\mathcal{D}} = [\varepsilon, \eta] \ , \tag{3.2}$$

$$[x, y]_{\mathcal{D}} = [x, y]^* \ , \tag{3.3}$$

$$[\varepsilon, x]_{\mathcal{D}} = ad^*(\varepsilon)x - ad^*(x)\varepsilon \ . \tag{3.4}$$

must be a Lie algebra. In the last formula (3.4) $ad^*(\varepsilon)$ is the usual ad^*-operator for the Lie algebra $\mathcal{G}$ acting on $\mathcal{G}^*$. The symbol $ad^*(x)$ corresponds to the coadjoint action of the Lie algebra $\mathcal{G}^*$ on its dual space $\mathcal{G}$.

The only thing we have to check is the Jacobi identity for the commutator $[,]_{\mathcal{D}}$. If it is satisfied, we call the pair $(\mathcal{G}, \mathcal{G}^*)$ Lie bialgebra. Algebra $\mathcal{D}$ is called Drinfeld double. It has the nondegenerate scalar product $<,>_{\mathcal{D}}$:

$$< (\varepsilon, x), (\eta, y) >_{\mathcal{D}} = < y, \varepsilon > + < x, \eta > \ , \tag{3.5}$$

where in the r.h.s. $<,>$ is the canonical pairing of $\mathcal{G}$ and $\mathcal{G}^*$. It is easy to see that

$$< \mathcal{G}, \mathcal{G} >_{\mathcal{D}} = 0 \ , \quad < \mathcal{G}^*, \mathcal{G}^* >_{\mathcal{D}} = 0 \ . \tag{3.6}$$

In other words, $\mathcal{G}$ and $\mathcal{G}^*$ are isotropic subspaces in $\mathcal{D}$ with respect to the form $<,>_{\mathcal{D}}$. We call the form $<,>_{\mathcal{D}}$ on the algebra $\mathcal{D}$ standard product in $\mathcal{D}$.

We shall need two operators P and P^* acting in $\mathcal{D}$. P is defined as a projector onto the subspace $\mathcal{G}$:

$$P(x + \varepsilon) = \varepsilon \ . \tag{3.7}$$

The operator P^* is its conjugate with respect to form (3.5). It appears to be a projector onto the subspace $\mathcal{G}^*$:

$$P^*(x + \varepsilon) = x \ . \tag{3.8}$$

The standard product in $\mathcal{D}$ enables us to define the canonical isomorphism $J : \mathcal{D}^* \longrightarrow \mathcal{D}$ by means of the formula

$$< J(a^*), b >_{\mathcal{D}} = < a^*, b > \ , \tag{3.9}$$

where a^* is an element of $\mathcal{D}^*$ and b belongs to $\mathcal{D}$. In the r.h.s. we use the canonical pairing of $\mathcal{D}$ and $\mathcal{D}^*$. The standard product can be defined on the space $\mathcal{D}^*$:

$$< a^*, b^* >_{\mathcal{D}^*} = < J(a^*), J(b^*) >_{\mathcal{D}} \ , \tag{3.10}$$

where a^* and b^* belong to $\mathcal{D}^*$. The scalar product $<, >_{\mathcal{D}}$ is invariant with respect to the commutator in $\mathcal{D}$:

$$< [a, b], c >_{\mathcal{D}} + < b, [a, c] >_{\mathcal{D}} = 0 \ . \tag{3.11}$$

It is easy to check that the operator J converts ad^* into ad:

$$J ad^*(a) J^{-1} = ad(a) \ . \tag{3.12}$$

Using the standard scalar product in $\mathcal{D}$, one can construct elements r and r^* in $\mathcal{D} \otimes \mathcal{D}$ which correspond to the operators P and P^*:

$$< a \otimes b, r >_{\mathcal{D} \otimes \mathcal{D}} = < a, Pb >_{\mathcal{D}} \ , \tag{3.13}$$

$$< a \otimes b, r^* >_{\mathcal{D} \otimes \mathcal{D}} = - < a, P^* b >_{\mathcal{D}} \ . \tag{3.14}$$

In terms of dual bases $\{\varepsilon^a\}$ and $\{l_a\}$ in $\mathcal{G}$ and $\mathcal{G}^*$

$$r = \sum_a \varepsilon^a \otimes l_a \ , \quad r^* = -\sum_a l_a \otimes \varepsilon^a \ . \tag{3.15}$$

The Lie algebra $\mathcal{D}$ may be used to construct the Lie group D. We suppose that D exists (for example, for finite dimensional algebras it is granted by the Lie theorem) and we choose it to be connected. Originally the double is defined as a connected and simply connected group. However, we may use any connected group D corresponding to Lie algebra $\mathcal{D}$. Property (3.12) can be generalized for Ad and Ad^*:

$$J Ad^*(d) J^{-1} = Ad(d) \ , \tag{3.16}$$

where d is an element of D.

Let us denote by G and G^* the subgroups in D corresponding to subalgebras g and g^* in $\mathcal{D}$. In the vicinity D_0 of the unit element of D the following two decompositions are applicable:

$$d = gg^* = h^*h \ , \tag{3.17}$$

where d is an element of D, coordinates g, h belong to the subgroup G, coordinates g^*, h^* belong to the subgroup G^*. In general, the subset D_0 does not cover the whole group D. However, it is open and dense. In the further consideration we restrict ourselves to the cell D_0 in D and send the reader to [1] for complete description.

Now we turn to the description of the Poisson brackets on the manifold D. Double D admits two natural Poisson structures. First of them was proposed by Drinfeld [13]. For two functions f and h on D the Drinfeld bracket is equal to

$$\{f, h\} = < \nabla_L f \otimes \nabla_L h, r > - < \nabla_R f \otimes \nabla_R h, r > \ , \tag{3.18}$$

where $<, >$ is the canonical pairing between $\mathcal{D} \otimes \mathcal{D}$ and $\mathcal{D}^* \otimes \mathcal{D}^*$. Poisson bracket (3.18) defines a structure of a Poisson-Lie group on D. However, the most important for us is the second Poisson structure on D suggested by Semenov-Tian-Shansky [14]:

$$\{f, h\} = -(< \nabla_L f \otimes \nabla_L h, r > + < \nabla_R f \otimes \nabla_R h, r^* >) \ . \tag{3.19}$$

The manifold D equipped with bracket (3.19) is called Heisenberg double or D_+. It is a natural analogue of T^*G in the Poisson-Lie case. When g^* is abelian, $G^* = g^*$ and $D_+ = T^*G$. If the double D is a matrix group, we can rewrite the basic formula (3.19) in the following form:

$$\{d^1, d^2\} = -(rd^1d^2 + d^1d^2r^*) \ , \tag{3.20}$$

where $d^1 = d \otimes I$, $d^2 = I \otimes d$.

For concrete calculations let us choose the left identification of the tangent space to D with $\mathcal{D}$. We can rewrite the Poisson bracket (3.19) in terms of left derivatives ∇_L:

$$\{f, h\}(d) = -(< \nabla_L f \otimes \nabla_L h, r > + < Ad^*(d^{-1})\nabla_L f \otimes Ad^*(d^{-1})\nabla_L h, r^* >) =$$
$$= - < \nabla_L f \otimes \nabla_L h, r + Ad(d) \otimes Ad(d) r^* > \ . \tag{3.21}$$

Here we use relation (2.15) between left and right derivatives on a group.

Given a hamiltonian h one can produce the hamiltonian vector field v_h so that the formula

$$< \delta f, v_h > = \{h, f\} \tag{3.22}$$

holds for any function f. Using (3.21), (3.22) we can reconstruct the field v_h:

$$v_h = < \nabla_L h, r + Ad(d) \otimes Ad(d) r^* >_2 \ . \tag{3.23}$$

Having identified $\mathcal{D}$ and $\mathcal{D}^*$ by means of the operator J, we can rewrite the r.h.s. of (3.23) as follows:

$$v_h\big|_d = \mathcal{P}\delta h = (P - Ad(d)P^* Ad(d^{-1}))J(\nabla_L h(d)) \ , \tag{3.24}$$

where $\mathcal{P}$ acts in $\mathcal{D}$:

$$\mathcal{P} = P - Ad(d)P^* Ad(d^{-1}) \ . \tag{3.25}$$

It is called Poisson operator.

The problem which appears immediately in the theory of D_+ is the possible degeneracy of Poisson structure (3.19) in some points of D. Stratification of D_+ into the set of symplectic leaves is described in [1]. Here we need only a simple fact about this stratification:

Lemma 1 *The subset*

$$D_0 = GG^* \cap G^* G \tag{3.26}$$

is a symplectic leave in D with respect to the Poisson bracket (3.19).

It means that the bracket (3.19) is actually nondegenerate on D_0.

3.2 Symplectic structure of the Heisenberg double

The subject of this subsection is to find an efficient description of the symplectic form Θ on D_0. Let us introduce two sets of coordinates on D_0:

$$d = gg^* = h^* h. \tag{3.27}$$

In terms of (g, g^*) and (h^*, h) we can write down the answer for Θ.

Theorem 2 *The symplectic form Θ on D_0 can be represented as follows:*

$$\Theta = \frac{1}{2}(<\theta_{h^*} \overset{\wedge}{,} \theta_g> + <\mu_{g^*} \overset{\wedge}{,} \mu_h>) \ . \tag{3.28}$$

In the formula (3.28) $\theta_g, \theta_{h^*}, \mu_h, \mu_{g^*}$ are Maurer-Cartan forms on G and G^*. The pairing $<,>$ is applied to their values, which can be treated as elements of $\mathcal{g}$ and $\mathcal{g}^*$ embedded to $\mathcal{D} = \mathcal{g} + \mathcal{g}^*$. So we can use $<,>_\mathcal{D}$ as well as $<,>$.

Proof of Theorem 2

The strategy of the proof is quite straightforward. We consider Poisson bracket (3.19) on the symplectic leaf D_0. If we use dual bases $\{e_a\}$ and $\{e^a\}$ ($a = 1, \ldots, n = dimD$) of right-invariant vector fields and one-forms on D, the formula (3.19) acquires the following form:

$$\{f, h\}(d) = - <\nabla_L f \otimes \nabla_L h, r + Ad(d) \otimes Ad(d) r^* >=$$

$$= - \sum_{a,b=1}^{n} <\nabla_L f, e_a><\nabla_L h, e_b><e^a, \mathcal{P}Je^b> \ . \tag{3.29}$$

The last multiplier in (3.29) is Poisson matrix corresponding to the bracket (3.19):

$$\mathcal{P}^{ab} = < e^a, \mathcal{P}Je^b > \quad . \tag{3.30}$$

Here $\mathcal{P}$ is the same as in (3.25). It is ensured by *Lemma 1* that
the matrix $\mathcal{P}^{ab}$ is nondegenerate. The symplectic form Θ can be represented as follows (see subsection 2.1):

$$\Theta = \sum_{a,b=1}^{n} \Theta_{ab} e^a \otimes e^b \quad , \tag{3.31}$$

where the matrix Θ satisfies the following condition:

$$\sum_{c=1}^{n} \Theta_{ac} \mathcal{P}^{cb} = \delta_a^b \quad . \tag{3.32}$$

So what we need is inverse matrix $\mathcal{P}^{-1}$ for $\mathcal{P}^{ab}$. To this end let us introduce two operators $\mathcal{P}_1$ and $\mathcal{P}_2$:

$$\mathcal{P}_1 = P + Ad(d)P^* \quad , \tag{3.33}$$

$$\mathcal{P}_2 = P^* - Ad(d)P \quad . \tag{3.34}$$

P may be decomposed in two ways, using $\mathcal{P}_1$ and $\mathcal{P}_2$:

$$P = \mathcal{P}_1 \mathcal{P}_2^* = -\mathcal{P}_2 \mathcal{P}_1^* \quad . \tag{3.35}$$

The definition of $\mathcal{P}_1$ and $\mathcal{P}_2$ permit us to write down the answer for Θ_{ab}:

$$\Theta_{ab} = < e_a, \Theta e_b >_{\mathcal{D}} \quad , \quad \Theta = P\mathcal{P}_1^{-1} - P^*\mathcal{P}_2^{-1} \quad . \tag{3.36}$$

We must check condition (3.32):

$$\delta_a^b = \sum_{c=1}^{n} \Theta_{ac} \mathcal{P}^{cb} =$$

$$= \sum_{c=1}^{n} < e_a, \Theta e_c >< e^c, \mathcal{P}J(e^b) >= \tag{3.37}$$

$$= < e_a, \Theta \mathcal{P}J(e^b) >_{\mathcal{D}} \quad .$$

The product $\Theta \mathcal{P}$ can be easily calculated using (3.35), (3.36):

$$\Theta P = P\mathcal{P}_1^{-1} \mathcal{P}_1 \mathcal{P}_2^* + P^* \mathcal{P}_2^{-1} \mathcal{P}_2 \mathcal{P}_1^* =$$

$$= P(P - P^* Ad(d^{-1})) + P^*(P^* + P Ad(d^{-1})) = \tag{3.38}$$

$$= P + P^* = I \quad .$$

So, the answer is

$$< e_a, \Theta \mathcal{P}J(e^b) >_{\mathcal{D}} = < e^b, e_a >= \delta_a^b \tag{3.39}$$

as it is required by (3.32).

We can rewrite formula (3.36) in more invariant way:

$$\Theta = < \theta_d \overset{\otimes}{,} \Theta \theta_d >_{\mathcal{D}} \ , \tag{3.40}$$

where θ_d is the Maurer-Cartan on D. Expression (3.36) for the operator Θ still includes inverse operators $\mathcal{P}_{1,2}^{-1}$ implying that some equations must be solved. To this end we represent the Maurer-Cartan form θ_d in two different ways:

$$\theta_d = \theta_g + Ad(d)\mu_{g^*} \ , \tag{3.41}$$

$$\theta_d = \theta_{h^*} + Ad(d)\mu_h \ . \tag{3.42}$$

Representations (3.41), (3.42) allow us to calculate $\mathcal{P}_{1,2}^{-1} p^* \theta_d$ explicitly:

$$\mathcal{P}_1^{-1} \theta_d = \theta_g + \mu_{g^*} \ , \tag{3.43}$$

$$\mathcal{P}_2^{-1} \theta_d = \theta_{h^*} - \mu_h \ . \tag{3.44}$$

Putting together (3.36), (3.40), (3.43) and (3.44), we obtain the following formula for the symplectic form:

$$\Theta = < (\theta_g + Ad(d)\mu_{g^*}) \overset{\otimes}{,} \theta_g >_{\mathcal{D}} - < (\theta_{h^*} + Ad(d)\mu_h) \overset{\otimes}{,} \theta_{h^*} >_{\mathcal{D}} =$$
$$= < Ad(d)\mu_{g^*} \overset{\otimes}{,} \theta_g >_{\mathcal{D}} - < Ad(d)\mu_h \overset{\otimes}{,} \theta_{h^*} >_{\mathcal{D}} \ . \tag{3.45}$$

Actually, the form (3.45) is antisymmetric. To make it evident, let us consider the identity

$$< p^* \theta_d \overset{\otimes}{,} p^* \theta_d >_{\mathcal{D}} =$$
$$= < Ad(d)\mu_{g^*} \overset{\otimes}{,} \theta_g >_{\mathcal{D}} + < \theta_g \overset{\otimes}{,} Ad(d)\mu_{g^*} >_{\mathcal{D}} = \tag{3.46}$$
$$= < Ad(d)\mu_h \overset{\otimes}{,} \theta_{h^*} >_{\mathcal{D}} + < \theta_{h^*} \overset{\otimes}{,} Ad(d)\mu_h >_{\mathcal{D}} \ .$$

Or, equivalently,

$$< Ad(d)\mu_{g^*} \overset{\otimes}{,} \theta_g >_{\mathcal{D}} - < Ad(d)\mu_h \overset{\otimes}{,} \theta_{h^*} >_{\mathcal{D}} =$$
$$= - < \theta_g \overset{\otimes}{,} Ad(d)\mu_{g^*} >_{\mathcal{D}} + < \theta_{h^*} \overset{\otimes}{,} Ad(d)\mu_h >_{\mathcal{D}} \ . \tag{3.47}$$

Applying 3.47 to make (3.45) manifestly antisymmetric, one gets:

$$\Theta = \frac{1}{2}(< Ad(d)\mu_{g^*} \overset{\wedge}{,} \theta_g >_{\mathcal{D}} + < \theta_{h^*} \overset{\wedge}{,} Ad(d)\mu_h >_{\mathcal{D}}) \ . \tag{3.48}$$

Using representation of d in terms of (g, g^*) and (h^*, h), it is easy to check that formula (3.48) coincides with

$$\Theta = -\frac{1}{2}(< \mu_g \overset{\wedge}{,} \theta_{g^*} >_{\mathcal{D}} + < \theta_h \overset{\wedge}{,} \mu_{h^*} >_{\mathcal{D}}) \ . \tag{3.49}$$

To obtain formula (3.28) one can use (3.41), (3.42):

$$\theta_d = \theta_g + Ad(d)\mu_{g^*} = \theta_{h^*} + Ad(d)\mu_h \ . \tag{3.50}$$

Or, equivalently,

$$\theta_g - Ad(d)\mu_h = \theta_{h^*} - Ad(d)\mu_{g^*} \quad . \tag{3.51}$$

Due to antisymmetry we have

$$< (\theta_g - Ad(d)\mu_h) \overset{\wedge}{,} (\theta_{h^*} - Ad(d)\mu_{g^*}) >_D = 0 \quad . \tag{3.52}$$

Therefore,

$$\frac{1}{2}(< \theta_{h^*} \overset{\wedge}{,} \theta_g >_D + < \mu_{g^*} \overset{\wedge}{,} \mu_h >_D) =$$
$$= \frac{1}{2}(< Ad(d)\mu_{g^*} \overset{\wedge}{,} \theta_g >_D + < \theta_{h^*} \overset{\wedge}{,} Ad(d)\mu_h >_D) = \Theta \quad , \tag{3.53}$$

which coincides with (3.28).

One can easily check that the r.h.s. of formula (3.28) does represent the pull-back of some two-form on D_0.

It is known from general Poisson theory that

$$\delta\Theta = 0 \quad , \tag{3.54}$$

but it is interesting to check that form (3.28) is closed by direct calculations. Rewriting equation (3.51) we get:

$$\theta_g - \theta_{h^*} = Ad(d)\mu_h - Ad(d)\mu_{g^*} \quad . \tag{3.55}$$

Taking the cube of the last equation we get:

$$< \theta_g \overset{\wedge}{,} \theta_g \wedge \theta_g >_D - < \theta_{h^*} \overset{\wedge}{,} \theta_{h^*} \wedge \theta_{h^*} >_D +$$
$$+3 < \theta_g \overset{\wedge}{,} \theta_{h^*} \wedge \theta_{h^*} >_D -3 < \theta_g \wedge \theta_g \overset{\wedge}{,} \theta_{h^*} >_D =$$
$$= < \mu_h \overset{\wedge}{,} \mu_h \wedge \mu_h >_D - < \mu_{g^*} \overset{\wedge}{,} \mu_{g^*} \wedge \mu_{g^*} >_D + \tag{3.56}$$
$$+3 < \mu_h \overset{\wedge}{,} \mu_{g^*} \wedge \mu_{g^*} >_D -3 < \mu_h \wedge \mu_h \overset{\wedge}{,} \mu_{g^*} >_D \quad .$$

As $\theta_g \wedge \theta_g = \frac{1}{2}[\theta_g \overset{\wedge}{,} \theta_g]$ and $\mu_h \wedge \mu_h = \frac{1}{2}[\mu_h \overset{\wedge}{,} \mu_h]$ take values in g, $\theta_{h^*} \wedge \theta_{h^*} = \frac{1}{2}[\theta_{h^*} \overset{\wedge}{,} \theta_{h^*}]$ and $\mu_{g^*} \wedge \mu_{g^*} = \frac{1}{2}[\mu_{g^*} \overset{\wedge}{,} \mu_{g^*}]$ take values in g^* we may use the pairing $<,>_D$ for them. Moreover, as both g and g^* are isotropic subspaces in D, we rewrite (3.56) as follows:

$$< \theta_g \overset{\wedge}{,} \theta_{h^*} \wedge \theta_{h^*} >_D - < \theta_g \wedge \theta_g \overset{\wedge}{,} \theta_{h^*} >_D -$$
$$- < \mu_h \overset{\wedge}{,} \mu_{g^*} \wedge \mu_{g^*} >_D + < \mu_h \wedge \mu_h \overset{\wedge}{,} \mu_{g^*} >_D = 0 \quad . \tag{3.57}$$

We remind that $\delta\theta_g = \theta_g \wedge \theta_g$ and $\delta\mu_g = -\mu_g \wedge \mu_g$. Thus,

$$\delta\Theta = - < \delta\theta_g \overset{\wedge}{,} \theta_{h^*} >_D + < \theta_g \overset{\wedge}{,} \delta\theta_{h^*} >_D -$$
$$- < \delta\mu_h \overset{\wedge}{,} \mu_{g^*} >_D + < \mu_h \overset{\wedge}{,} \delta\mu_{g^*} >_D = 0 \quad . \tag{3.58}$$

Now it is interesting to consider the classical limit of our theory to recover the standard answer for T^*G. There is no deformation parameter in bracket (3.19) but it may be introduced by hand:

$$\{f, h\}_\gamma = \gamma\{f, h\} \quad . \tag{3.59}$$

For the new bracket (3.59) we have the symplectic form:

$$\Theta^\gamma = \frac{1}{\gamma}\Theta \ . \tag{3.60}$$

To recover coordinates on T^*G one have to parametrize a vicinity of the unit element in the group G^* by means of the exponential map:

$$g^* = \exp(\gamma m) \ , \tag{3.61}$$

$$h^* = \exp(\gamma l) \ , \tag{3.62}$$

where m and l belong to $\mathfrak{g}^*$. Coordinates m and l are adjusted in such a way that they have finite values after the limit procedure. When γ tends to zero, the formula

$$d = gg^* = h^*h \tag{3.63}$$

leads to the following relations:

$$g = h \ , \quad l = Ad^*(g)m \ . \tag{3.64}$$

Expanding the form Θ^γ into the series in γ we keep only the constant term (singularity γ^{-1} disappears from the answer because the corresponding two-form is identically equal to zero). The answer is the following:

$$\Theta^\gamma = \frac{1}{2}(< \delta m \stackrel{\wedge}{,} \mu_g > + < \delta l \stackrel{\wedge}{,} \theta_g >) \tag{3.65}$$

and it recovers classical answer (see subsection 2.1). Deriving formula (3.65), we use the expansions for the Maurer-Cartan forms on G^*:

$$\theta_{g^*} = \gamma\delta m + O(\gamma^2) \ , \tag{3.66}$$

$$\mu_{h^*} = \gamma\delta l + O(\gamma^2) \ . \tag{3.67}$$

We have considered general properties of the symplectic structure on the main cell D_0 of the Heisenberg double D_+. Our next aim is the Poisson-Lie analogue of the theory of coadjoint orbits. The necessary technical tools will be introduced
in the next subsection.

3.3 Dual pairs

One of powerful tools in Hamiltonian mechanics is the language of dual pairs. Let X be a symplectic space. Obviously, it carries a nondegenerate Poisson structures.

Definition 1 *A pair of Poisson mappings*

$$\mu : X \to Y,$$
$$\nu : X \to Z \tag{3.68}$$

is called a dual pair if

$$\{\{f,h\} = 0, \forall f = \tilde{f} \circ \mu, \tilde{f} : Y \to C\} \Leftrightarrow \{\exists \tilde{h} : Z \to C, h = \tilde{h} \circ \nu\}. \qquad (3.69)$$

In other words, any function lifted from Y is in involution with any function lifted from Z and moreover, if some function commute with any function lifted from Y it means that it is lifted from Z.

The standard source of dual pairs is Hamiltonian reduction. If we have a Hamiltonian action of a group G on a symplectic manifold X, the following pair of projections is dual:

$$\mu : X \to \mathcal{G}^*,$$
$$\nu : X \to X/G. \qquad (3.70)$$

Here the mapping μ is the momentum mapping from the manifold X to the space dual to the Lie algebra $\mathcal{G}$.

Dual pairs provide the method to classify symplectic leaves in the Poisson spaces Y and Z. For any point $y \in Y$ the subspace $\nu(\mu^{-1}(y))$ is a symplectic leaf in Z. It carries nondegenerate symplectic structure. The same is true in the other direction. Take any point $z \in Z$, then the subspace $\mu(\nu^{-1}(z))$ is a symplectic leaf in Y. Actually, in this paper we don't need the full machinery of dual pairs. Only one simple fact will be of importance for us.

Lemma 2 *Let the pair of mappings (μ, ν) (3.68) be a dual pair. Under these conditions the restriction of the symplectic form Ω on X to the subspace $\mu^{-1}(y)$ coincides with the pull back of the symplectic form ω_y on the symplectic leave $\nu(\mu^{-1}(y))$ along the projection ν:*

$$\Omega \mid_{\mu^{-1}(y)} = \nu^* \omega_y. \qquad (3.71)$$

This lemma relates the symplectic structure of the reduced phase space with the symplectic structure of the global space X which is usually much simpler.

3.4 Theory of orbits

In this subsection we describe reductions of the Heisenberg double D_+ which lead to Poisson-Lie analogues of coadjoint orbits.

The coordinates g, g^*, h, h^* introduced in subsection 3.2 will be quite convenient for this purpose. Let us remark that the relation

$$gg^* = h^* h \qquad (3.72)$$

may be used to define the action of G on G^*

$$g : g^* \to g^{*'}(g, g^*) = h^*. \qquad (3.73)$$

This action usually appears in literature with the name dressing transformation [14].

The decomposition

$$d = gg^* = h^*h \tag{3.74}$$

induces Poisson structures on the groups G and G^*. Indeed, let us consider for example the realization of the group $G^* : G_L^* \approx D/G$. This formula is not quite correct because the decomposition $D \approx G^*G$ is not global. However, Poisson and symplectic structures are local objects and we can ignore this subtlety.

We have used the notation G_L^* to indicate that we treat G^* as a special quotient of D.

Functions on G_L^* may be regarded as functions on D invariant with respect to right action of G:

$$f(dg) = f(d) . \tag{3.75}$$

The right derivative $\nabla_R f$ is orthogonal to g for functions on G_L^*:

$$< \nabla_R f, g > = 0 . \tag{3.76}$$

For a pair of invariant functions f and h the second term in the formula (3.19) vanishes because $r^* \in g^* \otimes g$. The first term is an invariant function because the left derivative ∇_L preserves the condition (3.75). So we conclude that the Poisson bracket

$$\{f, h\} = - < \nabla_L f \otimes \nabla_L h, r > \tag{3.77}$$

is well-defined on invariant functions and hence it can be treated as a Poisson bracket on G_L^*. This bracket is consistent with the group multiplication in G^* so that the group G^* equipped with such Poisson bracket becomes a Poisson-Lie group. The same is true for the other three quotients $G_R^* = G \setminus D$, $G_R = G^* \setminus D$ and $G_L = D/G^*$. The purpose of this subsection is to study the stratification of the space G_R^* into symplectic leaves and describe the corresponding symplectic forms on them.

It is instructive to consider the classical limit, when g^* and h^* are very close to the identity. Then formula (3.73) transforms into the coadjoint action of G on g^*:

$$g^* = I + \gamma l + \dots , \tag{3.78}$$

$$h^* = I + \gamma l' + \dots , \tag{3.79}$$

$$l' = Ad^*(g)l . \tag{3.80}$$

We denote the transformations (3.73) by AD^* to remind their relation to the coadjoint action:

$$h^*(g, g^*) = AD^*(g)g^* . \tag{3.81}$$

In order to describe symplectic leaves in G^* let us consider the following pair of Poisson mappings:

$$
\begin{array}{ccc}
 & D_0 & \\
\swarrow & & \searrow \\
G_L^* & & G_R^* \ .
\end{array}
\tag{3.82}
$$

This pair is a dual pair [14],[15].

Let us apply the general prescription of the previous subsection to the dual pair (3.83). In order to find a symplectic leaf in G_R^* one should pick up some element $h^*G \in D/G$, consider its preimage in D and project it into $G \backslash D$. It is easy to see that we get an orbit of dressing transformations

$$\mathcal{O}_{h^*} = \{g^* \in G^*, \quad g^* = AD(g^{-1})h^*\}. \tag{3.83}$$

The definition (3.83) introduces at the same time the projection p from G to $\mathcal{O}_{h^*}$:

$$p: g \to g^* = AD(g^{-1})h^*. \tag{3.84}$$

So, the orbits of dressing transformations coincide with symplectic leaves in G^*. Our next task is to evaluate the corresponding symplectic forms. Due to *Lemma 2* the pull back of

symplectic form on the orbit to its preimage in D coincides with the restriction of the symplectic form (3.49) on D_0 to this preimage. As h^* is set to be equal to constant, the first term in (3.49) disappears and we end up with the following formula for the symplectic form ϑ on the orbit:

$$p^*\vartheta = \frac{1}{2} < \theta_{g^*}, \mu_g > . \tag{3.85}$$

To consider the classical limit we can introduce a deformation parameter into the formula (3.85):

$$p^*\vartheta_\gamma = \frac{1}{2\gamma} < \theta_{g^*}, \mu_g > . \tag{3.86}$$

In this way one can recover the classical Kirillov form (2.28) as we did it for T^*G in subsection 3.2.

3.5 Example: simple group

In this subsection we rewrite formulae for symplectic forms on D_+ and orbits of dressing transformations for the case of G being a simple Lie group. We begin with form (3.85) on the orbit. In order to make the expression for this form more transparent we need more detailed information about the group G^*. Let us introduce two Borel subgroups B_+ and B_- in the group G. In the case of $G = SL(n)$ these are subgroups of upper-triangular and lower-triangular matrices correspondingly. For both B_+ and B_- one can define a canonical projection to the Cartan subgroup in G. For $SL(n)$ the projection picks up a diagonal part of upper- or lower-triangular matrix. If we denote elements of B_+ or B_- by big letters, then the corresponding small letters always denote the diagonal parts. The group G^* is defined as follows [14]:

$$G^* = \{(L_+, L_-) \in B_+ \times B_-, \quad l_+l_- = I\}. \tag{3.87}$$

Multiplication in G^* is component-wise:

$$(L_+, L_-)(M_+, M_-) = (L_+M_+, L_-M_-). \tag{3.88}$$

There is a natural mapping α from G^* to G which is given by Gauss decomposition formula:

$$\alpha : (L_+, L_-) \to L = L_+ L_-^{-1}. \tag{3.89}$$

The group structures of G and G^* are different and the mapping α is not a group homomorphism. However, we shall see in Section 4 that it may be useful to replace the requirements of group homomorphism by some weaker conditions. The mapping (3.89) provides an identification of the spaces g and g^*. Then the pairing $<,>$ may be replaced by the invariant form Tr on g.

It is remarkable that for the element L the dressing action simplifies and acquires the form of group conjugations:

$$AD(g)L = gLg^{-1}. \tag{3.90}$$

Let us choose the orbit of dressing transformations which contains a Cartan matrix C:

$$L = AD(g^{-1})C = g^{-1}Cg = L_+ L_-^{-1}. \tag{3.91}$$

Here we specify the definition (3.83) for the case of simple group G. In the notations (3.91) the symplectic form (3.85) may be represented as:

$$\begin{aligned}
\vartheta(g, C) &= \frac{1}{2}Tr(\delta L_+ L_+^{-1} - \delta L_- L_-^{-1}) \wedge g^{-1}\delta g = \\
&= \frac{1}{2}Tr\{C\delta g g^{-1} \wedge C^{-1}\delta g g^{-1} + L_+^{-1}\delta L_+ \wedge L_-^{-1}\delta L_-\}.
\end{aligned} \tag{3.92}$$

The second line may be obtained from the first by straightforward but lengthy calculation.

Now we have an efficient formula for symplectic forms on the orbits and the symplectic form on D_+ is in order. As we learn from formula (3.49), the symplectic form on D_+ consists of two terms. Each term resembles the symplectic form on the orbit of dressing transformations. Let us make this statement more precise. In the simple case one can rewrite the relation (3.72) as follows:

$$L' = gL^{-1}g^{-1}. \tag{3.93}$$

Here L' represents an analogue of right momentum in D_+. We have inverted matrix L in order to get similar Poisson brackets for L and L'. Following the pattern of the dressing orbits, we introduce the diagonal matrix C which consists of common eigenvalues of L and L'^{-1}:

$$\begin{aligned}
L &= u^{-1}Cu, \\
L' &= v^{-1}C^{-1}v.
\end{aligned} \tag{3.94}$$

The group variable g may be represented as a ratio of u and v:

$$g = v^{-1}u. \tag{3.95}$$

In the notations (3.94,3.95) the form Θ looks as

$$\Theta(u,v,C) = \vartheta(u,C) + \vartheta(v,C^{-1}) + Tr\delta CC^{-1} \wedge (\delta uu^{-1} - \delta vv^{-1}). \tag{3.96}$$

The last term in (3.96) corresponds to the fact that the diagonal matrix C is dynamical in D_+. Speaking about the dressing orbits we have no analogue of this term because there C is constant.

In the next Section we shall be considering the symplectic form on the moduli space of flat connections on a Riemann surface with marked points. We shall find that the orbit symplectic structure ϑ may be naturally assigned to a marked point and the form Θ to a handle. Formula (3.96) demonstrates that in some sense one handle is equivalent to two marked points.

4 Symplectic structure of the moduli space

This section is devoted to symplectic geometry of the Chern-Simons theory. As we discussed in Introduction, this theory is defined by the canonical symplectic structure on the moduli space of flat connections on a Riemann surface. Surprisingly, this symplectic structure may be expressed in terms of Poisson-Lie symplectic forms introduced in the previous Section.

4.1 Chern - Simons model

The purpose of this subsection is to provide some physical motivations for study of the moduli space of flat connections starting from the Chern-Simons theory. We follow the approach of [6].

The Chern-Simons theory is a gauge theory in 3 dimensions (in principle the CS term exists in any odd dimension). It is defined by the action principle

$$CS(A) = -Tr \int_M (AdA + \frac{2}{3}A^3). \tag{4.1}$$

Here M is a 3-dimensional (3D) manifold, the gauge field A takes values in some simple Lie algebra $\mathcal{G}$

$$A = A_i^a t^a dx_i. \tag{4.2}$$

The generators t^a form a basis in $\mathcal{G}$ and satisfy the commutation relations

$$[t^a, t^b] = f_c^{ab} t^c. \tag{4.3}$$

We concentrate on the very special case of the CS theory. Suppose that the manifold M locally looks like a cylinder $\Sigma \times R$ (Cartesian product of a Riemann surface Σ and a segment of the real line). In this case we may interpret the theory in terms of Hamiltonian mechanics. We choose the direction parallel to the real line R to be the time direction. Two space–like components of the gauge field A become dynamical variables and we often denote by A the two component gauge field on the Riemann surface Σ. As usual, the time-component A_0 is a Lagrangian multiplier. After the change of variables the action (4.1) acquires the form

$$CS(A) = Tr \int (A\partial_0 A - 2A_0 F)dt, \qquad (4.4)$$

where the first term is a short action $\int pdq$ and the second term introduces a first class constraint

$$F = dA + A^2 = 0. \qquad (4.5)$$

The first term in (4.4) determines the Poisson brackets of dynamical variables. In particular, the Poisson bracket of the constraints (4.5) may be easily calculated:

$$\{F^a(z_1), F^b(z_2)\} = f_c^{ab} F^c(z_1)\delta^{(2)}(z_1 - z_2). \qquad (4.6)$$

As one expects, the constraints (4.5) generate gauge transformations

$$A^g = g^{-1}Ag + g^{-1}dg. \qquad (4.7)$$

Thus, the phase space in the Hamiltonian CS theory is a quotient of the space $\mathfrak{S}$ of flat connections (4.5) over the gauge group ΣG (4.7). We see that the moduli space (we shall often refer to the moduli space of flat connections as to the moduli space) appears to be a phase space of the CS theory on the cylinder. The action principle (4.4) provides canonical Poisson brackets on the moduli space. The efficient description of this Poisson bracket was given in [16].

We continue our brief survey of the CS theory by consideration of possible observables. The CS model enjoys two important symmetries: gauge symmetry and the symmetry with respect to diffeomorphisms. The reparametrization symmetry appears due to the geometric nature of the action (4.1) which is written in terms of differential forms and automatically invariant with respect to diffeomorphisms of the manifold M. It is natural to require that the observables in the CS model respect the invariant properties of the theory. Some observables of this type may be constructed starting from the following data. Let us choose the closed contour Γ in M and a representation I of the algebra $\mathcal{G}$. Apparently the following functional of the gauge field A

$$W_I(\Gamma) = Tr_I Pexp(\int_\Gamma A^I) \qquad (4.8)$$

is invariant with respect to both gauge and reparametrization symmetries. Usually the contour Γ and also the expression (4.8) are called a Wilson line and a Wilson line observable. The connection A^I is equal to

$$A^I = A^a T_I^a, \qquad (4.9)$$

where matrices T_I^a represent the algebra $\mathcal{G}$ in the representation I.

In the Hamiltonian formulation we may choose two special classes of Wilson lines— vertical and horizontal.

We call a Wilson line horizontal if it lies on an equal time surface. The observable corresponding to a horizontal Wilson line is a functional of two-dimensional gauge field and after quantization it becomes a physical operator. It is important to stress that Wilson lines do not cover the whole set of observables in the CS model.

The Wilson line is called vertical if the contour Γ is parallel to the time axis. In Hamiltonian picture we do not actually control the fact that vertical Wilson lines are closed. They come from the past through reality and disappear in the future. The vertical Wilson line is characterized by the representation I and the point z where it intersects the Riemann surface Σ. The choice of the time axis produces a big difference in the role of horizontal and vertical Wilson lines in the theory. Vertical Wilson lines *do not* correspond to observables in the Hamiltonian formulation. Instead they change the Hamiltonian system (4.4) so that both short action and the constraint get modified.

Using the formula (2.51), one may treat the CS correlator with n vertical Wilson lines inserted

$$Z_k(I_1, \ldots, I_n) = \int DA e^{\frac{ik}{4\pi} CS(A)} W_{I_1} \ldots W_{I_n} \qquad (4.10)$$

as an expression where the gauge field is still classical, whereas some modes corresponding to the matrices T_I are already quantized. The original functional integral would be

$$Z = \int DA Dg_1 \ldots Dg_n e^{iS^{tot}}. \qquad (4.11)$$

The action S^{tot} is defined by the formula

$$S^{tot} = \frac{k}{4\pi} CS(A) + \sum_{i=1}^{n} (S_{I_i}(v_i) + Tr \int dt A_0(z_i) T_i). \qquad (4.12)$$

Here the first term coincides with the standard Chern-Simons action, the second term consists of two parts. The first part collects auxiliary orbit actions for each Wilson line, the second part represents contributions of the Wilson lines into the CS partition function (4.10).

We have reformulated the Hamiltonian Chern-Simons model with vertical Wilson lines as a theory of the 2D gauge field A interacting with a set of finite dimensional systems with coordinates T_i localized at the points z_i. As in the case of the pure CS theory, the Hamiltonian (4.12) is equal to zero. The action of the modified system may be rewritten as

$$S^{tot} = Tr(\frac{k}{4\pi} \int A \partial_0 A + \sum_{i=1}^{n} D_i \partial_0 v_i v_i^{-1}) + Tr \int A_0 (\sum_{i=1}^{n} T_i \delta(z - z_i) - \frac{k}{2\pi} F).$$

$$(4.13)$$

The first term in (4.13) is a short action of the Hamiltonian system. It is responsible for the Poisson brackets of dynamical variables. The second term gives the modified constraint

$$\Phi(z) = \sum_{i=1}^{n} T_i \delta(z - z_i) - \frac{k}{2\pi} F = 0. \qquad (4.14)$$

Let us remark that after quantization the formula (4.14) is still true if we shift the central charge k in the standard way $k \rightarrow k + h$ (h is the dual Coxeter number of the algebra $\mathcal{G}$). Actually, the shift of the parameter k is of the same nature as a shift of the highest weight in formula (2.50).

The constraints (4.14) satisfy the same algebra (4.6) as in the pure CS theory. They generate gauge transformations for the gauge field A and conjugations for the variables T_i:

$$A^g = g^{-1} A g + g^{-1} dg, \quad T_i^g = g(z_i)^{-1} T_i g(z_i). \qquad (4.15)$$

The analogue of flatness condition (4.14) together with the modified gauge transformations (4.15) lead to the definition of the moduli space of flat connections on a Riemann surface with marked points (see the next subsection).

So the moduli space of flat connections emerges naturally as a phase space in the Chern-Simons theory. The rest of the paper is devoted to the analysis of the symplectic structure of this space.

4.2 Definition of the symplectic structure on the moduli space

Let Σ be a Riemann surface of genus g with n marked points. Consider a connection A on Σ taking values in a simple Lie algebra $\mathcal{g}$. The canonical symplectic structure [17] on the space $\mathcal{A}$ of all smooth connections may be read from the action (4.4)

$$\Omega_{\mathcal{A}} = \frac{k}{4\pi} Tr \int_{\Sigma} \delta A \wedge \delta A. \qquad (4.16)$$

The form (4.16) is obviously nondegenerate and invariant with respect to the action of the gauge group G_{Σ}:

$$A^g = g^{-1} A g + g^{-1} dg. \qquad (4.17)$$

We denote the exterior derivative on the Riemann surface by d, whereas the exterior derivative on the space of connections, moduli space or elsewhere is always δ. The action (4.17) is actually Hamiltonian and the corresponding momentum mapping is given (up to a multiplier) by the curvature:

$$\mu(A) = -\frac{k}{2\pi} F;$$
$$F = dA + A^2. \qquad (4.18)$$

Let us start with a case when there is no marked points.

Definition 2 *The space of flat connections $\Im_g$ on a Riemann surface of genus g is defined as a zero level surface of the momentum mapping (4.18):*

$$F(z) = 0. \tag{4.19}$$

Definition 3 *The moduli space of flat connections is a quotient of the space of flat connections $\Im_g$ over the gauge group action (4.17):*

$$\mathcal{M}_g = \Im_g / G_\Sigma. \tag{4.20}$$

The curvature being the momentum mapping for the gauge group, the moduli space may be obtained by Hamiltonian reduction from the space of smooth connections. General theory of Hamiltonian reduction [9],[18] ensures that the moduli space carries canonical nondegenerate symplectic structure induced from the symplectic structure (4.16) on $\mathcal{A}$.

Now we turn to more sofisticated case of the Riemann surface with marked points.

We have a coadjoint orbit assigned to each marked point z_i. As the gauge field A may develop a singularity in the vicinity of a marked point we have to choose a class of connections different from smooth connections on Σ. To this end we introduce a notion of decoration.

Definition 4 *A decorated Riemann surface with n marked points is a Riemann surface and a set of coadjoint orbits $\mathcal{O}_1, \ldots, \mathcal{O}_n$ assigned to the marked points $z_1, \ldots, z_n$.*

In order to explain this definition let us introduce the local coordinate ϕ_i in the small neighborhood of the marked point z_i so that

$$\oint_{S_i} d\phi_i = 2\pi. \tag{4.21}$$

Here S_i is a closed contour which surrounds the marked point. Apparently, the coordinate ϕ_i measures the angle in the neighborhood of z_i. On the surface with marked points we admit connections which have singularities of the form

$$A(z)_{z \sim z_i} = A_i d(\frac{\phi_i}{2\pi}) + \tilde{A}(z), \tag{4.22}$$

where A_i are constant coefficients and $\tilde{A}(z)$ is a smooth connection. We call the coefficients A_i singular parts of A.

Definition 5 *The space of connections $\mathcal{A}_{g,n}$ on a decorated Riemann surface with marked points is defined by the requirement that the singular parts of the connection belong to the coadjoint orbits assigned to the corresponding marked points:*

$$\frac{2\pi}{k} A_i \in \mathcal{O}_i. \tag{4.23}$$

It is remarkable that the symplectic structure (4.16) may be used for the space $\mathcal{A}_{g,n}$ as well. It is convenient to introduce one more symplectic space which is the direct product of $\mathcal{A}_{g,n}$ and its collection of coadjoint orbits:

$$\mathcal{A}_{g,n}^{tot} = \mathcal{A}_{g,n} \times \mathcal{O}_1 \times \ldots \times \mathcal{O}_n. \tag{4.24}$$

It carries the symplectic structure

$$\Omega_{\mathcal{A}}^{tot} = \Omega_{\mathcal{A}} + \sum_i^n \varpi_i, \tag{4.25}$$

The action of the gauge group may be defined on the space $\mathcal{A}_{g,n}^{tot}$ as follows:

$$A^g = g^{-1}Ag + g^{-1}dg :$$
$$T_i^g = g(z_i)^{-1}T_i g(z_i), \quad v_i^g = v_i g(z_i). \tag{4.26}$$

As we see, the modified gauge transformations are combined from the standard gauge transformations (4.17) and orbit conjugations (2.44). The momentum mapping is given by the coefficient before A_0 in the action (4.13):

$$\mu(z) = \sum_i^n T_i \delta(z - z_i) - \frac{k}{2\pi} F(z). \tag{4.27}$$

It is easy to see that the definition of $\mathcal{A}_{g,n}$ ensures that there is a lot of solutions of the zero level conditions.

Definition 6 *The space of flat connections on a decorated Riemann surface $\mathfrak{S}_{g,n}$ is defined as a space of solutions of the following equation which replaces the zero curvature condition:*

$$\mu(z) = 0. \tag{4.28}$$

Let us choose a loop S_i surrounding the marked point z_i. One can define the monodromy matrix (or parallel transport) M_i along this way. It is easy to check that if A and $\{T_i\}$ satisfy (4.28), the monodromy matrix M_i belongs to the conjugancy class of the exponent of D_i

$$M_i = u_i^{-1} exp(\frac{2\pi}{k} D_i) u_i. \tag{4.29}$$

Definition 7 *The moduli space of flat connections on a Riemann surface of genus g with n marked points $\mathcal{M}_{g,n}$ is defined as a quotient of the space of flat connection on a decorated Riemann surface over the gauge group action (4.26):*

$$\mathcal{M}_{g,n} = \mathfrak{S}_{g,n}/G_\Sigma. \tag{4.30}$$

It is important that the moduli space $\mathcal{M}_{g,n}$ is obtained by Hamiltonian reduction from the symplectic space $\mathcal{A}_{g,n}^{tot}$. This procedure provides the nondegenerate symplectic form on $\mathcal{M}_{g,n}$ which is the main object of this paper.

4.3 Combinatorial description of the symplectic structure on the moduli space

As it was explained in subsection 3.3, the symplectic structure on the reduced phase space obtained by Hamiltonian reduction from some symplectic space is easy to describe. More exactly, we get the reduced phase space as a projection of some constant momentum surface to the quotient of the global phase space over the group action. The pull-back of the reduced symplectic form to the constant momentum surface is equal to the restriction of the global symplectic form to the same subspace.

The moduli space of flat connections plays the role of the reduced phase space the global space being the space of smooth connections with the symplectic form (4.16). So our main concern is to restrict the form (4.16) to the space of flat connections efficiently. To this end it is convenient to introduce some intermediate finite dimensional space between the moduli space and the space of flat connections which admits an efficient parametrization.

Let us choose a point P on the Riemann surface which does not coincide with marked points z_i. One can define a subgroup of the gauge group $G_\Sigma(P)$ by the requirement:

$$G_\Sigma(P) = \{g \in G_\Sigma, \quad g(P) = I\}. \tag{4.31}$$

The quotient space

$$\mathcal{M}_{g,n}(P) = \Im_{g,n}/G_\Sigma(P) \tag{4.32}$$

is already finite dimensional and admits efficient parametrization.

Let us draw a bunch of circles on the Riemann surface so that there is only one intersection point P. In this bunch we have two circles for each handle (corresponding to a- and b- cycles) and one circle for each marked point. We shall denote the circles corresponding to the i's handle by a_i and b_i $(i = 1, \ldots, g)$ and we shall use symbols m_i $(i = 1, \ldots, n)$ for the circles surrounding marked points. We assume that the circles on Σ are chosen in such a way that the only defining relation in $\pi_1(\Sigma_{g,n})$ looks as

$$m_1 \ldots m_n (a_1 b_1^{-1} a_1^{-1} b_1) \ldots (a_g b_g^{-1} a_g^{-1} b_g) = id. \tag{4.33}$$

To each circle we assign the corresponding monodromy matrix defined by the flat connection A. Let us denote these matrices by A_i, B_i and M_i for a-, b- and m-circles. The set of monodromy matrices provides coordinates on $\mathcal{M}_{g,n}$ and a representation of the fundamental group $\pi_1(\Sigma_{g,n})$. It implies the relation

$$M_1 \ldots M_n (A_1 B_1^{-1} A_1^{-1} B_1) \ldots (A_g B_g^{-1} A_g^{-1} B_g) = I \tag{4.34}$$

imposed on the values of A_i, B_i and M_i. Actually, monodromies M_i are not arbitrary. They belong to conjugacy classes $C_i(G)$ defined by

$$M_i = u_i^{-1} C_i u_i, \tag{4.35}$$

where

$$C_i = \exp\left(\frac{2\pi}{k}D_i\right). \tag{4.36}$$

So the space $\mathcal{M}_{g,n}(P)$ is a subspace in

$$\mathcal{F}_{g,n} = G^{2g} \times \prod_{i=1}^{n} C_i(G) \tag{4.37}$$

defined by the relation (4.34).

The original moduli space may be represented as a quotient of $\mathcal{M}_{g,n}$ over the residual gauge group which is isomorphic to the group G:

$$\mathcal{M}_{g,n} = \mathcal{M}_{g,n}(P)/G. \tag{4.38}$$

It is convenient to define some additional coordinates K_i on $\mathcal{F}_{g,n}$:

$$
\begin{aligned}
K_0 &= I, \\
K_i &= M_1 \ldots M_i, 1 \le i \le n \\
K_{n+2i-1} &= K_{n+2i-2}A_i, \\
K_{n+2i} &= K_{n+2i-1}B_i^{-1}A_i^{-1}B_i.
\end{aligned}
\tag{4.39}
$$

It follows from the equation (4.34) that

$$K_{n+2g} = K_0 = I. \tag{4.40}$$

Unfortunately, coordinates A, B, M and K are not sufficient for analysis of the symplectic form on the moduli space and we have to introduce a new space $\tilde{\mathcal{F}}$:

$$\tilde{\mathcal{F}} = G^{n+2g} \times H^{n+g}. \tag{4.41}$$

Here H is a Cartan subgroup of G. $\tilde{\mathcal{F}}$ may be parametrized by matrices $u_i, i = 1, \ldots, n + 2g$ from the group G and by Cartan elements $C_i, i = 1, \ldots, n + g$. We define a projection from $\tilde{\mathcal{F}}$ to $\mathcal{F}$ by the formulae:

$$
\begin{aligned}
M_i &= u_i^{-1}C_i u_i, \\
A_i &= u_{n+2i-1}^{-1}C_{n+i}u_{n+2i-1}, \\
B_i &= u_{n+2i}u_{n+2i-1}^{-1}.
\end{aligned}
\tag{4.42}
$$

Let us call $\tilde{\mathcal{M}}_{g,n}(P)$ the preimage of $\mathcal{M}_{g,n}(P)$ in $\tilde{\mathcal{F}}$.

After this lengthy preparations we are ready to formulate the main result of this subsection.

Theorem 3 *The pull-back of the canonical symplectic form on $\mathcal{M}_{g,n}$ to $\tilde{\mathcal{M}}_{g,n}(P)$ coincides with the restriction of the following two-form defined on $\tilde{\mathcal{F}}$:*

$$\Omega_{\mathcal{F}} = \frac{k}{4\pi} Tr \left[\sum_{i=1}^{n+2g} \delta u_i u_i^{-1} C_i \wedge \delta u_i u_i^{-1} C_i^{-1} - \sum_{i=1}^{n+2g} \delta K_i K_i^{-1} \wedge \delta K_{i-1} K_{i-1}^{-1} + \right.$$

$$\left. + \sum_{i=1}^{g} \delta C_{n+i} C_{n+i}^{-1} \wedge (\delta u_{n+2i} u_{n+2i}^{-1} - \delta u_{n+2i-1} u_{n+2i-1}^{-1}) \right] \quad (4.43)$$

The rest of the subsection is devoted to proof of *Theorem 3*.

Proof.

Let us cut the surface along every circle a_i, b_i, m_i. We get $n+1$ disconnected parts. The first n are similar. Each of them is a neighborhood of the marked point with the cycle m_i as a boundary. We denote these disjoint parts by P_i. The last one is a polygon. There is no marked points inside and the boundary is composed of a-,b-, and m-cycles as it is prescribed by formula (4.33). We denote the polygon by P_0.

Being restricted to P_0 a flat connection A becomes trivial:

$$A\mid_{P_0} = g_0^{-1} dg_0. \quad (4.44)$$

For any other part P_i we get a bit more complicated expression:

$$A\mid_{P_i} = \frac{1}{k} g_i^{-1} D_i g_i d\phi_i + g_i^{-1} dg_i. \quad (4.45)$$

We remind that D_i is a diagonal matrix which characterizes the orbit attached to the marked point z_i. There is a set of consistency conditions which tells that the connection described by formulae (4.44,4.45) is actually smooth on the Riemann surface everywhere except the marked points. It means that when one approaches the cuts from two sides, one always gets the same value of A. To be explicit, let us consider the m-cycle which surrounds the marked point z_i. Comparison of equations (4.44,4.45) gives:

$$g_0^{-1} dg_0 \mid_{m_i} = (\frac{1}{k} g_i^{-1} D_i g_i d\phi_i + g_i^{-1} dg_i) \mid_{m_i}. \quad (4.46)$$

This equation may be easily solved:

$$g_0 \mid_{m_i} = N M g_i \mid_{m_i}, \quad (4.47)$$

where N is an arbitrary constant matrix and M is equal to

$$M(\phi_i) = \exp\left(\frac{1}{k} D_i \phi_i\right). \quad (4.48)$$

Now we turn to consistency conditions which arise when one considers a- or b-cycles. In this case both sides of the cut belong to the polygon P_0. Let us denote the restrictions of g_0 on the cut sides by g' and g''. So we have:

$$g'^{-1}dg' = g''^{-1}dg''. \tag{4.49}$$

We conclude that the matrices g' and g'' may differ only by a constant left multiplier:

$$g'' = Ng'. \tag{4.50}$$

By now we considered connection A in the region of the surface where it is flat. However, it is not true at the marked points. We calculate the curvature in the region P_i and get a δ-function singularity:

$$F(z)\,|_{P_i} = \frac{2\pi}{k}g_i^{-1}D_i g_i \delta(z - z_i). \tag{4.51}$$

Equations (4.51,4.27,4.28) imply that the value $g_i(z_i)$ coincides with the matrix v_i:

$$g_i(z_i) = v_i. \tag{4.52}$$

Let us remind that v_i diagonalizes the matrix T_i attached to the marked point z_i by definition of the decorated Riemann surface.

Now we are prepared to consider the symplectic structure on the space of flat connections. First, let us rewrite the definition (4.25) in the following way:

$$\Omega^{tot} = \omega_0 + \sum_{i=1}^{n} \omega_i, \tag{4.53}$$

where the summands correspond to different parts of the Riemann surface:

$$\omega_0 = \frac{k}{4\pi}Tr \int_{P_0} \delta A \wedge \delta A,$$

$$\omega_i = \frac{k}{4\pi}Tr \int_{P_i} \delta A \wedge \delta A + \varpi_i. \tag{4.54}$$

The next step must be to substitute (4.44,4.45) into formulae (4.54). The following lemma provides an appropriate technical tool for this operation.

Lemma 3 *Let A be a g-valued connection defined in the region P of the Riemann surface Σ. Suppose that*

$$A = g^{-1}Bg + g^{-1}dg. \tag{4.55}$$

Then the canonical symplectic form

$$\omega_P = Tr \int_P \delta A \wedge \delta A \tag{4.56}$$

may be rewritten as

$$\omega_P = Tr \int_P \{\delta B \wedge \delta B + 2\delta[F_B \delta g g^{-1}]\} + Tr \int_{\partial P} \{\delta g g^{-1} d(\delta g g^{-1}) - \delta[B \delta g g^{-1}]\}, \tag{4.57}$$

where F_B is a curvature of the connection B

$$F_B = dB - B^2. \tag{4.58}$$

One can prove *Lemma 3* by straightforward calculation.

Let us apply *Lemma 3* to the polygon P_0. In this case $B = 0$ and the answer reduces to

$$\omega_0 = \frac{k}{4\pi} Tr \int_{\partial P_0} \delta g_0 g_0^{-1} d(\delta g_0 g_0^{-1}). \tag{4.59}$$

The boundary of the polygon ∂P_0 consists of $n + 4g$ cycles (4.33). So actually we have $n + 4g$ contour integrals in the r.h.s. of (4.59).

Now we use formula (4.57) to rewrite symplectic structures ω_i:

$$\omega_i = \frac{k}{4\pi} Tr \int_{\partial P_I} \{\delta g_i g_i^{-1} d(\delta g_i g_i^{-1}) - \frac{2\pi}{k} \delta[D_i \delta g_i g_i^{-1}]\} -$$
$$-Tr \int_{P_I} \delta\{D_i \delta g_i g_i^{-1}\} \delta(z - z_i) + Tr D_i (\delta v_i v_i^{-1})^2. \tag{4.60}$$

The last term in (4.60) represents Kirillov form attached to the marked point z_i. Taking into account relation (4.52) we discover that this term together with the third term in (4.60) cancel each other.

At this point it is convenient to denote the values of g_0 at the corners of the polygon. We enumerate the corners by the index $i = 0, \dots, n + 4g - 1$ so that the end-points of the cycle m_i are labeled by $i - 1$ and i. One can easily read from formula (4.33) the enumeration of the ends of a- and b-cycles (see Fig. 1). For example, the end-points of a_i are labeled by $n + 4(i - 1)$ and $n + 4(i - 1) + 1$, whereas the end-points of a_i^{-1} entering in the same word are labeled by $n + 4(i - 1) + 2$ and $n + 4(i - 1) + 3$. We denote the value of g_0 at the i's corner by h_i.

Monodromies A_i, B_i and M_i may be expressed in terms of h_i as

$$M_i = h_{i-1}^{-1} h_i, \tag{4.61}$$

$$A_i = h_{n+4(i-1)}^{-1} h_{n+4(i-1)+1} = h_{n+4(i-1)+3}^{-1} h_{4(i-1)+2}, \tag{4.62}$$

$$B_i = h_{n+4(i-1)+1}^{-1} h_{n+4(i-1)+2} = h_{n+4i}^{-1} h_{4(i-1)+3}. \tag{4.63}$$

Let us remark that without loss of generality we can choose g_0 in such a way that its value h_0 is equal to unit element in G. After that some of the corner values h_i may be identified with K_i;

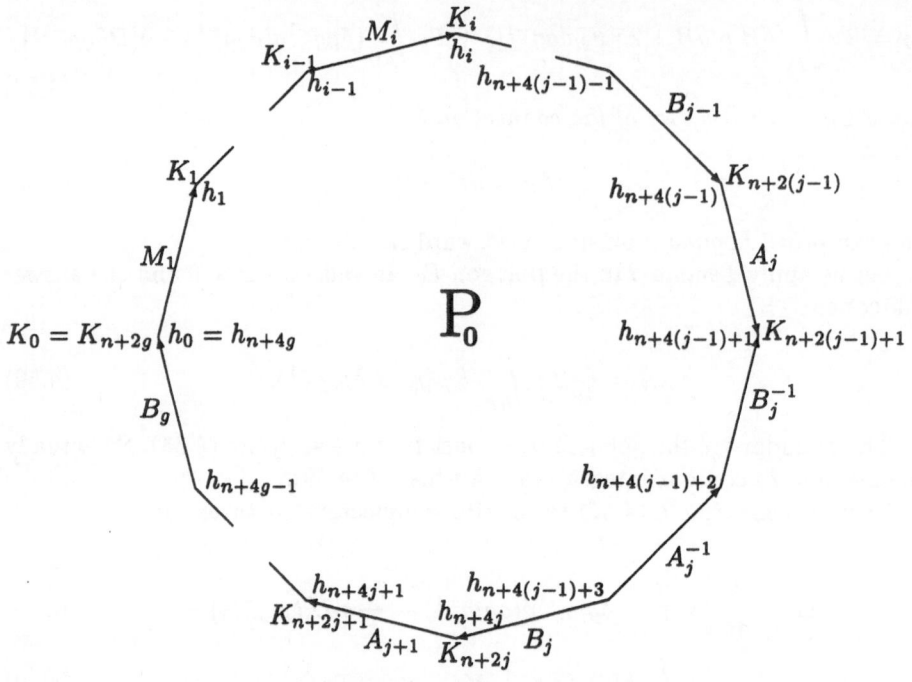

Figure 1

$$K_i = \begin{cases} h_i & \text{for } 1 \le i \le n \\ h_{2i-n-1} & \text{for } (i-n) \text{ odd} \\ h_{2i-n} & \text{for } (i-n) \text{ even} \end{cases} \qquad (4.64)$$

Our strategy is to adjust notations to the description of Poisson-Lie symplectic forms (see subsection 3.2). Using formula (4.47) one can diagonalize M_i

$$M_i = u_i^{-1} C_i u_i. \qquad (4.65)$$

Here u_i is the value of the variable g_i at the point P.

Let us rewrite formula (4.59) in the following way:

$$\omega_0 = \sum_{i=1}^{n} \varphi_i + \sum_{i=1}^{g} \psi_i. \qquad (4.66)$$

Here φ_i is a contribution corresponding to the marked point:

$$\varphi_i = \frac{k}{4\pi} Tr \int_{m_i} \delta g_0 g_0^{-1} d(\delta g_0 g_0^{-1}), \qquad (4.67)$$

and ψ_i is a contribution of the handle:

$$\psi_i = \frac{k}{4\pi} Tr \int_{a_i b_i^{-1} a_i^{-1} b_i} \delta g_0 g_0^{-1} d(\delta g_0 g_0^{-1}). \tag{4.68}$$

First, we are going to evaluate the total contribution of the given M-cycle which is equal to a sum of two terms:

$$\Omega_i = \omega_i + \varphi_i. \tag{4.69}$$

Actually, each summand in (4.69) includes an integral over the m-cycle. However, this sum of integrals is an integral of exact form and it depends only on some finite number of boundary values. This situation is typical and will repeat when we consider a contribution of a handle.

Lemma 4 *The form ω_i depends only on finite number of parameters and may be written as*

$$\omega_i = \frac{k}{4\pi} Tr[C_i \delta u_i u_i^{-1} \wedge C_i^{-1} \delta u_i u_i^{-1} - \delta K_i K_i^{-1} \wedge \delta K_{i-1} K_{i-1}^{-1}]. \tag{4.70}$$

To prove *Lemma 4* one should substitute formula (4.47) into expression for φ_i, integrate by parts and compare the result with the expression for ω_i. The integrals in φ_i and ω_i cancel each other and after rearrangements the boundary terms reproduce formula (4.70).

Now we turn to the contribution of a handle ψ_i into the symplectic form on the moduli space. One can see that each a-cycle and each b-cycle enter twice into expression (4.66). These two contributions correspond to two sides of the cut. As usual, the result simplifies if we combine the contributions of two cut sides together.

Lemma 5 *Let g', g'' be two mappings from the segment $[x_1, x_2]$ into the group G with boundary values $g'_{1,2}, g''_{1,2}$. Suppose that these mappings differ by the x-independent left multiplier*

$$g'' = Ng'. \tag{4.71}$$

Then the following equality holds:

$$\Omega_{[x_1,x_2]} = Tr \int_{x_1}^{x_2} \delta g'' g''^{-1} d(\delta g'' g''^{-1}) - Tr \int_{x_1}^{x_2} \delta g' g'^{-1} d(\delta g' g'^{-1}) =$$
$$= Tr(g_1'^{-1} \delta g_1' \wedge g_1''^{-1} \delta g_1'' - g_2'^{-1} \delta g_2' \wedge g_2''^{-1} \delta g_2''). \tag{4.72}$$

Proof is straightforward.

Let us parametrize A_i and B_i as in (4.43):

$$A_i = u_{n+2i-1}^{-1} C_{n+i} u_{n+2i-1}, \quad u_{n+2i} = B_i u_{n+2i-1}. \tag{4.73}$$

One of the motivations for such notations is the following identity:

$$B_i^{-1} A_i^{-1} B_i = u_{n+2i}^{-1} C_{n+i}^{-1} u_{n+2i}. \tag{4.74}$$

In principle, one can introduce the following uniformal variables

$$
\begin{aligned}
M_{n+2i-1} &= A_i = u_{n+2i-1}^{-1} C_{n+i} u_{n+2i-1}, \\
M_{n+2i} &= B_i^{-1} A_i^{-1} B_i = u_{n+2i}^{-1} C_{n+i}^{-1} u_{n+2i}.
\end{aligned}
\tag{4.75}
$$

so that the defining relation (4.34) looks as

$$M_1 \ldots M_n M_{n+1} \ldots M_{n+2g} = I. \tag{4.76}$$

In these variables we treat handles and marked points in the same way. Roughly speaking, one handle produces two marked points which have the inverse values of C: $C_1 = C_{n+i}$, $C_2 = C_{n+i}^{-1}$. It resembles the relation between the double D_+ and two orbits of dressing transformations (see subsection 3.4). Using the definition of M (4.75) we can clarify the definition of K_i:

$$K_i = M_1 \ldots M_i. \tag{4.77}$$

Now we turn to the contribution ψ_i of a handle into symplectic form (4.66).

Lemma 6 *The handle contribution into symplectic form depends only on the values of g_0 at the end-points of the corresponding a- and b-cycles and may be written as*

$$
\begin{aligned}
\psi_i \;=\; & \frac{k}{4\pi} Tr[C_{n+i} \delta u_{n+2i-1} u_{n+2i-1}^{-1} \wedge C_{n+i}^{-1} \delta u_{n+2i-1} u_{n+2i-1}^{-1} - \\
& - \; \delta K_{n+2i-1} K_{n+2i-1}^{-1} \wedge \delta K_{n+2(i-1)} K_{n+2(i-1)}^{-1} + \\
& + \; C_{n+i}^{-1} \delta u_{n+2i} u_{n+2i}^{-1} \wedge C_{n+i} \delta u_{n+2i} u_{n+2i}^{-1} - \\
& - \; \delta K_{n+2i} K_{n+2i}^{-1} \wedge \delta K_{n+2i-1} K_{n+2i-1}^{-1} + \\
& + \; \delta C_{n+i} C_{n+i}^{-1} \wedge (\delta u_{n+2i-1} u_{n+2i-1}^{-1} - \delta u_{n+2i} u_{n+2i}^{-1})].
\end{aligned}
\tag{4.78}
$$

If we take into account *Lemma 5*, the proof of *Lemma 6* becomes straightforward but long calculation. Let us remark that the terrible formula (4.78) contains two copies of the marked point contribution (4.70) with parameters C_{n+i} and C_{n+i}^{-1}. The last term includes $\delta C_{n+i} C_{n+i}^{-1}$ and coincides with the corresponding additional term in formula (3.96) for the symplectic form on the double D_+.

Summarizing *Lemma 4* and *Lemma 6* we get the proof of *Theorem 3* completed.

4.4 Equivalence to Poisson-Lie symplectic structure

Formula (4.43) contains cross-terms with different indices i. In this subsection we represent the canonical symplectic structure as a direct sum of several terms.

Using the results of Section 3, each term may be identified with either Kirillov form for the Poisson-Lie group G^* or symplectic form on the Heisenberg double D_+ of the Poisson-Lie group G. To achieve this result we have to make a change of variables. The new set of variables is designed to "decouple" contributions of different handles and marked points.

The following remark is important for understanding of the construction of decoupled variables. Monodromy matrices M_i, A_i and B_i are elements of the group G. In accordance with this fact we use G-multiplication to define the variables K_i (4.77) and to constraint monodromies (4.34). On the other hand, natural variables for description of orbits of dressing transformations or double D_+ must belong to G^*. In subsection 3.5 we defined the mapping $\alpha : G^* \to G$. Unfortunately, α is not a group homomorphism. So, we would face difficulties applying α to identities (4.77,4.34). This is a motivation to introduce a notion of a weak group homomorphism.

Definition 8 *Let G and G' be two groups. A set of mappings*

$$\alpha^{(n)} : G^n \to G'^n \qquad (4.79)$$

is called a weak homomorphism if the following diagram is commutative for any i:

$$
\begin{array}{ccc}
G^n & \xrightarrow{\alpha^{(n)}} & G'^n \\
\mathbf{m}_i \downarrow & & \mathbf{m}'_i \downarrow \\
G^{n-1} & \xrightarrow{\alpha^{(n-1)}} & G'^{n-1} \, .
\end{array}
\qquad (4.80)
$$

Here $\mathbf{m}_i$ and $\mathbf{m}'_i$ are multiplication mappings in G and G' correspondingly which map the product of n copies of the group into the product of $n - 1$ copies:

$$
\begin{aligned}
\mathbf{m}_i &: (g_1, \ldots, g_i, g_{i+1}, \ldots, g_n) \to (g_1, \ldots, g_i g_{i+1}, \ldots, g_n) : \\
\mathbf{m}'_i &: (g'_1, \ldots, g'_i, g'_{i+1}, \ldots, g'_n) \to (g'_1, \ldots, g'_i g'_{i+1}, \ldots, g'_n) :
\end{aligned}
\qquad (4.81)
$$

The mapping α (3.89) may be considered as a first mapping of a weak homomorphism from G^* to G. To define the other mappings $\alpha^{(n)}$ we introduce the products

$$K_\pm(i) = L_\pm(1) \ldots L_\pm(i). \qquad (4.82)$$

The action of $\alpha^{(n)}$ looks as follows. A tuple $(L_+(i), L_-(i)) \in G^*, i = 1, \ldots n$ is mapped into the tuple $M_i \in G, i = 1, \ldots n$:

$$M_i = K_-(i-1)L_i K_-(i-1)^{-1}. \qquad (4.83)$$

Here L_i is the image of the pair $(L_+(i), L_-(i))$ under the action of α:

$$L_i = L_+(i)L_-(i)^{-1}. \tag{4.84}$$

One can easily check that the set of mappings (4.83) satisfies the requirements of a weak homomorphism.

The next step is to implement the definition (4.83) to the space $\tilde{\mathcal{F}}$. Let us introduce a set of variables on $\tilde{\mathcal{F}}$ which consists of $v_i, i = 1, \ldots, n + 2g$ taking values in G and $C_i', i = 1, \ldots, n + g$ taking values in H. In addition we introduce the elements of G^*:

$$
\begin{aligned}
L_i &= v_i^{-1} C_i' v_i & for 1 \leq i \leq n; \\
L_{n+2i-1} &= v_{n+2i-1} C_{n+i}' v_{n+2i-1} & for 1 \leq i \leq g; \\
L_{n+2i} &= v_{n+2i} C_{n+i}'^{-1} v_{n+2i} & for 1 \leq i \leq g.
\end{aligned}
\tag{4.85}
$$

together with their Gauss components (3.89). So, we have natural variables to describe n copies of the orbit of dressing transformations in G^* and g copies of the Heisenberg double. The canonical symplectic form on this object is equal to the sum of symplectic forms for each copy of the orbit (3.92) and each copy of double (3.96):

$$\Omega_{PL} = \sum_{i=1}^{n} \vartheta(u_i, C_i') + \sum_{i=1}^{g} \Theta(u_{n+2i-1}, u_{n+2i}, C_{n+i}'). \tag{4.86}$$

Let us compare the forms (4.43) and (4.86). Motivated by the definition (4.83) we introduce the mapping $\sigma : \tilde{\mathcal{F}} \to \tilde{\mathcal{F}}$ defined by the relations:

$$u_i = v_i K_-^{-1}(i-1), \quad C_i = C_i'. \tag{4.87}$$

Here $K_-(i)$ are defined as in (4.82). It is easy to see that the mapping σ induces the mapping $\alpha^{(n+2g)}$ from the set of pairs $(L_+(i), L_+(i))$ into the set of monodromies M_i. It is guaranteed by the definition of weak homomorphism that G-product in the relation (4.34) is now replaced by G^*-product:

$$K_\pm(n + 2g) = L_\pm(1) \ldots L_\pm(n + 2g) = I. \tag{4.88}$$

Equation (4.88) defines the preimage of $\tilde{\mathcal{M}}_{g,n}$ in $\tilde{\mathcal{F}}$ with respect to the mapping σ. It is worth mentioning that the matrices K_i from the previous subsection may be represented as

$$K_i = K_+(i)K_-(i)^{-1}. \tag{4.89}$$

This also a consequence of the definition of weak homomorphism. Indeed, K_i has been defined as a product in G of the first i monodromies. Formula (4.82) defines a product in G^* of i first elements $(L_+(i), L_+(i))$. Using the basic property of weak homomorphism $(i-1)$ times we check (4.89).

The mapping σ provides a possibility to compare two-forms $\Omega_{\mathcal{F}}$ and Ω_{PL}.

Lemma 7 *The two-forms $\Omega_{\mathcal{F}}$ is proportional to the pull-back of the form Ω_{PL} along the mapping σ:*

$$\Omega_F = \frac{k}{4\pi}\sigma^*(\Omega_{PL}). \tag{4.90}$$

Lemma 7 may be proved by straightforward calculation. *Theorem 3* and *Lemma 7* imply the following theorem which is the main result of this paper.

Theorem 4 *Being restricted to the subset (4.88), the direct sum of n copies of Kirillov symplectic form on the orbit of dressing transformations in G^* and g copies of the canonical form on the Heisenberg double of the group G coincides up to a scalar multiplier with the pull-back of the canonical symplectic form on the moduli space of flat connections on the Riemann surface of genus g with n marked points.*

5 Conclusions

We have started in Section 4 from the correlator of the Chern-Simons theory on the cylinder with inserted vertical Wilson lines. This system may be represented as a 3D gauge field interacting to a finite number degrees of freedom living on the Wilson lines. As there is no Hamiltonian, the system is completely defined by the symplectic form on the phase space. We proved that this symplectic form may be decomposed into the direct sum of Poisson-Lie symplectic structures subject to constraint (4.34). So, the functional integral for the correlator (4.10) may be rewritten as:

$$Z_k(I_1,\dots,I_n) = \int \prod_{i=1}^{n+2g} \mathcal{D}u_i \prod_{i=n+1}^{n+g} \mathcal{D}C_i \times$$

$$\times e^{\frac{k+h}{2\pi}\int \delta^{-1}\{\sum_{i=1}^{n}\vartheta(u_i,C_i)+\sum_{i=1}^{g}\Theta(u_{n+2i-1},u_{n+2i},C_{n+i})\}}\delta(M_1\dots B_g). \tag{5.91}$$

Here we took into account the standard shift $k \to k + h$ in the Chern-Simons action and used the symbol δ^{-1} as in Wess-Zumino action.. If we compare expression (5.91) with original formula (4.10), we find that the gauge field disappeared and the Wilson line insertions got modified. One can say that the Chern-Simons theory in the bulk quantizes the group variables living on the Wilson lines. In addition to the modified Wilson lines one finds in the partition function (5.91) the finite number of degrees of freedom which carry topological information about genus of the Riemann surface.

If we turn to operator approach, each Wilson line multiplier in (5.91) presents a deformed analogue of the orbit quantum mechanics considered in subsection 2.2. It is natural to expect that quantization leads to the Hilbert space which coincides with the space of certain irreducible representation of the quantum group with the highest weight w given by the formula (2.50). Each multiplier corresponding to a handle gives a regular representation $\Re$ of the same quantum group. The parts of the system corresponding to the different summands in the action (5.91) are related only by the constraint (4.34). It prescribes that the Hilbert space of the whole system is equal to the space of invariants in the tensor product.

$$\mathcal{H} = Inv_q(I_1 \otimes \ldots I_n \otimes \mathfrak{R}^{\otimes g}). \tag{5.92}$$

in agreement with the known results.

More detailed information about operator formalism and the corresponding representation theory may be found in [19], [20]. It would be interesting to work out the functional integral (5.91) by direct calculation and generalize this procedure for 3D topologies different from the cylinder.

Acknowledgements

We are grateful to the organizers of the 3d Baltic Rim Student Seminar for a possibility to present our results during the conference. Special thanks to A.J.Niemi for perfect conditions in Uppsala where this work was completed.

References

[1] A.Yu.Alekseev, A.Z.Malkin, *Symplectic structures associated to Lie-Poisson groups*, preprint PAR-LPTHE 93-08, UUITP 5/1993, HEP-TH 9303038, to be published in Comm. Math. Phys.

[2] A.Yu.Alekseev, A.Z.Malkin, *Symplectic structure of the moduli space of flat connections on a Riemann surface*, Hep-th 9312004, submitted to Comm. Math. Phys.

[3] E.Witten, *Quantum Field Theory and the Jones Polynomial*, Comm. Math. Phys. **121** (1989) p.351.

[4] A.Gerasimov, *Localization in GWZW and Verlinde formula*, Hep-th 9305090.

[5] M.Blau, G.Thompson, *Derivation of the Verlinde Formula From Chern-Simons Theory And The G/G Model*, Nucl. Phys. **B** 408, 1993, p.345.

[6] S.Elitzur, G.Moore, A.Schwimmer, N.Seiberg, *Remarks on the canonical quantization of the Chern-Simons-Witten theory*, Nucl. Phys. **B 326** (1989) p.108.

[7] A.Yu.Alekseev, L.D.Faddeev, S.L.Shatashvili, *Quantization of symplectic orbits of compact Lie groups by means of the functional integral*, Journ. Geom. Phys., vol.5, no.3, 1989, pp.391-406.

[8] K.Gawedzki, F.Falceto, *On quantum group symmetries of conformal field theories*, Preprint IHES/P/91/59, September 1991.

[9] V.I.Arnold, *Mathematical methods of classical mechanics*. Springer-Verlag 1980.

[10] A.A.Kirillov, *Elements of the theory of representations*, Springer-Verlag 1976.

[11] H.B.Nielsen, D.Rohrlich, *A path integral to quantize spin*, Nucl. Phys. **B** 299, 1988, p. 471.

[12] A.J.Niemi, P.Pasanen, *Orbit geometry, group representations and topological quantum field theories*, Phys. Lett. **B** 253, 1991, pp. 349-356.

[13] V.G.Drinfeld, *Quantum groups*, in Proc. ICM, MSRI, Berkeley, 1986, p.798.

[14] M.A.Semenov-Tian-Shansky, *Dressing transformations and Poisson-Lie group actions*, In: Publ. RIMS, Kyoto University 21, no.6, 1985, p.1237.

[15] J.H.Lu, A.Weinstein,

Poisson-Lie groups, dressing transformations and Bruhat decompositions, J.Diff.Geom., v.31, 1990, p.501.

[16] V.V.Fock, A.A.Rosly *Poisson structure on the moduli space of flat connections on Riemann surfaces and r-matrix*, preprint ITEP 72-92, June 1992, Moscow.

[17] M.Atiyah, R.Bott, *The Yang-Mills equations over a Riemann surface*, Phil. Trans. R. Soc A **308** (1982), p.523.

[18] A.Weinstein, *The local structure of Poisson manifolds*, J. Diff. Geom., v.18, n.3, pp.523-557. (1983)

[19] A.Yu.Alekseev, *Integrability in the Hamiltonian Chern-Simons theory*, Hep-th 9311074.

[20] A.Yu.Alekseev, H.Grosse, V.Schomerus, *Combinatorial quantization of the Hamiltonian Chern-Simons theory*, to appear.

Quantization of Field Theories Generalizing Gravity-Yang-Mills Systems on the Cylinder

P. Schaller[*] and T. Strobl[†]

Inst. f. Theor. Physik, Techn. Univ. Vienna,
Wiedner Hauptstr. 8-10, A-1040 Wien, Austria

1   Introduction

Non abelian gauge theories as well as several models of gravity on two dimensional space time manifolds have become an active field of research in recent years [1],[2],[3],[4],[5],[6]. Both types of theories are closely related to each other. This is best illustrated by the fact that the Lagrangian of a gravity theory with vanishing torsion and constant curvature (Jackiw Teitelboim model [7]) can be rewritten as the one of a nonabelian gauge theory with vanishing field strength (BF-theory) [8].

Indeed, as shown in section 2 of the present article pure gravity and gauge theories in two dimensions may be seen as special cases of a more general class of models each of which is characterized by an antisymmetric tensorfield on a finite dimensional target space L (more precisely by a Poisson structure on L). A general scheme for the quantization of these models in a Hamiltonian formulation (restricting the topology of the space time manifold to the one of a cylinder) is presented. The heart of this scheme is the reinterpretation of the constraints as horizontality conditions on $U(1)$-bundles over loop spaces. In the special case of non abelian gauge theories the manifolds underlying the loop spaces are the coadjoint orbits (the orbits generated by the gauge group on the dual of its Lie algebra) equipped with the standard symplectic structure [9],[10]. In any case the result is a finite dimensional quantum mechanical system generically including discrete degrees of freedom of topological origin.

The considerations in section 2 are rather formal and abstract. Explicit examples are given in section 3, including several theories of quantum gravity.

---
[*]email: schaller@email.tuwien.ac.at,
[†]tstrobl@email.tuwien.ac.at

The reader may find it helpful to have a look at these examples while reading section 2.

The constraints generate those symmetry transformations only which are connected to the unity. The effect of large symmetry transformations is studied in section 4 at the example of gauge theories based on the $so(2,1)$ Lie algebra. We argue that the quantum theory obtained by the implementation of the constraints corresponds to an $\widetilde{SL}(2,\mathbf{R})$ gauge theory ($\tilde{}$ denoting the universal covering). For an $SO(2,1)$ gauge theory the implementation of big gauge transformations yields a one parameter family of unitarily inequivalent quantum theories. The results obtained are checked for both theories by investigating the topological structure of the reduced phase spaces. The latter are compared to the reduced phase space (RPS) of the Jackiw Teitelboim model on a cylinder characterized by the same action. Inequivalences found are traced to the fact that the action of the constraints generates diffeomorphisms only for space time manifolds with nondegenerate metric and thus connect gravitationally inequivalent solutions of the equations of motion.

In section 5 we investigate R^2-gravity coupled to an $SU(2)$-Yang Mills theory. The Hilbert space and the operators corresponding to a set of independent Dirac observables are constructed explicitly. As in any quantum theory of gravity the Dirac observables are space-time independent and the Hamiltonian vanishes on physical quantum states. Strategies to resolve this apparent 'problem of (space-)time' [11] are developed at the example of the reparametrization invariant nonrelativistic free particle. Realizing these strategies in the gravity-Yang-Mills system, one finds some partial confirmation of them through the fact that a gravity flat limit reproduces the usual $SU(2)$ quantum dynamics.

The material covered in this work is based on two talks delivered in Helsinki and St Petersburg (cf. [12],[13],[14]). To allow for a comprehensive treatment, more recent developments have been included as well (cf. also [15]).

2   The General Formalism

The action S of a non abelian gauge theory with a finite dimensional semisimple gauge group is given by

$$S = \int \langle F, *F \rangle, \quad F = dA + A \wedge A. \tag{1}$$

Here $\langle .,. \rangle$ denotes the Killing metric on the Lie algebra of the gauge group, $A = A_\mu dx^\mu$ a Lie algebra valued one form (gauge connection), and the Hodge dual $*$ is to be taken with respect to a fixed metric on the space time manifold. If the latter is a cylinder $S^1 \times \mathbf{R}$, one may parametrize it by a coordinate $x^0 \in \mathbf{R}$ and a 2π periodic coordinate x^1. For the Hamiltonian formulation one may choose x^0 as the evolution parameter of the Hamiltonian system. One then finds the zero component A_0 of the gauge connection to play the role of a Lagrange multiplier giving rise to the system of first class constraints (Gauss law constraints)

$$\partial B(x^1) + ad^*_{A_1(x^1)} B(x^1) \approx 0, \tag{2}$$

where B is the momentum conjugate to the one component A_1 of the gauge connection and takes its values in the dual space of the Lie algebra. The symbol ad^* denotes the coadjoint action of the Lie algebra on its dual space and ∂ the derivative with respect to x^1.

All the physical systems considered in this paper have a Hamiltonian structure generalizing the one of the non abelian gauge theories. They are obtained by modifying the constraints (2) due to

$$G_i(x) = \partial B_i(x) + v_{ij}(B)A_1^j(x) \approx 0 \quad x \in S^1 \tag{3}$$

where B takes values in a linear space L, A_1 takes values in the dual space L^*, and $v : L \to L \wedge L$ is a map from L into the space of antisymmetric tensors over L. [Indices refer to an arbitrary basis in L and the dual basis in L^*. Summation over pairs of upper and lower indices is assumed. Throughout this section we will abbreviate x^1 by x, as done already in (3). The fundamental Poisson brackets are given by $\{A_1^i(x), B_j(y)\} = \delta_j^i \delta(x - y)$]. The Gauss law of the nonabelian gauge theory is recovered from (3) if L^* is identified with the Lie algebra of the gauge group and v is chosen due to $v_{ij}(B) = f_{ij}{}^k B_k$, where $f_{ij}{}^k$ denote the structure constants of the Lie algebra.

For general v, (3) will not define a system of first class constraints. Calculating the commutator of two constraints, we find (3) to be first class, iff

$$\frac{\partial v_{ij}}{\partial B_k} v_{kl} + cycl(i, j, l) = 0. \tag{4}$$

This is precisely the condition for v to generate a Poisson structure on L.[1]. The constraint algebra reads

$$\{G_i(x), G_j(y)\} = \delta(x - y)\frac{\partial v_{ij}}{\partial B_k}G_k(x). \tag{5}$$

The aim of the rest of this section is to investigate the quantization of the system under the restriction (4).

To quantize the system in a momentum representation we consider quantum wave functions as complex valued functionals on the space Γ_L of smooth parametrized loops in L:

$$\Gamma_L = \{\mathcal{B} : S^1 \to L, x \to B(x)\}. \tag{6}$$

Following the Dirac procedure [17], we consider the kernel of the quantum constraints

$$\hat{G}_i(x)\Psi[\mathcal{B}] = \left(\partial B_i(x) + i\hbar v_{ij}(B)\frac{\delta}{\delta B_j(x)}\right)\Psi[\mathcal{B}] = 0 \tag{7}$$

as the space $\mathcal{H}$ of physical states.

Let us consider two simple examples: For $v \equiv 0$ the support of wave functions in $\mathcal{H}$ is restricted to constant loops. Thus there is a natural identification of $\mathcal{H}$

---

[1]Cf., e.g., [16]; cf. also the article of A. Alekseev and A. Malkin in the present Lecture Notes as well as [15]

with the space of complex valued functions on L. If $v_{ij}(B)$ is a constant invertible matrix, we may rewrite (7) according to

$$\left((v^{-1})^{ij}\partial B_j(x) - \frac{\hbar}{i}\frac{\delta}{\delta B_i(x)} \right) \Psi[\mathcal{B}] = 0 \qquad (8)$$

and the physical wave functions have the form

$$\Psi[\mathcal{B}] = c\,\exp\left(\frac{i}{2\hbar}\oint B_i(x)(v^{-1})^{ij}\partial B_j(x)dx \right), \quad c \in C. \qquad (9)$$

So in this case $\mathcal{H}$ can be identified with the complex plane.

In general, v is neither trivial nor nondegenerate. For (4) the vector fields

$$V_i = v_{ij}(B)\frac{\partial}{\partial B_j} \qquad (10)$$

are in involution. Thus they generate an integral surface[2] (symplectic leave) I_{B_0} through any point $B_0 \in L$. Denote by $J = \{I_{B_0}, B_0 \in L\}$ the space of these integral surfaces. In $B_0 \in L$ the tangent vectors V_i span a subspace S_{B_0} of the tangent space $T_{B_0}(L)$. Given a cotangent vector $w = w^i dB_i \in T^*_{B_0}(L)$ in the kernel of S_{B_0} (i.e. $v_{ij}(B_0)w^j = 0$), we may use the antisymmetry of v to find $(B(x_0) := B_0)$

$$w^i \hat{G}_i(x_0)\Psi = w^i \partial B_i(x_0)\Psi[\mathcal{B}] = 0. \qquad (11)$$

Thus in any point B_0 the tangent vector ∂B along a loop $\mathcal{B} \in \Gamma_L$ is tangential to S_{B_0}, if $\Psi[\mathcal{B}] \neq 0$. In other words: The support of Ψ is restricted to loops, which are entirely contained in some integral surface $I \in J$.

Given a fixed element $I_0 \in J$, let us denote by Γ_{I_0} the space of loops on I_0. The vector fields

$$\mathcal{V}_i(x) = v_{ij}(B(x))\frac{\delta}{\delta B_j(x)} \qquad (12)$$

form an overcomplete basis in the tangent space over Γ_{I_0}. The ansatz $\Psi|_{\Gamma_{I_0}} = \exp\Phi$ for the restriction of Ψ to Γ_{I_0} allows to rewrite the constraint equation according to

$$\mathcal{A} = \frac{\hbar}{i}d\Phi \qquad (13)$$

where $\mathcal{A}$ denotes the one form on Γ_{I_0} given implicitly by

$$\mathcal{A}(\mathcal{V}_i(x)) = \partial B_i(x), \qquad (14)$$

and d denotes the exterior derivative on Γ_{I_0}. The above ansatz is general, if we exclude the trivial solution $\Psi|_{\Gamma_{I_0}} \equiv 0$. Locally eq. (13) is integrable, iff $\mathcal{A}$ is closed. With the general identity

$$d\mathcal{A}(\mathcal{V}_i, \mathcal{V}_j) = \mathcal{V}_i\left(\mathcal{A}(\mathcal{V}_j)\right) - \mathcal{V}_j\left(\mathcal{A}(\mathcal{V}_i)\right) + \mathcal{A}([\mathcal{V}_i, \mathcal{V}_j]) \qquad (15)$$

---

[2]possibly with singularities

and (4) we can indeed verify $d\mathcal{A} = 0$, if the constraints are first class. Still, there could be global obstructions to the integrability of (13), if the first homotopy group of the underlying space, $\Pi_1(\Gamma_{I_0})$, is nontrivial. At this point one should note, however, that $\mathcal{A}$ need not be exact, as Φ is determined by Ψ up to transitions $\Phi \to \Phi + i2\pi n$, $n \in \mathbf{Z}$, only. Therefore Ψ is well defined, iff $\mathcal{A}$ is integral, i.e. iff ($h \equiv 2\pi\hbar$)

$$\int_\gamma \mathcal{A} = nh, \quad n \in \mathbf{Z} \tag{16}$$

for any (noncontractible) closed loop γ representing an element of $\Pi_1(\Gamma_{I_0})$. This condition yields a restriction on the support of Ψ to a (possibly discrete) subset $\bar{J}$ of J. For I_0 in this subset $\bar{J}$, Ψ is determined up to a multiplicative integration constant on any connected component of Γ_{I_0}. (The space $\Pi_0(\Gamma_{I_0})$ of connected components of Γ_{I_0} is in one to one correspondence with the first homotopy group $\Pi_1(I_0)$, [18]). Let

$$\mathcal{I} = \cup_{I \in \bar{J}} \Pi_0(\Gamma_I). \tag{17}$$

Then $\mathcal{H}$ is identified naturally with space of complex valued functions on $\mathcal{I}$.

There is also a less abstract description of $\mathcal{H}$: Denote by $\{Q_{(\alpha)}, \alpha = 1, ..., r\}$ a maximal set of independent functions on L invariant under the action of the vector fields V_i. We will denote those subspaces of L, where the $Q_{(\alpha)}$ are constant, as their level surfaces M_Q. If the connected components of M_Q are elements of J (this is the generic situation in many examples, c.f. next section), the wave functions can be written as

$$\Psi[\mathcal{B}] = \tilde{\Psi}(Q_{(r)}, m_Q, n_Q) \exp\left(\frac{i}{\hbar} \int \mathcal{A}\right), \tag{18}$$

where the discrete parameters m_Q, n_Q characterize the zeroth and first homotopy group of the level surfaces described above. (18) yields the physical wavefunctions in terms of the variables $B(x)$ and thus allows to describe the action of quantum operators in $\mathcal{H}$.

An alternative formulation of the integrality condition (16) is provided by the relation between one forms on a loop space Γ_M and two forms on the underlying space M: Any path γ in Γ_M corresponds to a one parameter family of loops in M spanning a two dimensional surface $\sigma(\gamma)$. To any closed loop in Γ_M the corresponding surface in M is closed. Thus any two form ω on M generates a one form α on Γ_M via

$$\int_\gamma \alpha = \int_{\sigma(\gamma)} \omega \tag{19}$$

and α is closed and integral, iff ω is closed and integral. (The latter means that the integral of ω over any closed surface is an integer multiple of $2\pi\hbar$).

Of course, not every one form on Γ_M can be described in this way. In our case, however, the one form $\mathcal{A}$ on Γ_I, $I \in \bar{J}$ is generated by a two form Ω on I characterized by its contraction with the vector fields (10):

$$\Omega(V_i, V_j) = v_{ij}. \tag{20}$$

To prove this let us choose a path $\gamma \in \Gamma_I$ parametrized by a parameter $\tau \in [0,1]$. Any point in γ corresponds to a loop $B(x)$. Thus γ induces a map $S^1 \times [0,1] \to L : (x,\tau) \to B(x,\tau)$. Denote by $\dot{B}$ the tangent vector in the tangent space of I corresponding to the derivative of this map with respect to τ. $\dot{B}$ as well as ∂B can be written as linear combination of the V_i:

$$\dot{B}(x,\tau) = \epsilon^i(x,\tau)V_i , \quad \partial B(x,\tau) = \mu^i(x,\tau)V_i \Rightarrow \partial B_i = \mu^i v_{ij}. \tag{21}$$

The corresponding vectors in the tangent space of Γ_I are now given by

$$\dot{\mathcal{B}} = \int_x \epsilon^i(x,\tau)\mathcal{V}_i(x) , \quad \partial \mathcal{B} = \int_x \mu^i(x,\tau)\mathcal{V}_i(x) . \tag{22}$$

With (14) we have

$$
\begin{aligned}
\int_\gamma \mathcal{A} &= \int \mathcal{A}(\dot{\mathcal{B}})d\tau = \int \epsilon^i \partial B_i dx d\tau = \\
&= \int \epsilon^i v_{ij} \mu^j dx d\tau = \int \Omega(\partial \mathcal{B}, \dot{\mathcal{B}})dx d\tau = \int_{\sigma(\gamma)} \Omega .
\end{aligned}
\tag{23}
$$

Thus our assertion is proven. So (16) is equivalent to the condition

$$I \in \bar{J} \quad \Leftrightarrow \quad \Omega \text{ is integral on } I. \tag{24}$$

Ω is an integral symplectic form and thus gives I the structure of an integrable space. Furthermore, Ω is invariant under the flow of the vector fields V_i (i.e. $\mathcal{L}_{V_i}\Omega = 0$, where $\mathcal{L}$ denotes the Lie derivative). So the vector fields V_i are locally Hamiltonian with respect to the symplectic form Ω on I.

Let us illustrate the formalism by the example of non abelian gauge theories (cf. also [6]). There L is the dual space g^* of the Lie algebra g of the gauge group. Condition (4) becomes the Jacoby identity. J is the space of coadjoint orbits, i.e. the space of orbits generated by the action of the gauge group in g^*. The vector fields V_i are the vector fields on the coadjoint orbits associated with the coadjoint action of the generators of the Lie algebra and $\{Q_{(\alpha)}\}$ is the set of Casimirs on g. Ω is the standard symplectic form on the coadjoint orbits of a Lie group as introduced in [9]. The coadjoint orbits are quantizable spaces, iff this symplectic form is integral and their quantization yields the unitary irreducible representations of g. This observation establishes a connection of the momentum representation to the configuration space representation of quantum mechanics for non abelian gauge theories on a cylinder: In the configuration space representation, where wave functions are functionals on the space of gauge connections, the physical wave functions (i.e. the kernel of the constraints) can be identified with the functions on the space of unitary irreducible representations of g [1].

The generalization of our considerations to the case, where the B_i are local coordinates on a nonlinear space, is straightforward. In this case $v \in T(L) \wedge T(L)$ is an skew symmetric two tensor over the tangent bundle and the v_{ij} are the components of v with respect to the coordinate basis in $T(L)$. The formulation (7) of the constraints can be made coordinate independent:

$$i_{df} [\partial B(x) + i\hbar v(B(x))] \Psi[\mathcal{B}] = 0 \quad \forall f : L \to R. \tag{25}$$

For $I \in \bar{J}$ and $B_0 \in I$ denote by $v|_S(B_0)$ the restriction of v to $S(B) = T_B(I)$. We then have $\Omega = (v|_S)^{-1}$. To prove this let us choose coordinates $\{c_1, ..., c_s, Q_{(1)}, ..., Q_{(r)}; s+r = dimL\}$ on L. In these coordinates we have

$$v = \sum_{\alpha=1}^{s} v_{\alpha\beta} \frac{\delta}{\delta c_\alpha} \frac{\delta}{\delta c_\beta} \tag{26}$$

as the $Q_{(\alpha)}$ are constant on I. Now (20) immediately implies

$$\Omega = (v^{-1})^{\alpha\beta} dc_\alpha dc_\beta. \tag{27}$$

3 Explicit Examples

The constraints (3) are induced by the action

$$\int [B_i dA^i + \frac{1}{2} v_{ij}(B) A^i \wedge A^j] \tag{28}$$

where A is an L^* valued one-form and the zero-form B is a map into the corresponding dual space L. The action (28) is already of first order, i.e. Hamiltonian: A_1 and B are seen to be canonical conjugates and A_0 enforces the constraints (3).

The simplest nontrivial examples for the considerations of the previous section can be formulated in a three dimensional target space L. For the rest of this section we will thus stick to such a space. In three dimensions any antisymmetric two-tensor v can be rewritten according to

$$v_{ij}(B) = \varepsilon_{ijk} u^k(B) \tag{29}$$

for some u^i, ε_{ijk} being the standard antisymmetric ε-tensor with $\varepsilon_{123} = 1$. Indices may be raised and lowered by means of the metric $\kappa_{ij} = diag(\pm 1, 1, 1)$. The ϵ_{ij}^k can be thought of as being the structure constants of $L^* := so(3)$ or $L^* := so(2, 1)$ in an appropriate basis $\{T_i\}$. If u is the identity map, (28) takes the form $\int B_i F^i$, which is the weak coupling limit of (1).

There are several further possibilities to satisfy the generalized Jacobi identity (4). One is provided by $u^k = u^k(B_k)$. Another choice, of more interest for the following, is given by

$$u_a = B_a, u_3 = u_3((B)^2, B_3), \quad (B)^2 \equiv B_a B^a, \quad a \in \{1, 2\}. \tag{30}$$

Rewriting A in the latter case as

$$A = e^a T_a + \omega T_3, \tag{31}$$

the action (28) takes the form

$$S_G = \int [B_a(de^a - \varepsilon^a{}_b \omega \wedge e^b) + B_3 d\omega + u_3(e^1 \wedge e^2)], \tag{32}$$

with $\varepsilon_{ab} = \varepsilon_{ab3}$. Much of our interest in (32) stems from the fact that this action can be reinterpreted as the action of a gravitational theory: Viewing e^a as a zweibein and ω as a spin connection, the term following B_a is identified with the torsion two-form De^a, whereas $d\omega$ becomes the curvature two-form. B_a and B_3 are vector and scalar valued functions, respectively, living on the two-dimensional manifold characterized by the metric $g = e^a e^b \kappa_{ab}$. The latter is of Euclidean or Minkowski type, corresponding to the respective signature of the frame metric $\kappa_{ab} = diag(\pm 1, 1)$. Eliminating the B fields by means of their equations of motion for some special choices of u_3, we can establish the equivalence of S_G with purely geometrical actions of two dimensional gravity. E.g., $u_3 = (1/4\gamma)(B_3)^2 - \lambda + \frac{\alpha}{2}(B)^2$ with $\alpha \neq 0$ is easily seen to lead to 2D gravity with torsion [5]

$$S_G^{KV} = - \int [\pm \gamma d\omega \wedge *d\omega \pm \frac{1}{2\alpha} De^a \wedge *De_a + \lambda\varepsilon], \qquad (33)$$

where $\varepsilon \equiv e^1 \wedge e^2$ is the metric induced volume form or ε-tensor and '$*$' denotes the Hodge dual operation ($*\varepsilon = \pm 1$). (33) is the most general Lagrangian yielding second order differential equations for e^a and ω in two dimensions. With $\alpha = 0$ the same choice for u_3 leads to the similar action of torsionless R^2 gravity.

To find the set J of integral surfaces I, on which physical wave functionals $\Psi[B]$ have their support, let us for a moment return to an arbitrary dimensional target space L and a Poisson structure v of the form $v = f_{ij}{}^k u_k(B)(\partial/\partial B_i) \wedge (\partial/\partial B_j)$ where the f's are the structure constants of some Lie algebra of rank r; (29) may be regarded as a special case of this. If $C(u) = C^{ijk\cdots} u_i u_j u_k \cdots$ denotes one of the r independent Casimirs,

$$w = \frac{\partial C(u)}{\partial u_i} dB_i \qquad (34)$$

will be annihilated by v. In the case of a non abelian gauge theory $u_i = B_i$ and $w = dC(B)$ so that according to (11) the (physical) wave functions Ψ have support only on loops B with constant values of the Casimirs, as has been noted already at the end of the previous section. In the present case of (29) the only independent Casimir is the Killing metric κ_{ij}, so that (11) with (34, 30) becomes

$$(B^a \partial B_a + u_3 \partial B_3) \Psi = 0. \qquad (35)$$

For reasons of calculational simplicity, let us specify u_3 to

$$u_3 = U(B_3) + \frac{\alpha}{2}(B)^2. \qquad (36)$$

Multiplying now (35) by the integrating factor $2\exp(\alpha B_3)$, we obtain

$$\partial Q \, \Psi = 0, \qquad Q = (B)^2 \exp(\alpha B_3) + 2 \int^{B_3} U(y) \exp(\alpha y) dy. \qquad (37)$$

Thus generically the level surfaces M_Q generated by $Q = const$, and thus (generically) also the integral surfaces $I \in J$, will be two-dimensional for the considered class of examples (32). In the following we will specify the potential u_3 to study these surfaces in more detail.

Prototypes are provided by the $SO(3)$ and $SO(2,1)$ BF-theories based on the $so(3)$ and $so(2,1)$ algebra, respectively, resulting from $u_3 = B_3$.[3] Fixed values of the Casimir $Q \equiv B^i B_i$ yield the coadjoint orbits of the groups, i.e. in the former case spheres for $Q > 0$ and the origin for $Q = 0$ and in the latter case one sheet hyperboloids for $Q > 0$, two sheet hyperboloids for $Q < 0$, and the light cone in the target space for $Q = 0$. In the compact case, we see that all the level surfaces, which coincide with the integral surfaces, are two dimensional except for the zero dimensional origin. The spheres are connected and simply connected, but have $\Pi_2 = \mathbf{Z}$. According to our general considerations of section 2 we therefore know that the spectrum of Q becomes discrete. This (i.e. eq. (16) or (24)) was a necessary and sufficient condition for the integrability of the horizontality condition (7). (The determination of this spectrum shall be taken up at the end of this section, after having further analyzed the topological structure of the integral surfaces).

In the noncompact $sl(2, \mathbf{R})$ case Π_2 is always trivial and the spectrum of Q remains continous. For $Q < 0$, however, the level surfaces M_Q consist of two (simply connected) parts, thus corresponding to two different integral surfaces I of the vector fields V_i defined in (10). For $Q > 0$ $\Pi_0(M_Q)$ is trivial, but $\Pi_1(M_Q) = \mathbf{Z}$; thus the integral surfaces I of V_i coincide with the level surfaces M_Q in this case, but loops $\mathcal{B}$ with different winding number around the target space hyperboloid are not smoothly connected to each other in the space of loops on I $(\Pi_0(\Gamma_I) = \Pi_1(I) = \mathbf{Z})$; thus for any winding number of $\mathcal{B} \sim B(x^1)$ we can prescribe an independent initial value for the solution of the first order differential equation (7). This illustrates the necessity for the two quantum numbers m_Q and n_Q within (18). Actually, they correspond also to invariant Dirac observables, if we allow the latter to become discontinuous: Clearly

$$m_Q := \Theta(-Q)\Theta(B_1), \qquad n_Q := \Theta(Q) \oint \partial \phi dx^1 \qquad (38)$$

where Θ is the Heaviside step function and ϕ is the angle variable of polar coordinates in the (B_2, B_3)-plane, are also Dirac observalbes in this extended sense, independent from the continuous invariant Q.

The $Q = 0$ level surface plays a special role: Since V vanishes at the origin, the latter is an integral surface by itself and splits $Q = 0$ into three parts. (Note that (7) constrains the wave functionals to have support only on loops not passing through a target space point where V_i vanishes; thus this splitting transfers consistently to the spaces of loops on the integral surfaces). This implies also that Π_1 becomes nontrivial for the future and the past target space light

---

[3]Actually it is rather the BF-theories of the corresponding universal covering groups which have been quantized so far; the further steps necessary to quantize an $SO(2,1)$-BF-theory will be discussed in the following section.

cones. Allowing also for invariant distributions, we can uniquely describe the integral surfaces of $\mathcal{V}_i \equiv V_i[B(x)]$, i.e. the space $\mathcal{I}$ of eq. (17) (with $\bar{J} = J$ here), by means of Dirac observables:[4] Defining $\Theta(0) := 1$ we have to add merely $\delta[B_i]$ to Q, m_Q, and n_Q so as to get a complete set of independent commuting Dirac observables for the $sl(2, \mathbf{R})$-BF-theory. The space J of integral surfaces I as well as the space of loops on it, $\cup_{I \in J} \Gamma_I$, are, however, not Hausdorff at $Q = 0$. As a consequence there might arise some ambiguity in glueing together the orbit spaces with $Q \neq 0$, an issue which is certainly closely related to the determination of an inner product.

Summing up the sl_2 case, we find the physical wave functionals to effectively become functions (possibly also generalized ones) on the space of the above Dirac observables, the corresponding spectra remain classical, and the phase factor becomes essentially superfluous, one can get rid of it by changing the basis in the U(1) quantum bundle.

Concerning the question of the inner product, let us remark only that on large parts of the phase spaces of any of the models (32) with (36) and Minkowski signature, the variable conjugate to Q can be written as $(e^{\pm} =: (e^2 \pm e^1)/\sqrt{2})$

$$P = -\frac{1}{2} \oint \exp(-\alpha B_3) \frac{e_1^{-}}{B_+} dx^1 \approx -\frac{1}{2} \oint \exp(-\alpha B_3) \frac{e_1^{+}}{B_-} dx^1. \qquad (39)$$

Pulling through the phase factor of (18), which in local target space coordinates takes the form

$$\exp\left(-\frac{i}{\hbar} \oint \ln|B_+| \partial B_3 dx^1\right) \sim \exp\left(\frac{i}{\hbar} \oint \ln|B_-| \partial B_3 dx^1\right) \qquad (40)$$

the Dirac observable P acts via $(\hbar/i)(d/dQ)$ on $\tilde{\Psi}$. Requiring that it will become a Hermitean operator severely restricts the measure of the inner product, but, in the case that $\tilde{\Psi}$ depends also nontrivially on quantum numbers m or n, this does not determine the inner product entirely. It is not quite clear, if one should require the 'Dirac observables' m_Q and n_Q, introduced above for the sl_2-theory, to become hermitean as well. In this case the corresponding eigenspaces would be orthogonal.

Next let us find the space of integral surfaces for R^2-gravity, i.e. for (32) with potential $u_3 = -(B_3)^2 - \lambda$, yielding $Q((B)^2, B_3) = (B)^2 - 2(B_3)^3/3 - 2\lambda B_3$. For $\lambda > 0$, $Q = const$ allows to determine B_3 uniquely as a function of $(B)^2$. Thus the resulting surfaces in the target space are diffeomorphic to a plane so that there is no quantization of the classical spectrum of Q and there are also no additional quantum numbers within the wave functions (18). So for $\lambda > 0$ the resulting Hilbert space is the one of an ordinary particle on a line.

For $\lambda = 0$ the situation is similar, only that the value $Q = 0$ (critical value) plays a similarly exceptional role as in the BF case: one gets a conic singularity of the plane at $(B)^2 = B_3 = 0$ for Euclidean signature ($k_{ab} = \delta_{ab}$), and for Minkowski signature ($k_{ab} = diag(-1, 1)$) additionally a non Hausdorff structure (of J) at this point.

---

[4] The integral surfaces of V_i are characterized by the same quantities except for n_Q.

For $\lambda < 0$ there are two critical values of Q: $Q_{<(>)} \equiv Q(0, \pm\sqrt{-\lambda}) = \pm 8(-\lambda)^{(3/2)}/3$, the values $\pm\sqrt{-\lambda}$ of B_3 corresponding to the zeros of u_3 resp. U. For $Q \in (-\infty, Q_<) \cup (Q_>, \infty)$ the resulting surfaces are again manifolds with trivial topology. For $Q \in (Q_<, Q_>)$ and Euclidean signature we get two disconnected surfaces of the topology of a plane and a sphere, respectively. Thus the continuous spectrum $Q \in \mathbf{R}$ has a twofold degeneracy for some specific values of $Q \in (Q_<, Q_>)$. For $Q \in (Q_<, Q_>)$ and Minkowskian signature the level surfaces M_Q are connected and of trivial second homotopy; however, there are two fundamental noncontractible loops, the winding numbers of which give rise to a quantum number $n_Q \in \mathbf{Z}$.

To analyse the situation for general potential U, it is helpful to use a $(B)^2$ over B_3 diagram. Any fixed value of Q induces a curve C_Q in this diagram. The intersections of C_Q with the B_3 axis are most crucial for the topology of M_Q. Let us first consider the Euclidean case, where only non-negative values of $(B)^2$ are admissible: Any part of C_Q (in the positive of $(B)^2$) between two successive intersections with the B_3 axis leads to a spherical M_Q, any part of C_Q with $(B)^2 \geq 0$ and exactly one point of vanishing $(B)^2$ on it yields a 'plane', and a C_Q with no such points or intersections results in a cylindrical M_Q (or an empty M_Q for strictly negative $(B)^2$, as, e.g., in the $so(3)$-example for $Q < 0$). Changes of the topology of M_Q (along the choice of Q) can happen only at sliding intersections of C_Q with the B_3 axis; the latter are possible only at $B_3 = \beta_c$, $U(\beta_c) = 0$, and thus only for the 'critical values' $Q_c = Q(0, \beta_c)$ of Q. The critical points $(B_1, B_2, B_3) = (0, 0, \beta_c)$ (and only these) are then fixed points of the vector fields V_i and constitute an (zero dimensional) integral surface by itself. For Minkowski signature the transition from C_Q to M_Q is a bit more cumbersome. The result is, however, quite simple: If C_Q contains no points $(B)^2 = 0$, M_Q consists of two disconnected 'planes'; if C_Q contains l points of (nonsliding) intersections with the B_3 axis, it has $l - 1$ fundamental non-contractible loops. At the critical values $Q = Q_c$ (sliding intersections) we again have fixed points $(0, 0, \beta)$, and the set J of integral surfaces becomes non Hausdorff there.

For both signatures the fixed points correspond also to the distributional solutions $\delta[B_a]\delta[B_3 - \beta_c]$ of the quantum constraints and might be implemented via a point measure in the inner product. (A somewhat special case arises when choosing $u_3 \equiv 0$, describing 'flat gravity' on the cylinder, where the set of β_c becomes uncountable and needs a separate treatment). Aside from these fixed point solutions the wave functions have the form (18). We further observe that in our class of examples (32) the integral surfaces have a non trivial second homotopy only for Euclidean signature and that non trivial Π_1 implies trivial Π_2 and vice versa.

The discrete part of the spectrum of the Dirac observable Q is obtained most easily via the two-form Ω of section 2. According to (27) it is the inverse of v restricted to the integral surfaces, which are (deformed) spheres in the case under study. By construction $v(dQ, \cdot) = 0$. Furthermore, due to (29) and (30) $v(dB_3, \cdot)$ is independent of the potential u_3. Thus it will be convenient to calculate the inverse of $v|_{M_Q}$ in coordinates Q, B_3 and e.g. $\varphi = \arctan(B_2/B_1)$; these cover the spheres up to the poles at $B_a = 0$. Since $v(dB_3, d\varphi) = 1$ we obtain

$$\Omega = dB_3 \wedge d\varphi. \tag{41}$$

Integrating this two-form over the considered 'sphere', the integrability condition (16) becomes (cf. eqs. (24, (19))

$$B_{3,max}(Q) - B_{3,min}(Q) = n\hbar, \qquad n \in \mathbf{N} \tag{42}$$

where $B_{3,max}, B_{3,min}$ denote the values of B_3 at the poles. Given a curve C_Q introduced above, it is then easy to decide if this value of Q allows for a spherical integral surface or not. For the case of $u_3 = B_3$, (41) becomes the rotation invariant Kostant-Souriau form $\Omega = r\sin\vartheta d\vartheta d\varphi = (\varepsilon^{ijk} B_i dB_j dB_k/r^2)|_{r=\sqrt{Q}}$, where (r, ϑ, φ) denote spherical target space coordinates, and the quantization condition (42) can be expressed also explicitly in terms of $Q = r^2$, namely as $Q = n^2\hbar^2/4$, $n \in \mathbf{N}$. If we add to this $Q = 0$, corresponding to the distributional solution located at $\mathcal{B} = 0$, this spectrum coincides precisely with the one obtained in the connection representation [2].

4 Large Gauge Transformations and Metric Non-Degeneracy

The previous two sections have been devoted to the analysis of the models under consideration in a Hamiltonian formulation, where the symmetries of the system are expressed in terms of first class constraints. There are, however, some subtle points connected with this approach:

- The constraints are the generators of infinitesimal symmetry transformations. Large gauge transformations (i.e. symmetry transformations not connected to the unity) cannot be generated by infinitesimal transformations and thus they are not determined by the constraints.

- In the gravity theories presented in the previous sections the zero components of the zweibein and the spin connection played the role of Lagrange multiplier fields. We eliminated them from the phase space as unphysical degrees of freedom. But in a theory of gravity one usually requires the metric of the space time manifold to be nondegenerate (i.e. $\det g \neq 0$ everywhere). Obviously it is difficult to realize this condition after eliminating the zero components of the zweibein from the phase space. Even if we allow for a degenerate metric, the problem is not solved: The constraints (3) with (30) generate the symmetries of the gravity theory only for $\det g \neq 0$ and turn out to connect gravitationally inequivalent solutions separated in the phase space by regions with a degenerate metric.

In the present section we will illustrate the importance of these points by considering concrete examples. Our analysis will include the explicit calculation of the reduced phase space (i.e. the space of solutions of the equations of motion modulo the symmetries of the model) for gauge and gravity theories based on the $sl(2, R)$ Lie algebra. All of the theories considered are characterized by the same Lagrangian

$$\int \langle B, F \rangle \qquad (43)$$

and thus a naive calculation of the constraints yields equivalent Hamiltonian systems. Nevertheless we will find that the reduced phase spaces differ as the symmetry contents of the models differ.[5]

In the first example let us regard the action (43) as the one of a $PSL(2, R)$ gauge theory. $PSL(2, R)$ is the group obtained from $SL(2, R)$ by the identification $1 \sim -1$ and is isomorphic to $SO_e(2, 1)$, the component connected with the unity of $SO(2, 1)$. Thus its Lie algebra is given by

$$[T_i, T_j] = \varepsilon_{ij}{}^k T_k, \qquad (44)$$

where the last index in the ε-tensor has been raised by means of the Killing metric $\kappa_{ij} = diag(-1, 1, 1)$. A possible matrix representation of (44) is provided by the real matrices $T_1 = i\sigma_2/2$, $T_2 = -\sigma_1/2$, and $T_3 = -\sigma_3/2$, where the σ_i are the Pauli matrices. From this one finds $\kappa_{ij} = 2tr(T_i T_j)$ so that, e.g., the Dirac observable $Q = B_i B^i$ introduced in eq. (37) can be expressed alternatively as $Q = 2tr B^2 = -4det B$ $(B \equiv B_i T^i)$.

The group $\mathcal{G}$ of symmetry transformations is the group of smooth mappings from the cylinder into $PSL(2, \mathbf{R})$:[6]

$$\mathcal{G}_{PSL(2,R)} = \{g : S^1 \times \mathbf{R} \to PSL(2, \mathbf{R})\} \qquad (45)$$

The equations of motion,

$$F = 0, \qquad dB + [A, B] = 0, \qquad (46)$$

yield the connection to be flat and the Lagrange multiplier field B to be covariantly constant. Up to gauge transformations a flat connection A on a cylinder is determined by its monodromy $M_A = \mathcal{P} \exp \oint A \in PSL(2, \mathbf{R})$ generating parallel transport around the cylinder ($\mathcal{P}$ denotes path ordering and the integration runs over a closed curve C winding around the cylinder once). As the exponential map is surjective on $PSL(2, R)$, any monodromy matrix can be generated by a connection of the form $A = A_1 dx^1$ where A_1 is constant:

$$A = \begin{pmatrix} z & y+t \\ y-t & -z \end{pmatrix} dx^1, \quad t, y, z \in \mathbf{R} \qquad (47)$$

Constant gauge transformations act on A via the adjoint action leaving the determinant $t^2 - y^2 - z^2$ invariant and may be interpreted as Lorentz transformations in the three dimensional Minkowski space (t, y, z). Hyperbolic, elliptic

---

[5]This is similar to the inequivalence of the symmetry generators (d/dq) and $q(d/dq)$ on a line even when disregarding $q = 0$ [12].

[6]There are no nontrivial principal G-bundles on a cylindrical base manifold, iff the chosen structure (gauge) group G is connected.

and parabolic elements, respectively, in the Lie algebra correspond to spacelike, timelike, and lightlike vectors, respectively, in this Minkowski space. By Lorentz transformations in the (t, y, z) plane they can be brought into the form:

$$A^{hyp} = \begin{pmatrix} 0 & \alpha \\ \alpha & 0 \end{pmatrix} dx^1, \quad A^{ell} = \begin{pmatrix} 0 & \vartheta \\ -\vartheta & 0 \end{pmatrix} dx^1,$$

$$A^{par} = \begin{pmatrix} 0 & 0 \\ \pm 1 & 0 \end{pmatrix} dx^1 \tag{48}$$

with $\alpha, \vartheta \in \mathbf{R}$ and the identification $\alpha \sim -\alpha$. Exponentiation yields the monodromy matrices

$$M_{A^{hyp}} = \begin{pmatrix} \cosh 2\pi\alpha & \sinh 2\pi\alpha \\ \sinh 2\pi\alpha & \cosh 2\pi\alpha \end{pmatrix}, \quad M_{A^{ell}} = \begin{pmatrix} \cos 2\pi\vartheta & \sin 2\pi\vartheta \\ -\sin 2\pi\vartheta & \cos 2\pi\vartheta \end{pmatrix},$$

$$M_{A^{par}} = \begin{pmatrix} 1 & 0 \\ \pm 2\pi & 1 \end{pmatrix}. \tag{49}$$

inducing the further identification $\vartheta \sim \vartheta + 1/2$ in the elliptic sector (remember $\oint dx^1 = 2\pi$ and $1 \sim -1$). The integration of the second eq. (46) gives $B(x^0, x^1) = B(x^0, x^1 + 2\pi) = M_A B(x^0, x^1) M_A{}^{-1}$ and thus choosing a connection from (48) $B(x)$ has to commute with the corresponding monodromy matrix and consequently with the connection itself. Using (46) again one finds $B(x)$ to be constant. We obtain:

$$B^{hyp} = \begin{pmatrix} 0 & c_1 \\ c_1 & 0 \end{pmatrix}, \quad B^{ell} = \begin{pmatrix} 0 & c_2 \\ -c_2 & 0 \end{pmatrix},$$

$$B^{par} = \begin{pmatrix} 0 & 0 \\ c_3 & 0 \end{pmatrix}, \quad c_i \in \mathbf{R}. \tag{50}$$

In the case $A = 0$ (corresponding to $\alpha = 0$ or $\vartheta = 0$, respectively, in (48)) $B(x)$ is constant, too, but it is not restricted by its commutator with the monodromy matrix. It is, however, subject to constant gauge transformations, as they leave $A = 0$ invariant. Considerations similar to those above show that also in this case gauge representatives of the solutions are given by (50) with $c_3 = \pm 1$ and the identification $c_1 \sim -c_1$. (48) and (50) give a complete parametrization of the reduced phase space of the $PSL(2, \mathbf{R})$-gauge theory.

The group of gauge transformations $\mathcal{G}_{PSL(2, \mathbf{R})}$ as defined above is not connected; rather it consists of an infinite number of components not smoothly connected to each other: $\Pi_0(\mathcal{G}) = \Pi_1(PSL(2, \mathbf{R})) = \mathbf{Z}$. A complete set of representatives for the components of $\mathcal{G}_{PSL(2, \mathbf{R})}$ is given by

$$g_{(n)} = \begin{pmatrix} \cos(nx^1/2) & \sin(nx^1/2) \\ -\sin(nx^1/2) & \cos(nx^1/2) \end{pmatrix}, \quad n \in \mathbf{Z}. \tag{51}$$

Parametrizing the phase space as in (48) - (50) we also implemented these gauge transformations. The action of the group elements $g_{(n)}$ on the connections (48) gives in the hyperbolic sector

$$A^{hyp}_{(n)} = \begin{pmatrix} \alpha \sin(nx^1) & \alpha \cos(nx^1) + n/2 \\ \alpha \cos(nx^1) - n/2 & -\alpha \sin(nx^1) \end{pmatrix} dx^1$$

$$B^{hyp}_{(n)} = c_1 \begin{pmatrix} \sin(nx^1) & \cos(nx^1) \\ \cos(nx^1) & -\sin(nx^1) \end{pmatrix}. \tag{52}$$

An analogous result is obtained in the parabolic sector. In the elliptic sector the $g_{(n)}$ generate a transformation $\vartheta \to \vartheta + n/2$. They are responsible for the previous identification $\vartheta \sim \vartheta + 1/2$, which is removed now.

With this knowledge it is straightforward to find the RPS for an $SL(2, \mathbf{R})$ gauge theory: The action is the same as the one of the $PSL(2, \mathbf{R})$ theory. Gauge transformations of the type $g_{(2l+1)}$, $l \in \mathbf{Z}$ are not allowed, as we do not have the identification $1 \sim -1$. Consequently the hyperbolic sector of the RPS is parametrized by $(A^{hyp}_{(0)}, B^{hyp}_{(0)})$ and $(A^{hyp}_{(1)} B^{hyp}_{(1)})$. An analogous result holds for the parabolic sector. In the elliptic sector we have (A^{ell}, B^{ell}) but with the identification $\vartheta \sim \vartheta + 1$ rather than $\vartheta \sim \vartheta + 1/2$. In contrast to the $PSL(2, \mathbf{R})$-case there are elements of the RPS which cannot be represented by a constant connection. This is a consequence of the non surjectivity of the exponential map between Lie algebra and group in the case of $SL(2, \mathbf{R})$.

From the homotopical point of view $SL(2, \mathbf{R})$ is the double covering of $PSL(2, \mathbf{R})$. Analogously, excluding all gauge transformations not connected to the unity from $\mathcal{G}$ is equivalent to choosing the universal covering $\widetilde{SL}(2, \mathbf{R})$ as the gauge group of the theory. (Note that $\Pi_0(\mathcal{G}_{\widetilde{SL}(2,\mathbf{R})}) = \Pi_1(\widetilde{SL}(2, \mathbf{R})) = \{1\}$). Thus the RPS of $\widetilde{SL}(2, \mathbf{R})$ is parametrized by $(A^{hyp}_{(n)}, B^{hyp}_{(n)})$, $n \in \mathbf{Z}$, $\alpha \in \mathbf{R}$, the analogous solutions in the parabolic sector, and (A^{ell}, B^{ell}), $\vartheta \in \mathbf{R}$ (without any identification).

To see the significance of the difference between the $PSL(2, \mathbf{R})$ and the $\widetilde{SL}(2, \mathbf{R})$ gauge theory, let us have a look on the quantization of the models: In the first case the elliptic sector of the configuration space (i.e. the space of gauge inequivalent connections) is compact and thus we expect the possibility of unitarily inequivalent quantum theories with a discrete spectrum for the momentum operator (i.e. the Dirac observable $Q = -4 \det B$). We may compare this with the result obtained in the previous section: There we used the Gauss law constraints to realize gauge transformations in the quantum theory. As outlined above the constraints generate those gauge transformations only which are connected to the unity. So the quantum theory we obtained corresponds to the $\widetilde{SL}(2, \mathbf{R})$ gauge theory. Indeed a continuous spectrum for Q was found. Furthermore, the discrete parameter n_Q within the wave functions is also readily identified with the parameter n of the hyperbolic sector in the above parametrization of the RPS of the $\widetilde{SL}(2, \mathbf{R})$ theory.

To find the correct quantization of the $PSL(2, \mathbf{R})$ theory we have to implement large gauge transformations. To this end let us employ the exponential map in order to rewrite the Gauss law. Starting from an initial loop $\mathcal{B}_0$ in some coadjoint orbit any loop $\mathcal{B}$ may be written as $\mathcal{B} = g\mathcal{B}_0 g^{-1}$ for some $g = \exp X \in \mathcal{G}$,

$X : S^1 \to sl(2, \mathbf{R})$. (As we mentioned above the exponential map is surjective on $PSL(2, \mathbf{R})$). If g is connected to the unity, we also have $g(t) = \exp tX \in \mathcal{G}$ for $t \in [0, 1]$. The Gauss law can then be rewritten as

$$\oint \langle X, \partial(e^{tX} \mathcal{B}_0 e^{-tX}) \rangle dx^1 \, \Psi[\mathcal{B}] = i\hbar \frac{\partial}{\partial t} \Psi[e^{tX} \mathcal{B}_0 e^{-tX}]. \qquad (53)$$

With the identity

$$\oint dx^1 \langle X, \partial(e^{tX} \mathcal{B}_0 e^{-tX}) \rangle = -\oint dx^1 \langle e^{-tX} \partial X e^{tX}, \mathcal{B}_0 \rangle = -\oint dx^1 \frac{\partial}{\partial t} \langle e^{-tX} \partial e^{tX}, \mathcal{B}_0 \rangle \qquad (54)$$

integration over t leads to

$$\Psi[\mathcal{B}] = \Psi[\mathcal{B}_0] \exp\left(\oint \frac{i}{\hbar} \langle g^{-1} \partial g, \mathcal{B}_0 \rangle dx^1\right), \quad g \in \mathcal{G}_e. \qquad (55)$$

An alternative derivation of this exponantiated form of the Gauss law constraint is provided by a Fourier transformation of the gauge invariance property of the physical wave functionals in the connection representation [19]. By construction the wave functions calculated in the previous sections are the general solutions of eq. (55) for gauge transformations connected to the unity. To quantize the $PSL(2, \mathbf{R})$ theory we may rewrite $\mathcal{G}$ as the semidirect product of $\mathcal{G}_e$ (the component connected to the unity) and the zero'th homotopy group $\Pi_0(\mathcal{G}) = \mathbf{Z}$:

$$\mathcal{G} = \mathcal{G}_e \times_s \mathbf{Z} \qquad (56)$$

The most general incorporation of the second factor $\mathbf{Z}$ into the quantum theory will be to require the wave functionals to transform according to a unitary representation D_θ of $\mathbf{Z}$ characterized by an angle θ:

$$D_\theta(n) = \exp\left(\frac{i2\pi n\theta}{\hbar}\right). \qquad (57)$$

Taking together (55) and (57) we are thus lead to

$$\Psi[\mathcal{B}] \equiv \Psi[g\mathcal{B}_0 g^{-1}] = \Psi[\mathcal{B}_0] \exp\left[\frac{i}{\hbar}\left(\oint \langle g^{-1} \partial g, \mathcal{B}_0 \rangle dx^1 + 2\pi n\theta\right)\right] \qquad (58)$$

for g in the n-th component of $\mathcal{G}$.

It is easy to verify that (58) is compatible with group multiplication, i.e. if it holds for two gauge transformations g_1, g_2, it will also hold for the product $g_1 g_2$. The group $\mathbf{Z}$ is generated by one element which may be represented by $g_{(1)}$ as defined in (51). For this reason the wave functions obtained in the sections 2 and 3 will solve (58) for all $g \in \mathcal{G}$, if this identity holds for $g_{(1)}$. Furthermore (55) will hold for all loops $\mathcal{B}$ which may be written as $\mathcal{B} = g^{-1}\mathcal{B}_0 g$ for some $g \in \mathcal{G}_e$, if it holds for the loop $\mathcal{B}_0$. It is now obvious that in the hyperbolic sector $(Q > 0)$ the large gauge transformations simply relate the $\tilde{\Psi}(Q, n_Q)$, $(n_Q \equiv n)$, to each other for fixed value of Q and different values of n. In the elliptic sector

$(Q < 0)$ we may apply (58) to the constant loop $\mathcal{B}_0 = B^{ell}$ (B^{ell} as defined in (50)). Noting that $g_{(1)}$ commutes with B^{ell}, we find

$$\frac{1}{\hbar}\left(\frac{1}{2\pi}\oint \langle g_{(1)}^{-1}\partial g_{(1)}, B^{ell}\rangle dx^1 + \theta\right) = \frac{\theta - 2c_2}{\hbar} = \frac{\theta - sgn(B_1)\sqrt{-Q}}{\hbar} \in \mathbf{Z} \quad (59)$$

in which the signum function can be expressed also in terms of the quantum number m_Q of eq. (38) via $sgn(c_2) \equiv sgn(B_1) = 2m_Q - 1$. Thus in the elliptic sector the support of physical wavefunctions $\tilde{\Psi}(Q, m_Q)$ is restricted to $\sqrt{-Q} = (1 - 2m_Q)(l\hbar + \theta)$, $l \in \mathbf{Z}$. The quantum theories we obtain for different choices of θ are obviously unitarily inequivalent, as they generate different spectra of Q. This is precisely the result we expected.

At this point we want to mention that these results also hold for the $PSL(2, \mathbf{R})$ Yang-Mills theory. In this case Q plays the role of the Hamiltonian.

Via the identification (31) the action (43) together with (44) may also be regarded as the one of a gravity theory (Jackiw Teitelboim model). In this case the symmetry content of the model is given by diffeomorphisms and local Lorentz transformations (gravitational symmetries). This group consists of a finite number of components not smoothly connected to each other. They differ by x^0- and x^1-reflection on the space time manifold and by parity transformation and time reversal in the Lorentz bundle. So up to these transformations the symmetry content of the gravity theory seems to coincide with the one of the $\widetilde{SL}(2, \mathbf{R})$ gauge theory. So let us see, how the infinitesimal generators of the gravitational symmetries are identified with the Gauss law constraints (2) generating the $sl(2, \mathbf{R})$ algebra. To this end we may calculate the Hamiltonian density $\aleph$ ($H = \oint \aleph dx^1$) of the theory as the generator of diffeomorphisms in x^0-direction. We find $\aleph = -A_0^i G_i$, with $G_i = \partial B_i + \varepsilon_{ij}{}^k A^j B_k$. Analogously the generator of diffeomorphisms in x^1-direction is obviously given by $A_1^i G_i$. Noting that G_3 precisely generates local Lorentz transformations one concludes that identification of the $sl(2, \mathbf{R})$ generators and the infinitesimal generators of the gravitational symmetries crucially depends on the condition $\det e \neq 0$. Let us investigate the consequences of this observation for the RPS of the gravity theory: With the identifications (31) the solutions we used above to parametrize the RPS of the $\widetilde{SL}(2, \mathbf{R})$ gauge theory correspond to space time manifolds with $\det g = 0$. To any of these solutions, however, it is possible to find a gauge transformation yielding a solution corresponding to a nondegenerate space time metric. More precisely, this can be done in an infinite number of gravitationally inequivalent ways. E.g., in the elliptic sector, we might apply one of the following gauge transformations to A^{ell}

$$g_{[k]} = \begin{pmatrix} \cos\chi_k & \sin\chi_k \\ -\sin\chi_k & \cos\chi_k \end{pmatrix} \begin{pmatrix} 1 & b_k \\ 0 & 1 \end{pmatrix}$$

$$\chi_k = [\exp(x^0) + 2|\vartheta|]\sin(kx^1), \quad b_k = [\exp(x^0) + 2|\vartheta|]\cos(kx^1) \quad k \in \mathbf{N}. \quad (60)$$

We obtain

$$A_{[k]}^{ell} = \begin{pmatrix} b_k(\vartheta dx^1 + d\chi_k) & (1 + b_k{}^2)(\vartheta dx^1 + d\chi_k) + db_k \\ -(\vartheta dx^1 + d\chi_k) & -b_k(\vartheta dx^1 + d\chi_k) \end{pmatrix}. \quad (61)$$

The gauge transformations (60) are smoothly connected to the unity for arbitrary value of k as the χ_k are periodic functions in x^1. Nevertheless the solutions $A^{ell}_{[k]}$ are gravitationally inequivalent for different values of k. To prove this let us again choose a loop C running around the cylinder once. Under the restriction $\det g = -(\det e)^2 \neq 0$ the components of the zweibein $(e_0{}^+, e_1{}^+)$ induce a map $C \sim S^1 \to \mathbf{R}^2 \backslash \{0\}$ characterized by a winding number (not depending on the choice of C). Solutions with different winding numbers cannot be transformed into each other by gravitational symmetries, since they are separated by solutions with $\det e = 0$. (Also the discrete gravitational symmetry transformations mentioned above do not change the winding number). For different values of k the solutions (61) have different winding numbers, which proves our assertion.

This result generalizes to the other sectors of the theory: Solutions which are gauge equivalent in the $\widetilde{SL}(2, \mathbf{R})$ gauge theory are not equivalent in the gravity theory, if they have different winding number.

The winding number defined above is related to the kink number as defined in [20] by means of 'turn arounds' of the light cone along non contractible loops. More precisely, winding number k corresponds to kink number $2k$. (Odd kink numbers [20] characterize solutions which are not time orientable. Such solutions are not considered here).

The physical relevance of solutions with nontrivial winding number is not quite clear. They necessarily contain closed lightlike curves. There are, however, also solutions with trivial winding number containing closed lightlike curves. As outlined, in a conventional Hamiltonian treatment of the action (43) the constraints will generate infinitesimal gauge transformations rather than gravitational symmetry transformations. Thus on the Hamiltonian level the kink number will not appear in the parametrization of the reduced phase space, while, however, not all solutions with closed timelike curves can be excluded in this way. A similar situation occurs also when treating other models of two dimensional gravity contained in (32). It would be interesting to see, if the equivalence up to $\det g = 0$ of the Hamiltonian and Lagrangian formulation of four dimensional gravity leads to similarly inequivalent factoring spaces.

5 A Model for Quantum Gravity

In the previous sections we found that a large class of Hamiltonian systems, including gravitational ones, can be reduced to quantum systems of finitely many topological degrees of freedom. The question arises: Can such models serve as toy models for a quantum theory of four dimensional gravity? Indeed even in the absence of local degrees of freedom an illustrative treatment of some conceptual questions of quantum gravity is possible. Most prominent among these is the so called 'problem of time' [11], which we shall take up in this section for the example of R^2-gravity with Minkowski signature coupled to $SU(2)$ Yang Mills.

The Lagrangian of this system is

$$S = \int_{S^1 \times \mathbf{R}} \left[\frac{1}{8\beta^2} R_{ab} \wedge *R^{ab} + \frac{1}{4\gamma^2} tr(F \wedge *F) \right] \tag{62}$$

where the Hodge dual operation is, in contrast to (1), performed with the dynamical metric used to define also the torsionless curvature two-form R_{ab}, and the trace is taken, e.g., in the fundamental representation of $su(2)$ (the generators T_i, fulfilling (44) with $\kappa_{ij} = \delta_{ij}$, are then represented by $T_i = -i\sigma_i/2$, which yields $\kappa_{ij} = -2tr T_i T_j$ now). Rewriting (62) by means of Cartan variables ($\omega^a{}_b := -\varepsilon^a{}_b\omega, e^a$) in a Hamiltonian first order form, it becomes

$$S_H = \int_{S^1 \times \mathbf{R}} B_a De^a + B_3 d\omega + tr(EF) + [-\beta^2(B_3)^2 + \gamma^2 tr(E^2)]\varepsilon \qquad (63)$$

where we have chosen $E = E^i T_i$ to denote the 'electric fields' conjugate to the $SU(2)$-connection one-components A_1, and the B's are the conjugates to the spin connection ω_1 and the zweibein one-components $e_1{}^a \equiv (e_1{}^-, e_1{}^+)$. (Our conventions are $e^\pm = (e^2 \pm e^1)/\sqrt{2}$, yielding a light cone frame metric $\kappa_{+-} = 1$, whereas $\varepsilon = e^1 \wedge e^2 = e^- \wedge e^+$ so that $\varepsilon^{+-} = \varepsilon_{-+} = 1$). Obviously S_H is the sum of an $SU(2)$-EF-theory (43) (up to a factor -2) and an action S_G (32) with $u_3 = -\beta^2(B_3)^2 + \gamma^2 tr(E^2)$. In explicit terms the constraints following (naturally) from S_H are

$$G_a = \partial B_a + \varepsilon^b{}_a B_b \omega_1 + \varepsilon_{ab}[-\beta^2(B_3)^2 + \gamma^2 tr E^2]e_1{}^b, \qquad (64)$$
$$G_3 = \partial B_3 + \varepsilon_b{}^a B_a e_1{}^b, \qquad (65)$$

beside the unmodified $SU(2)$ Gauss law $G \approx 0$. We will not attempt to reformulate these constraints so as to possibly cure the global deficiencies of them with respect to diffeomorphisms noted at the end of the previous section. Instead we proceed with a straightforward quantization.

There are two independent Dirac observables as functions of the momenta ($q_{(s)} \equiv \oint Q_{(s)} dx^1/2\pi$)

$$q_{(1)} = \frac{-1}{\pi} \oint tr(E^2) dx^1 \equiv \frac{1}{2\pi} \oint E_i E_i dx^1$$

$$q_{(2)} = \frac{1}{2\pi} \oint [(B)^2 - \frac{2}{3}\beta^2(B_3)^3 + 2\gamma^2 tr(E^2)B_3] dx^1.$$

The corresponding level surfaces have topology $S^2 \times \mathbf{R}^2$ for $q_{(1)} \neq 0$ and $\mathbf{R}^2$ for $q_{(1)} = 0$.[7] This gives rise to the quantization condition (cf. end of sec. 3): $q_{(1)} = n^2/4, n \in \mathbf{N}_0$. Expanding the physical wave functionals in terms of eigenfunctions of $q_{(1)}$, the corresponding coefficients are

$$\exp\left(\frac{i}{\hbar} \oint (E_3 \partial\varphi \pm \ln B_\mp \, \partial B_3 dx^1\right) \tilde{\Psi}_n(q_{(2)}), \quad n = 2\sqrt{q_{(1)}} \in \mathbf{N}_0, \, q_{(2)} \in \mathbf{R},$$
$$(66)$$

where we have written the phase factor in some local target space coordinates with $\tan\varphi \equiv (E_2/E_1)$. The inner product with respect to $q_{(2)}$ is determined by

---

[7]Within the latter level surface the origin is an integral surface by itself. We will in the following disregard this small complication. – As suggested already through the chosen notation we assume β^2 and γ^2 to be non negative.

the hermiticity requirement on

$$p_{(2)} = -\frac{1}{2} \oint \frac{e_1{}^{\pm}}{B_{\mp}} dx^1, \tag{67}$$

the Dirac observable conjugate to $q_{(2)}$: as noted already in section 3, $p_{(2)}$ acts as the usual derivative operator on $\tilde{\Psi}_n$, thus leading to the ordinary Lebesgue measure $dq_{(2)}$.

We end up with the Hilbert space $\mathcal{H}$ of an effective two-point particle system with nontrivial phase space topology. As a basic set of operators acting in $\mathcal{H}$ we could use $q_{(2)}, p_{(2)}, q_{(1)}$, and $tr[\mathcal{P} \exp(\oint A_1 dx^1)]$. From the latter one may construct a ladder operator $l : n \rightarrow n + 1$.

All operators acting in $\mathcal{H}$ are thus found to be expressible in terms of $q_{(2)}, p_{(2)}$, and the number and ladder operators. However, we do not have an operator such as $g_{\mu\nu}(x^\mu)$. Following, furthermore, any textbook on elementary quantum mechanics, the next step in the quantization procedure would be to introduce an evolution parameter 'time', which we will call τ, and to require the wave functions to evolve in this parameter according to the Schroedinger equation. In the present case, however, the Hamiltonian following from (63) is a combination of the constraints,

$$H = - \oint [e_0{}^a G_a + \omega_0 G_3 + tr(A_0 \, G)], \tag{68}$$

so that the naive Schroedinger equation becomes meaningless.

Both of these items, the nonexistence of space-time dependent quantum operators as well as the apparent lack of dynamics, are correlated and they are not just a feature of the topological theory (62). Also in four dimensional gravity the quantum observables are some (not explicitly space-time dependent) holonomy equivalence classes and the Hamiltonian vanishes when acting on physical wave functions [21]. Diffeomorphisms are part of the symmetries of any gravity theory; as a consequence the Lie derivative into any 'spatial' direction can be found to equal the Hamiltonian vector field of some linear combination of the constraints (in our case $\mathcal{L}_1 = e_1{}^a G_a + \omega_1 G_3 + tr A_1 \, G$), whereas, on shell, x^0-diffeomorphisms will be generated by the Hamiltonian H. Thus, although 4D gravity has local degrees of freedom, any of its (uncountably many) Dirac observables will be also space-time independent.

To orientate ourselves as of how to introduce quantum dynamics within such a system, let us have recourse to the simple case of a nonrelativistic particle (NRP). As is well known, any Hamiltonian system can be reformulated in time reparametrization invariant terms. In the case of the NRP,

$$\int (p\frac{dq}{dt} - \frac{p^2}{2})dt = \int (p\dot{q} - \frac{p^2}{2}\dot{t})d\tau, \tag{69}$$

the equivalent system has canonical coordinates $(q, t; p, p_t)$ and the 'extended' Hamiltonian is proportional (via a Lagrange multiplier) to the constraint $C =$

$p^2/2 + p_t \approx 0$. Quantizing this system, e.g., in the coordinate representation, we observe that the implementation of the constraint $C\psi(q,t) = 0$ is equivalent to the Schroedinger equation of the original formulation, if one reinterpretes the canonical variable t as evolution parameter τ. Therefore, given this formulation of the NRP or similarly of any other system, the postulate of a Schroedinger equation within the transition from the classical to the quantum system becomes superfluous; rather it is already included within the Dirac quantization procedure in terms of a constraint equation.

The identification $t = \tau$ above can be looked upon also as a gauge condition with gauge parameter τ. This interpretation is helpful for the quantization of the parametrization invariant NRP in the momentum representation, in which case the space of physical wave functions is isomorphic to the space of functions of the Dirac observable p. The gauge condition $\bar{C} \equiv t - \tau = 0$ provides a perfect cross section for the flow of C. Thus it is possible to determine any phase space variable in terms of the Dirac observables p, $Q = q - pt$, as well as the gauge fixing parameter τ. Interpreting τ as a dynamical flow parameter 'time', the obtained evolution equations for p and q, transferred to the quantum level as $q(\tau) = i\hbar\, d/dp + \tau p$, $p(\tau) = p$, become equivalent to the Heisenberg evolution equations of the parametrized NRP.

The operator $q(\tau)$ above corresponds to a measuring device that determines the place of the particle at time τ. A measuring device that determines the time t at which the particle is at a given point $q = q_0$, on the other hand, corresponds to the alternative gauge condition $\tilde{C} \equiv q - q_0 = 0$. $\tilde{C}$ provides a good cross section only for $p \neq 0$. Ignoring this subtlety, e.g. by regarding only wave functions with support at $p \neq 0$, the (hermitian) quantum operator for such an experiment is $t(q_0) = -i\hbar\,[(1/p)d/dp - (1/2p^2)] + q_0/p$. In this second experimental setting Heisenberg's 'fourth uncertainty relation' between time t and energy $p^2/2 \sim -p_t$, usually motivated only heuristically, becomes a strict mathematical equation. We learn that different experimental settings are realized by means of different gauge conditions, and, at least in principle, vice versa.

The wave functions of (63) are basically functions of the Dirac observables, although part of the latter became discretized in the quantum theory. Transferring the ideas above to the gravity system, we should find gauge conditions to the constraints (64, 65). (It will not be necessary to gauge fix also G). As such we will choose

$$\partial B_+ = 0, \quad B_3 + \tau B_+ = 0, \quad e_1{}^- = 1. \tag{70}$$

It is somewhat cumbersome to convince oneself that this is indeed a good gauge condition. However, for $q_{(1)} \neq 0$ it provides even a globally well-defined cross section.[8] The gauge conditions together with the constraints allow to express

---

[8]One possibility to check the obtainability of (70) is to carefully analyse the Faddeev matrix, taking into account that due to $\oint \partial Q dx^1 \equiv 0$ and (11, 37) the gravity constraints are not completely linearly independent. This (infinite dimensional) matrix turns out to be non-degenerate, iff $B_+ \oint e_1{}^- dx^1 \neq 0$. For $q_{(1)} \neq 0$ any gauge orbit in the loop space contains a representative fulfilling this condition, which suffices to prove the assertion since the space of

all gravity phase space variables in terms of Dirac observables. In this way one obtains evolution equations such as

$$B_-(\tau) = -\frac{1}{2\pi}p_{(2)}q_{(2)} - \frac{\gamma^2}{2}q_{(1)}\tau - \frac{\beta^2\pi^2}{3(p_{(2)})^2}\tau^3, \quad B_+(\tau) = \frac{-\pi}{p_{(2)}}. \tag{71}$$

Antisymmetrizing this with respect to $q_{(2)}$ and $p_{(2)}$, (71) can be taken as an operator in the Hilbert space $\mathcal{H}$ defined above.[9] Similarly one finds $g_{11}(x^0) = 2e_1{}^+(x^0) = -p_{(2)}B_-(x^0)/\pi$, $(x^0 \equiv \tau)$, which now, up to operator ambiguities, becomes a well defined operator in our small quantum gravity theory, too.

Requiring that the τ-dependence of (70) is generated by the Hamiltonian H, the gauge conditions determine also the zero components of the zweibein and the spin connection. Actually, one zero mode of these Lagrange multiplier fields remains arbitrary as a result of the linear dependence of the constraints G_i (cf. also [13]). Requiring this zero mode to vanish as a further gauge condition, one finds $e_0{}^+ = 1$ and $e_0{}^- = \omega_0 = 0$. In other gauges the Lagrange multipliers can become also non trivial quantum operators. Furthermore, it is a special feature of the chosen gauge that the obtained operators are x^1-independent. (The existence of this gauge shows that $B_3 = const$ is an isometry or Killing direction of the metric). Again different choices of gauge conditions are interpreted as corresponding to different types of questions or measuring devices.

The alternative, at least for the paramtrization invariant NRP equivalent procedure to reintroduce time within the quantum theory was the direct implementation of the gauge in the wave functions. For this it was decisive that the initially chosen polarization of the wave functions contained the phase space variable subject to the gauge. To implement (70) analogously within the gravity theory under consideration, we Fourier transform (66), multiplied by $\delta[\partial Q_{(2)}]$, with respect to $B_-(x^1)$. The result is

$$\exp\left(\frac{i}{\hbar}\oint[E_3\partial\varphi + \frac{\partial B_+ B_3 + [\frac{\beta^2}{3}(B_3)^3 - \gamma^2tr(E^2)B_3]e_1{}^-}{B_+}]dx^1\right)$$

$$\Pi_{x^1}\left(\frac{const}{B_+}\right)\hat{\Psi}_n(p_{(2)}), \tag{72}$$

in which $\hat{\Psi}_n$ is the Fourier transform of the ordinary function $\tilde{\Psi}_n$. Eq. (72) certainly is in agreement with the general solution of the quantum constraints in a $(B_+, B_3, e_1{}^-, E)$ representation. In the gauge (70) the quantum wave functions take the form

$$\sum_n \exp\left[\frac{-i}{\hbar}\left(\frac{\gamma^2n^2}{8}\tau + \frac{\beta^2\pi^2}{3p_{(2)}^2}\tau^3\right)\right]c_n(p_{(2)})|n\rangle, \tag{73}$$

gauge orbits is connected in the case under study (no quantum number n_Q).

[9]The elementary procedure above coincides with the use of Dirac brackets for τ-dependent systems (in which case one extends the symplectic form by $d\tau \wedge dp_\tau$); this explains also B_- and B_+ do not commute anymore.

where $|n\rangle$ denotes the eigenfunctions of $q_{(1)}$ (inclusive $\exp[(i/\hbar)\oint E_3\partial\varphi dx^1]$) and we have reabsorbed the divergent factor of (72), being a function of $p_{(2)}$, into $c_n(p_{(2)})$.

At this point the case $\beta = 0$ is of special interest: for it S_H is seen to describe a Yang Mills theory coupled to a *flat* metric. Thus in some sense it is the parametrization (i.e. diffeomorphism) invariant formulation of the usual Yang Mills theory on the cylinder (with rigid Minkowski background metric). If we ignore the $p_{(2)}$ dependence of c_n for a moment, (73) with $\beta = 0$ indeed coincides with the time evolution generated by the (nonvanishing) Yang Mills Hamiltonian $-\gamma^2 \oint tr E^2 dx^1 \equiv \gamma^2\pi q_{(1)}$. This agreement gives support to the method used to derive (73).

The reason for the $p_{(2)}$-dependence of c_n is due to the fact that in the formulation (63) with $\beta = 0$ the metric induced circumference of the cylinder became a dynamical variable (on shell one has $p_{(2)} \propto \oint_{B_3=const} \sqrt{g_{11}}dx^1$). Within (70) one finds $-\oint G_+ \sim H$ to effectively implement the Schroedinger equation corresponding to (73). The effective Hamiltonian acting on $c_n|n\rangle$ is $-(\gamma^2/2)\oint tr E^2 dx^1 - \beta^2\pi^2\tau^2/p_{(2)}^2$. Thus generically the above procedure yields time dependent Hamiltonians (cf. also [13]).

The strategies developed at the example of a NRP to resolve the 'issue of time' within a quantum theory of gravity produced sensible results for the toy model (62). They, however, relied heavily on either the knowledge of all Dirac observables or on some specifically chosen polarization. To cope with the considerable technical difficulties of a quantum theory of four dimensional gravity, it might be worthwhile to extend the applicability of the method. One way to do so within our model is to allow for equivalence classes of wave functions coinciding at $\partial Q_{(2)} = 0$, the latter condition being enforced within the inner product [13]. In this way one can, e.g., implement the gauge condition $\partial e_1^- = 0$ as an operator condition in the B polarization of the wave functions as well, whereas a straightforward implementation of $\oint e_1^- = const$ seems again inadmissible.

References

[1] S.G. Rajeev, , Phys. Lett. B 212 (1988) 203. A. Migdal, Sov. Phys. Jept. *42* (1976) 413. K.S. Gupta, R.J. Henderson, S.G. Rajeev, O.T. Turgut, *Yang-Mills Theory on a Cylinder Coupled to Point Particles*, preprint UR-1327, ER-40685-777.

[2] J.E. Hetrick and Y. Hosotani, Phys. Lett. B 230 (1989) 88. E. Langmann and G.W. Semenoff, Phys. Lett. B 296 (1992) 117; B 303 (1993) 303. S. Shabanov, *Phys. Lett.* **318B** (1993) 323; *2D Yang-Mills theories, gauge orbit spaces, and the path integral quantization*, preprint T93/139, hep-th/9312160.

[3] J.E. Hetrick, *Canonical Quantization of Two Dimensional Gauge Fields*, preprint UvA-ITFA 93-15, hep-th/9305020. L. Chandar and E. Ercolessi,

Inequivalent Quatizations of Yang-Mills Theory on a Cylinder, preprint SU-4240-537.

[4] C.G. Callan, S.B. Giddings, J.A. Harvey, A. Strominger, *Phys. Rev.* D 45/4 (1992) R1005. H. Verlinde, in *The Sixth Marcel Grossmann Meeting on General Relativity*, edited by M. Sato (World Scientific, Singapore, 1992). D. Cangemi and R. Jackiw, *Ann. Phys.* (NY) **225**, (1993) 229. D. Christensen and R.B. Mann, Class. Quan. Grav. **9** (1992) 1. S.A. Hayward, *Cosmic Censorship in 2-dimensional dilaton gravity*, Max Planck Inst. preprint 1992. A. Mikovic, *Phys. Lett.* **B304** (1993) 70. H. Kawai and R. Nakayama, *Quantum R^2 Gravity in Two Dimensions*, preprint KEK 92-212.

[5] M. O. Katanaev and I. V. Volovich, *Phys. Lett.* 175B (1986) 413; M. O. Katanaev, *J. Math. Phys.* **32** (1991) 2483; *J. Math. Phys.* **34**/2 (1991) 700; W. Kummer, D.J. Schwarz, *Nucl.Phys.* **B382** (1992) 171; F. Haider, W. Kummer, *Int.Journ.Mod.Phys.* **9** (1994) 207; N. Ikeda and K.I. Izawa, *Prog.Theor.Phys.*, **89** (1993) 223

[6] D. Amati, S. Elitzur, E. Rabinovici, *On Induced Gravity in 2-d Topological Theories*, preprint hep-th/9212003.

[7] C. Teitelboim, *Phys. Lett.* **126B** (1983) 41; R. Jackiw, *1984 Quantum Theory of Gravity*, ed S. Christensen (Bristol: Hilger) p 403.

[8] T. Fukuyama and K. Kamimura, *Phys. Lett.* **160B** (1985) 259; K. Isler and C. Trugenberger, *Phys. Rev. Lett.* 63 (1989) 834; A. Chamsedine and D. Wyler, *Phys. Lett.* **228B** (1989) 75; *Nucl. Phys.* **340B** (1990) 595.

[9] Kostant, B. (1970) Quantization and unitary representations. In *Lectures in modern analysis III* (ed. C.T. Taam). Lecture notes in mathematics, Vol. 170. Springer, Berlin. Souriau, J.-M (1970) *Structure des systmes dynamiques* Dunod, Paris

[10] Kirillov, A.A. (1976) *Elements of the theory of representations* Springer, Berlin. Woodhouse, N.M.J. Geometric Quantization, second Ed. 1992, Clarendon Press, Oxford.

[11] C.J. Isham in *Recent Aspects of Quantum Fields*, ed. Gausterer et. al. LNP **396**, p.123, Springer Berlin Heidelberg 1991.

[12] P. Schaller and T. Strobl, *Diffeomorphisms versus Nonabelian Gauge Transformations: An Example of 1+1 Dimensional Gravity*, preprint TUW9325, hep-th/9401110, to be published in *Phys. Letts. B*.

[13] P. Schaller and T. Strobl, *Class.Quan.Grav.* 11 (1993) 331.

[14] T. Strobl, *Quantization and the Issue of Time for Various Two-Dimensional Models of Gravity*, hep-th/9308155, to be publ. in *J.Mod.Phys.* D, (Proceedings of Journees Relativistes).

[15] T. Strobl, *Dirac Quantization of Gravity-Yang-Mills Systems in 1+1 Dimensions*, preprint TUW9326, hep-th/9403121. P. Schaller and T. Strobl, *Poisson Structure Induced (Topological) Field Theories*, preprint TUW9403, hep-th/9405110.

[16] Y. Choquet-Bruhat, C. DeWitt-Morette, *Analysis, Manifolds and Physics, Part II: 92 Applications*, North-Holland Physics 1989.

[17] P.A.M. Dirac, *Lectures on Quantum Mechanics*, Yeshiva University, New York 1964. M. Henneaux and C. Teitelboim, *Quantisation of Gauge Systems*, Princeton University Press 1992.

[18] J.F. Adams, *Infinite Loop Spaces*, Annals of Mathamatics Studies 90, Princeton University Press, Princeton 1978.

[19] J. Goldstone and R. Jackiw, *Phys. Lett.* **74B** (1978) 81.

[20] D. Finkelstein and C. W. Misner, *Ann. Phys.* (NY) **6**, 230 (1959); cf. also K. A. Dunn, T. A. Harriott, J. G. Williams, *J. Math. Phys.* **33/4** (1992) 1437, where, however, the solutions in chapter V do not correspond to constant curvature solutions

[21] A. Ashtekar *Lectures on Non–Perturbative Canonical Gravity*, World Scientific, Singapore, 1991.

Wodzicki Residue and Anomalies of Current Algebras

Jouko Mickelsson

Royal Institute of Technology, Stockholm
e-mail jouko@ theophys.kth.se, tel. 46 8 790 7278

Abstract. The commutator anomalies (Schwinger terms) of current algebras in $3 + 1$ dimensions are computed in terms of the Wodzicki residue of pseudo-differential operators; the result can be written as a (twisted) Radul 2-cocycle for the Lie algebra of PSDO's. The construction of the (second quantized) current algebra is closely related to a geometric renormalization of the interaction Hamiltonian $H_I = j_\mu A^\mu$ in gauge theory.

1 Introduction

One of the problems one meets all the time in quantum field theory is that products of field operators at equal times in the same points in the physical space are ill-defined. The first renormalization which one always performs is the normal ordering of the operator products, i.e. a shift of the energy lowering operators to the right, those giving a vanishing contribution when acting on the Dirac vacuum. In fact, in many cases in 1+1 dimensional models the normal ordering is sufficient to produce well-defined second quantized operators (more precisely, operator valued distributions). One such a case is the algebra of local charges formed as certain quadratic expressions in field operators. In one space dimension the construction leads to affine Kac-Moody and related algebras.

In higher dimensions one needs further renormalizations in addition to the normal ordering. The aim of this talk is to explain the renormalizations needed for local charges in 3+1 space-time dimensions. More specifically, we shall study (chiral) fermions minimally coupled to external gauge fields. The basic idea in the present renormalization scheme is to conjugate the Gauss law generators by unitary operators, which are functions of the external gauge field, in the one-particle space such that the resulting conjugated operators can be quantized using the standard normal ordering prescription. As a by-product, we obtain a new geometric renormalization of the Dirac-Yang-Mills interaction hamiltonian.

In one space dimension there is a nontrivial 2-cocycle which defines a central extension of the Lie algebra of pseudodifferential operators, [KK]. This algebra can be identified as the quantum W_∞ algebra. It has as a subalgebra (when the coefficients of the PSDO's are taken in a simple Lie algebra) an affine algebra.

The Kravchenko-Khesin cocycle has been generalized by Radul to all dimensions, [R]. We shall show that a twisted form of the Radul cocycle, when applied to the renormalized local charges, gives an extension of the naive current algebra which is equal to the Mickelsson-Faddeev algebra, [M, F-Sh], [M2]. Thus the MF algebra is closely related to a multidimensional version of the W_∞ algebra.

2   Central Extension of Algebras of PSDO's

Let us first consider pseudodifferential operators in one dimension, on a circle. Asymptotically, a PSDO is a defined by a Laurent series

$$(2.1) \qquad a(x,p) = \sum_{k \leq n} a_k(x) p^k$$

where n is some integer and the a_k's are smooth functions on the circle. The momentum p is the symbol of the operator $-i\partial_x$. The product is defined as

$$(2.2) \qquad a * b = \sum_{k=0,1,2,\ldots} \frac{(-i)^k}{k!} \partial_p^k a(x,p) \partial_x^k b(x,p).$$

Note that each $(a*b)_j$ is a finite sum of products of derivatives in the coefficients a_i, b_i.

The algebra of PSDO's becomes a Lie algebra B under the commutator $[a, b] = a * b - b * a$.

The Adler-Manin residue of a PSDO is defined as

$$(2.3) \qquad Res(a) = \frac{1}{2\pi} \int \operatorname{tr} a_{-1} dx$$

where we have included the trace in order to allow a generalization to matrix valued PSDO's. The residue behaves like a trace on the algebra of PSDO's. It is obviously a linear functional and furthermore it satisfies

$$(2.4) \qquad Res([a, b]) = 0.$$

The function $\log(p)$ is not a PSDO (since its expansion contains arbitrarily high powers of p) but nevertheless one can define a Lie algebra 2-cocycle, the KK cocycle, by the formula

$$(2.5) \qquad c(a, b) = Res[log(p), a] * b.$$

This is because the commutator $[log(p), a]$ is a PSDO,

$$(2.6) \qquad [log(p), a] = \sum_{k=1,2,\ldots} -\frac{i^k}{k} \partial_x^k a(x,p) p^{-k}$$

The 2-cocycle property

$$c(a, [b, c]) + \text{cycl. permutations} = 0$$

is a simple consequence of (2.4).

In the case when a, b are zeroth order PSDO's, i.e. they are multiplication operators, the value of the KK cocycle is

$$(2.7) \qquad c(a, b) = \frac{i}{2\pi} \int \text{tr}\, a(x) b'(x) dx.$$

This is exactly the central term of an affine Lie algebra (when a, b take values in a simple Lie algebra).

The KK cocycle can be defined in any number of space dimensions, [R]. Thus one might wonder whether the higher dimensional Radul cocyles have anything to do with anomalies of current algebras. The physically relevant extensions of current algebras in higher dimensions are generally not central extensions but extensions by some abelian ideal. For this reason it is not immediately obvious what is the relevance of the Radul cocycle in higher dimensions. We shall clarify this matter in section 5.

3 The Wodzicki Residue

Let M be a compact manifold of dimension n. A PSDO on M, with coefficients in a vector bundle V over M, is locally given by a matrix valued symbol $a(x, p)$. Here x is a local coordinate on M and p is a fiber coordinate in the cotangent bundle T^*M. The product rule for symbols is determined from the definition of the operator A acting on sections of V.

$$(3.1) \qquad (A\psi)(x) = \frac{1}{(2\pi)^{n/2}} \int e^{-ix \cdot p} a(x, p) \hat{\psi}(p) d^n p$$

where $\hat{\psi}$ is the Fourier transform of the section ψ. Thus the symbol of the product AB is

$$(3.2) \qquad (a * b)(x, p) = \frac{1}{(2\pi)^n} \int e^{i(x-y) \cdot (p-q)} a(x, q) b(y, p) d^n y d^n q.$$

The adjoint of A (in the Hilbert space of square-integrable sections, the measure defined by a Riemannian metric on M) is in general a complicated expression in terms of the symbol a. We shall give the formula only in the euclidean case:

$$(3.3) \qquad A^* \sim a^* + \Omega a^* + \frac{1}{2!} \Omega^2 a^* + \ldots$$

where

$$\Omega = -i \sum_j \partial_{x_j} \partial_{p_j}$$

and a^* is the matrix adjoint of the matrix valued symbol a.

Let a_{-k} be a symbol of integral order $-k$ in n space dimensions. Then the trace is asymptotically

$$\text{tr} a_{-k} = \int d^n x \int d^n p \, \text{tr} \, a_{-k}(x,p) \sim \int d^n x \int |p|^{-k} |p|^{n-1} d|p|$$

from which follows that a_{-k} is of trace-class if and only if $-k \leq -n-1$. Similarly, p_{-k} is Hilbert-Schmidt if and only if the integer $-2k \leq -n-1$.

Most of the time we are interested only on the asymptotic behaviour of the symbols a for large momenta p. We assume that a PSDO has an asymptotic expansion of the form

(3.4)
$$a = a_k + a_{k-1} + a_{k-2} \ldots$$

where each $a_k = a_k(x,p)$ is a smooth function of x and of $p \neq 0$, a_k is homogeneous of degree k in the momenta, $\frac{a_k}{|p|^{k+1}} \to 0$ as $|p| \to \infty$. One can show that the asymptotic expansion of the symbol $a * b$ can be written as (compare with (2.2))

(3.5)
$$a * b = \sum_m \frac{(-i)^{|m|}}{m!} \left(\partial_{p_1}^{m_1} \ldots \partial_{p_n}^{m_n} a(x,p) \right) \left(\partial_{x_1}^{m_1} \ldots \partial_{x_n}^{m_n} b(x,p) \right)$$

where the sum is over all multi-indices $m = (m_1, \ldots, m_n)$, $|m| = m_1 + \cdots + m_n$, and $m! = m_1! \ldots m_n!$, [H].

The Wodzicki residue [W] of a PSDO a is defined as a linear functional which depends only on the component a_{-n},

(3.6)
$$Res(a) = \frac{1}{(2\pi)^n} \int_{|p|=1} \text{tr} a_{-n}(x,p) \eta (d\eta)^{n-1}$$

where $\eta = \sum p_k dx_k$ and $d\eta = \sum dp_k dx_k$ is the symplectic 2-form on the cotangent bundle T^*M, $n > 1$. In the case $n = 1$ this is *almost* the Adler-Manin residue, [Ad], [Ma]. The difference is the following. If $a_1 = \frac{\alpha(x)}{p}$ then the resudue is zero, because the unit sphere in momentum space consists of two points ± 1 and the momentum space integral is

$$\int_{|p|=1} \frac{\alpha(x)}{p} p dx = (\alpha(x)|_{p=+1} - \alpha(x)|_{p=-1}) dx = 0.$$

However, if $a_{-1} = \frac{\alpha(x)}{|p|}$ then the integral becomes the sum $\alpha(x) + \alpha(x) = 2\alpha(x)$. Thus we can write

$$Res_{AM}(a) = \frac{1}{2} Res_W(\epsilon a)$$

where $\epsilon = p/|p|$. Usually one redefines the Wodzicki residue in one dimension so that it agrees precisely with the Adler-Manin residue, [W].

The Radul cocycle in n dimensions is defined as

(3.7) $$c(a, b) = Res([log|p|, a] * b).$$

For multiplication operators the Radul cocycle vanishes in dimensions higher than one. The structure of the Lie group defined by the Radul cocycle has recently been studied in [KV].

4 Renormalized Currents as PSDO's

Consider a system of quantized fermions in external vector potentials A in $3 + 1$ space-time dimensions. A takes values in $\mathbf{g}$, a finite-dimensional Lie algebra. We set up a hamiltonian formalism for second quantization: we consider field (anti)commutation relations at a fixed time $t = 0$.

In Schrödinger picture the wave functions are functions $\phi(A)$ of the potential, with values in a fermionic Fock space $\mathcal{F}$.

We use the temporal gauge $A_0 = 0$ and we consider only time-independent gauge transformations.

The (free) Dirac field ψ satisfies the CAR algebra

(4.1) $$\psi^*_{ia}(x)\psi_{jb}(y) + \psi_{jb}(y)\psi^*_{ia}(x) = \delta_{ij}\delta_{ab}\delta(x - y)$$

where i, j are space-time indices and a, b are internal symmetry indices; the latter refer to a unitary representation of $\mathbf{g}$.

The free vacuum is characterized by the property

(4.2) $$\psi(u)|0> = 0 = \psi^*(v)|v> \quad \text{for } u \in \mathcal{H}_+, v \in \mathcal{H}_-$$

where $\mathcal{H}_\pm$ are the positive and negative energy subspaces of the one particle Hilbert space $\mathcal{H}$ and

$$\psi(u) = \int \psi_{ia}(x)u_{ia}(x)d^3x.$$

In the perturbation theory based on Dyson expansion one writes the scattering amplitudes in terms of expressions like

(4.3) $$< 0|H_I(t_1)H_I(t_2)\ldots H_I(t_n)|0 > .$$

These are integrated over the times $t_1 > t_2 \cdots > t_n$. Divergencies for the scattering amplitudes occur since the interaction hamiltonian

(4.4) $$H_I(x) = \psi^*(x)\gamma_0\gamma^k T_a \psi(x)A^a_k(x)$$

involves products of field operators at the same point and does not lead to a well-defined operator distribution in Fock space. We need a renormalization even after normal ordering. (In $1 + 1$ dimensions a normal ordering is sufficient!)

More precisely, the technical reason for divergencies is the following. Let ϵ be the sign of the free hamiltonian $H_0 = \gamma_0 \gamma^k \nabla_k$ in the one-particle representation. Then the off-diagonal blocks $[\epsilon, H_I]$ are Hilbert-Schmidt only when the space-time dimension is at most 2. As discussed in [Ar], the Hilbert-Schmidt property is both necessary and sufficient for quantization of operators of the type

$$H_I = \sum X_{nm} a_n^* a_m$$

where the a_n's satisfy the CAR relations

$$a_n^* a_m + a_m a_n^* = \delta_{nm}$$

The X_{nm}'s are the matrix elements of H_I in a basis of eigenvectors of H_0 (compactify the physical space to get a discrete basis). Note that the norm of the state $H_I|0>$ is given by

$$\|H_I|0>\|^2 = \sum_{E_n > 0, E_m < 0} |X_{nm}|^2,$$

where E_n's are the energy levels for H_0. The finiteness of this norm is just the Hilbert-Schmidt property for one of the off-diagonal blocks of the one-particle operator.

The badly behaving interaction hamiltonian is renormalized as follows. For each A we construct a unitary operator T_A in the one-particle space such that

$$[\epsilon, T_A^{-1}(H_0 + H_I)T_A]$$

is Hilbert-Schmidt.

The strategy is to obtain $T_A = 1 + t_{-1} + t_{-2} \ldots$ as an expansion in homogeneous pseudodifferential operators t_k of degree $k = 0, -1, -2, \ldots$. We shall consider separately the left and right handed sectors in the space of 4-component fermions.

Let us consider the following operator acting on two-component fermions:

$$T_A = 1 - \frac{i}{4|p|^2}[p, A] - \frac{1}{32|p|^4}[p, A]^2 - \frac{1}{8}\left[\frac{\sigma_k}{|p|^2} - 2\frac{pp_k}{|p|^4}, \partial_k A\right]$$

(4.5)
$$- \frac{1}{8|p|^4}[p, A](A \cdot p) - \frac{1}{8|p|^4}(A \cdot p)[p, A] + O(-3)$$

We have used the following notation: $p = p_k \sigma_k$, $A = A_k \sigma_k$ The commutators in (4.5) are all ordinary commutators of Pauli matrices σ_k and of Lie algebra elements in $\mathbf{g}$. The Pauli matrices are normalized such that $\sigma_1 \sigma_2 = i\sigma_3$ and $\sigma_k^2 = 1$.

Since the physical space is assumed to be compact, there can be no infrared divergencies in the theory, so it sufficient to work with the asymptotic expansions of the operators in momentum space; the asymptotic expansion gives a complete picture of the ultraviolet properties of the theory. If one feels uneasy about the 'infrared singularity' at $p = 0$ of the terms in the asymptotic expansion one can always replace the inverse powers $1/|p|^n$ by some smooth functions which agree with the original for large values of $|p|$.

We can write $T_A^{-1}(H_0 + H_I)T_A = H_0 + W_A$. After a tedious computation we obtain

$$W_A = T_A^*(p + iA)T_A - p = \frac{ip}{|p|^2}A \cdot p$$

$$- \frac{1}{8}\left[p, \left[\frac{\sigma_k}{|p|^2} - 2\frac{pp_k}{|p|^4}, \partial_k A\right]\right] - \frac{\sigma_k}{4|p|^2}[p, \partial_k A]$$

$$(4.6) \qquad + \frac{i}{2|p|^2}\epsilon_{ljk}p_j[A_l, A_k] - \frac{p}{|p|^2}A_m A_m + \frac{p}{|p|^4}(A \cdot p)^2 + O(-2).$$

It is then a simple computation to show that $[\epsilon, W_A]$ is of degree -2, and therefore the operator is Hilbert-Schmidt. There is no magic in the derivation of the formula (4.5) for T_A. It is a simple recursive procedure. Writing

$$T_A = 1 + t_{-1} + t_{-2} + \ldots$$

in the asymptotic expansion, one gets

$$T_A^*(p + \alpha_0 + \alpha_{-1} + \ldots)T_A = p + \alpha_0' + \alpha_{-1}' + \ldots,$$

where

$$\alpha_0' = \alpha_0 + [p, t_{-1}]$$

$$(4.7) \qquad \alpha_{-1}' = \alpha_{-1} + [\alpha_0, t_{-1}] + pt_{-2} + (T_A^*)_{-2}p - i\sigma_k\partial_k t_{-1}.$$

Again, the commutators above are ordinary matrix commutators (and not *-commutators).

The Hilbert-Schmidt condition on $[\epsilon, W_A]$ is equivalent to the pair of equations

$$[\epsilon, \alpha_0'] = 0 \text{ and } [\epsilon, \alpha_{-1}'] - i(\partial_{p_k}\epsilon)(\partial_{x_k}\alpha_0') = 0.$$

Inserting from (4.7), the first equation is just a linear algebraic equation for the symbol t_{-1}. But then we can solve t_{-2} from the second equation above.

In the one-particle representation the infinitesimal gauge transformations X are acting on Schrödinger wave functions $\phi(A)$ as the operators $\mathcal{L}_X + X$, where the first part (Lie derivative) is the gauge action on vector potential A and the second is a multiplication operator acting on the value $\phi(A) \in \mathcal{H}$. After the conjugation by the A dependent operator T_A these become $\mathcal{L}_X + \theta(X; A)$ with

$$(4.8) \qquad \theta(X; A) = T_A^{-1}XT_A + T_A^{-1}(\mathcal{L}_X T_A).$$

By construction,

$$[\theta(X; A) + \mathcal{L}_X, \theta(Y; A) + \mathcal{L}_Y]$$
$$= [\theta(X; A), \theta(Y; A)] + \mathcal{L}_X \theta(Y; A) - \mathcal{L}_Y \theta(X; A) + [\mathcal{L}_X, \mathcal{L}_Y]$$
$$(4.9) \qquad = \theta([X, Y]; A) + \mathcal{L}_{[X,Y]}$$

That is, the functions $\theta(X; \cdot)$ form a 1-cocycle for the gauge action of $Map(M, \mathbf{g})$. If the function T_A is constructed as above, then $\theta(X; A)$ is in the Lie algebra of the *restricted unitary group* U_{res}, [M1]. The latter is the subgroup of $U(\mathcal{H})$ consisting of operators g such that $[\epsilon, g]$ is Hilbert-Schmidt.

Proof: Since $H'(A) = T_A^{-1}(H_0 + H_I(A))T_A = H_0 + W_A$ and $[\epsilon, W_A]$ is HS, we observe that the sign $\epsilon'(A)$ of the Hamiltonian $H'(A)$ differs from ϵ by a HS operator. Let g be a finite gauge transformation. It acts on Schrödinger wave functions by

$$(R(g)\phi)(A) = g \cdot \phi(g^{-1}Ag + g^{-1}dg)$$

But after the conjugation by T_A:

$$(R'(g)\phi)(A) = (T_{g \cdot A}^{-1} g T_A)\phi(g^{-1} \cdot A) \equiv \omega(g; A)\phi(g^{-1} \cdot A).$$

Thus

$$\begin{aligned}
[\epsilon, \omega(g; A)] &= \epsilon T_{g \cdot A}^{-1} g T_A - T_{g \cdot A}^{-1} g T_A \epsilon \\
&= (T_{g \cdot A}^{-1} \epsilon T_{g \cdot A}) g T_A - T_{g \cdot A}^{-1} g (T_A \epsilon T_A^{-1}) T_A \\
&= T_{g \cdot A}^{-1} \left((T_{g \cdot A} \epsilon T_{g \cdot A}^{-1}) g - g(T_A \epsilon T_A^{-1}) \right) T_A \\
&\equiv T_{g \cdot A}^{-1} \left(\epsilon(g \cdot A) g - g\epsilon(A) \right) T_A
\end{aligned}$$

where we have used the equivariantness of the family of Dirac operators,

$$gH(A)g^{-1} = H(g \cdot A).$$

Thus finite gauge transformations satisfy the HS condition; considering one-parameter subgroups one proves the HS condition for the generators.

The asymptotic expansion for (4.8) is

$$(4.10) \qquad \theta(X; A) = X + \frac{i}{4}\frac{[p, dX]}{|p|^2} + \theta_{-2} + O(-3)$$

with

$$\theta_{-2} = -\frac{1}{4}\frac{[\sigma_k, A]}{|p|^2}\partial_k X + \frac{1}{2}\frac{[p, A]}{|p|^4}p_k \partial_k X$$
$$+ \frac{1}{16}\frac{[p, A]}{|p|^4}[p, dX].$$

If X is any bounded bilinear quantity in the fermion creation and annihilation operators such that its off-diagonal blocks in the one-particle representation with respect to the energy polarization are HS, then the second quantized operator $\hat{X}$ is well-defined (after normal ordering) and

$$(4.11) \qquad [\hat{X}, \hat{Y}] = \widehat{[X, Y]} + c(X, Y),$$

where c is a Schwinger term, [L],

$$(4.12) \qquad c(Y, Y) = \frac{1}{4} \mathrm{tr} \epsilon [\epsilon, X][\epsilon, Y].$$

Example Multiplication operators in $1 + 1$ dimensions. Multiplication operators are PSDO's X of order zero with a p independent symbol $a(x)$. $\epsilon = \frac{p}{|p|}$ and

$$(4.13) \qquad [\epsilon, X] = -2i\delta(p)a'(x).$$

Now the trace (4.12) can be written as the conditionally converging trace

$$(4.14) \qquad c(X, Y) = \frac{1}{2} X[\epsilon, Y] = \frac{1}{2\pi i} \int \mathrm{tr}\, a(x) b'(x) dx.$$

This is the central term of an affine algebra.

In the $3 + 1$ dimensional case one just inserts from (4.8) to (4.12):

$$(4.15) \qquad c(X, Y; A) = \frac{1}{4} \mathrm{tr} \epsilon [\epsilon, \theta(X; A)][\epsilon, \theta(Y; A)].$$

Actually, we can write

$$(4.16) \qquad c(X, Y; A) = \frac{1}{2} \mathrm{tr}[\epsilon, \theta(X; A)]\theta(Y; A)$$

as a conditionally convergent trace: Compute first the traces for the finite- dimensional matrices, perform the momentum space integration over the spherical angles, next the integration over $|p|$, and finally integrate the star product of symbols in configuration space.

5 Computation of the Commutator Anomaly

The result of the computation starting from (4.16) is rather complicated expression involving terms of all orders in A. However, there is a great simplification if we are interested only on the cohomology class of the cocycle. A change in the renormalization $T_A \mapsto T'_A = T_A g_A$ of the gauge currents (with $g_A \in U_{res}$) leads to a modification

$$\theta(X; A) \mapsto \theta'(X; A) = g_A^{-1}\theta(X; A)g_A + g_A^{-1}\mathcal{L}_X g_A.$$

It is easy to check that

$$c' - c = (\delta\lambda)(X, Y; A) = \lambda([X, Y]; A) - \mathcal{L}_X\lambda(Y; A) + \mathcal{L}_Y\lambda(X; A),$$

where λ is the 1-cochain

$$\lambda(X; A) = \frac{1}{4}\mathrm{tr}\left[\epsilon, g_A^{-1}[X, g_A] + g_A^{-1}\mathcal{L}_X g_A - (\mathcal{L}_X g_A)g_A^{-1}\right]_+,$$

with $[A, B]_+ = AB + BA$.

We want to show that if we choose λ in a suitable way then the new cocycle c' takes the form

$$c'(X, Y; A) = \frac{i}{24\pi^2}\int \mathrm{tr} A[dX, dY].$$

Let us first define the regularized trace, with $\Lambda > 0$,

$$\mathrm{tr}_\Lambda R = \frac{1}{(2\pi)^3}\int_{|p|\leq\Lambda} \mathrm{tr}\, r(x, p)d^3x d^3p + \frac{1}{(2\pi)^3}\int_{|p|>\Lambda} \mathrm{tr}\left(r - \sum_{k=0}^{3} r_{-k}\right)d^3x d^3p$$

for a PSDO R of degree zero. Define

$$\lambda(X; A) = \frac{1}{2}\mathrm{tr}_\Lambda\, \epsilon\, \theta(X; A).$$

Then

$$(\delta\lambda)(X, Y; A) = \frac{1}{2}\mathrm{tr}_\Lambda\epsilon\,[\theta(X; A), \theta(Y; A)].$$

If the regularize trace were symmetric, this would be equal to the cocycle c. In order to show that $\mathrm{tr}(a*b) = \mathrm{tr}(b*a)$ for a pair of symbols one has to perform partial integration both in the momentum and coordinate variables. The coordinate integration does not cause any problems, since we assumed that the manifold is compact and without boundary. However, there can be boundary terms arising from the momentum space integration. Integrating a term of degree -3 leads to finite boundary contribution in partial integration: the integrand behaves like $|p|^{-2}$ at infinity, cancelling the factor $|p|^2$ coming from the integration measure. The boundary contributions from terms of degree less than 3 vanish at $|p| \to \infty$.

As we have seen, the terms in c which are not coboundaries of some 1-cochains must be of order -3. The terms of order greater than -3 vanish in (4.15) by the algebra of Pauli matrices and by the fact that the momentum space integration over spherical angles of a symbol of odd degree gives zero. According to the multiplication rule (3.5), with $X_k \equiv \theta(X; A)_k$,

$$S \equiv (\epsilon[\theta(X; A), \theta(Y; A)])_{-3} = \epsilon\,([X, Y_{-3}] + [X_{-1}, Y_{-2}]$$
$$+\{X, Y_{-2}\} + \{X_{-1}, Y_{-1}\} + \{X_{-2}, Y\} - (X \leftrightarrow Y)),$$

where we have denoted $\{a, b\} = -i\sum \partial_{p_k} a \partial_{x_k} b$, 'half Poisson bracket'. The commutators on the right are matrix commutators. The cutt-off trace of the

$\{X_{-1}, Y_{-1}\}$ term is seen to vanish after performing the spherical part of the momentum space integration. By partial integration we obtain

$$\operatorname{tr}_\Lambda S = \operatorname{tr}_\Lambda \left([\epsilon, X_{-1}]Y_{-2} + [\epsilon, X_{-2}]Y_{-1} - \{\epsilon, Y\}X_{-2} + \{\epsilon, X\}Y_{-2}\right)$$

$$+ \frac{i}{(2\pi)^3} \int_{|p|=1} \operatorname{tr} \epsilon Y_{-2}(p_k \partial_{x_k} X) \eta(d\eta)^2 - \frac{i}{(2\pi)^3} \int_{|p|=1} \operatorname{tr} \epsilon X_{-2}(p_k \partial_{x_k} Y) \eta(d\eta)^2$$

$$= \frac{1}{2}\operatorname{tr}_\Lambda \left(([\epsilon, \theta(X; A)] * \theta(Y; A))_{-3} - ([\epsilon, \theta(Y; A)] * \theta(X; A))_{-3}\right)$$

(5.1)

$$+ \operatorname{Res} \epsilon[log(|p|), \theta(X; A)] * \theta(Y; A).$$

It follows that

(5.2) $$c_\Lambda(X, Y; A) = \operatorname{Res} \epsilon[log(|p|), \theta(X; A)]\theta(Y; A).$$

(Here star product shoud be used). The residue is easely computed (as the spherical integrals in (5.1)). The result is

(5.3) $$c_\Lambda(X, Y; A) = \frac{i}{24\pi^2} \int_x \operatorname{tr} A[dX, dY],$$

which is the cocycle derived in [M, F-Sh] in a different context. Note that the final results (5.2) and (5.3) do not depend on the cut-off parameter Λ. This method of computing cocycles can be generalized to various directions, [LM].

The action functional in Connes noncommutative geometry model of Yang-Mills theory is also defined in terms of the Wodzicki residue, [C]. It would be interesting to see more precisely the relation between the present hamiltonian approach and the Lagrangian method of Connes.

Actually the formula (5.2) can be applied to the algebra W' of all PSDO's P (in three dimensions) obeying the condition

(5.4) $$deg\,[\epsilon, P] \leq -2$$

giving a twisted form of the Radul cocycle on the Lie algebra W',

(5.5) $$c(P, Q) = \operatorname{Res} \epsilon[log(|p|), P]Q$$

So we get a generalization of the W_∞ algebra as the central extension (defined by the 2-cocycle (5.5)) of the Lie algebra of restricted spinorial PSDO's in three dimensions. In n dimensions the condition (5.4) should be replaced by the requirement that the degree of the commutator is strictly less than $-\frac{1}{2}n$. We shall end the discussion by giving the proof that (5.5) is a 2-cocycle for the restricted Lie algebra W' is n dimensions. Denote

$$\operatorname{Res}' A = \operatorname{Res} \epsilon A.$$

Since $Res[A, B] = 0$ for any pair of PSDO's A, B we have

$$Res\, \epsilon[\epsilon, A][\epsilon, B] = Res(A\epsilon B - \epsilon AB - AB\epsilon + \epsilon A\epsilon B\epsilon)$$
$$= 2Res(A\epsilon B - \epsilon AB) = -2Res[\epsilon A]B = -2Res\epsilon[A, B].$$

If now $deg[\epsilon, A] < -n/2$ and $deg[\epsilon, B] < -n/2$ then $deg\epsilon[\epsilon, A][\epsilon, B] < -n$ and it follows that the residue vanishes. Thus

(5.6) $$Res'[A, B] = 0$$

for any $A, B \in W'$. Denote $\ell = log(|p|)$. By (3.5) and performing partial integration in momentum space one gets $Res[\ell, A] = 0$ for any PSDO A. Since ℓ commutes with ϵ, we have also

(5.7) $$Res'[\ell, A] = 0$$

although ℓ is not a PSDO. Define

$$\omega(A, B, C) = c(A, [B, C]) + c(B, [A, C]) + c(C, [A, B]).$$

Then

(5.8) $$\omega(A, B, C) = Res'([\ell, A][B, C] + [\ell, B][C, A] + [\ell, C][A, B]).$$

Using (5.6) and (5.7) we get

$$\omega(A, B, C) = Res'(A[[B, C], \ell] + A[[\ell, B], C] + A[[\ell, C], B]) = 0,$$

by Jacobi's identity, proving that c is indeed a 2-cocycle.

The formula (5.3) can be generalized to higher (odd) space dimensions in various ways. One approach is to use the cohomological descent equations starting from a Chern class in dimension $n + 3$, [M, F-Sh], [M2]. There exists another rather natural and much simpler generalization, which was found recently when studying p-brane symmetries, [CFNW]. The realization of that algebra as an algebra of PSDO's (of type W') is now under preparation.

References

[Ad] M. Adler, Invent. Math. **50**, 219 (1979).
[Ar] H. Araki in: *Contemporary Mathematics*, vol. 62, American Mathematical Society, Providence (1987).
[C] A. Connes, Commun. Math. Phys. **117**, 673 (1988).
[CFNW] M. Cederwall, G. Ferretti, B. Nilsson, and A. Westerberg, preprint ITP 93-37 Gothenburg; HEP-TH/940127.

[H] L. Hörmander: *The Analysis of Linear Partial Differential Operators III.* Springer-Verlag, Berlin (1985).

[KV] M. Kontsevich and S. Vishik, preprint Max-Planck-Institut für Mathematik, Bonn (April 1994); HEPTH 94 04 46.

[KK] O.S. Kravchenko and B.A. Khesin, Funct. Anal. Appl. **25**, 83 (1991).

[L] Lars-Erik Lundberg, Commun. Math. Phys. **50**, 103 (1976).

[LM] E. Langmann and J. Mickelsson, work under preparation.

[Ma] Yu.I. Manin, J. Sov. Mat. **11**, 1 (1979).

[M1] J. Mickelsson, Lett. Math. Phys. **28**, 97 (1993).

[M2] J. Mickelsson: *Current Algebras and Groups.* Plenum Press, London (1989).

[M,F-Sh] J. Mickelsson, Commun. Math. Phys. **97**, 361 (1985);
L. Faddeev and S. Shatasvili, Theoret. Math. Phys. **60**, 770 (1984).

[R] A.O. Radul, Funct. Anal. Appl. **25**, 25 (1991).

[W] M. Wodzicki: Noncommutative residue. In Lecture Notes in Mathematics 1289, ed. by Yu.I. Manin, Springer-Verlag.

Spacetime Locality of the Antifield Formalism: General Theorems Illustrated by Means of Examples

Marc Henneaux

Faculté des Sciences, Université Libre de Bruxelles, Campus Plaine
C.P. 231, B-1050 Bruxelles, Belgium.
and
Centro de Estudios Científicos de Santiago, Casilla 16443, Santiago
9, Chile

Abstract. Some general techniques and theorems on the spacetime locality of the antifield formalism are illustrated in the familiar cases of the free scalar field, electromagnetism and Yang-Mills theory. Common misconceptions in the field are corrected.

1   Introduction

The antifield-BRST formalism [1] provides a powerful approach to the quantization of gauge systems. Its geometric and algebraic features have been clarified in [2, 3, 4], where it was shown how the general BRST construction implements gauge invariance in cohomology. The crucial equation of the theory, namely the TTmaster equation", was in particular justified and derived from this point of view. A general exposition of these ideas with pedagogical emphasis may be found in [5].

A major feature of the theory is that the solution of the master equation is determined perturbatively as a power series in the antifields. As it has been shown in [2, 3], the rationale for introducing the antifields is that these provide a resolution of the algebra of functionals of on-shell field configurations. Namely, the antifields are there to implement the equations of motion when one passes to the BRST cohomology. The resolution associated with the antifields is called "Koszul-Tate" resolution, because it is patterned after a construction due to Koszul [6], supplemented, when the equations of motion are not independent, by the introduction of further variables killing unwanted homology along lines due to Tate [7]. The acyclicity of the Koszul-Tate differential in strictly positive

137

resolution degree is crucial for the existence of the higher order terms in the perturbative expansion of the solution of the master equation. [We assume some familiarity with the general ideas of the antifield formalism; we refer to [5] for a detailed exposition].

The analysis presented in [3] did not address the question of the spacetime locality of the construction. More precisely, it did not address the question as to whether the acyclicity of the Koszul-Tate differential in strictly positive resolution degree still holds in the space of local functionals. A few years ago, that question has been investigated and completely solved [8] (see also [5], chapters 12 and 17). The purpose of this paper is to make it clear how the approach developed in [8] works and does indeed solve the issue of locality by illustrating it in the familiar cases of the Klein-Gordon field, the electromagnetic field and the Yang-Mills field.

We shall analyse only the specific question of locality of the Koszul-Tate complex. The reference [5] contains a discussion as to why this complex is so useful in the quantization of gauge systems.

2 Definitions

Consider a field theory with field variables ϕ^i. We shall deal with both local functionals and local functions of ϕ^i. Local functions are functions of ϕ^i and a finite number of their derivatives, which may also involve the spacetime coordinates explicitly. So, a local function is given by

$$f(x^\mu, \phi^i, \partial_\mu \phi^i, ..., \partial_{\mu_1...\mu_k} \phi^i). \tag{1}$$

Local functionals are integrals of local functions. Hence,

$$F[\phi^i] = \int f(x^\mu, \phi^i, \partial_\mu \phi^i, ..., \partial_{\mu_1...\mu_k} \phi^i) d^n x \tag{2}$$

is a local functional.

The appropriate way to deal with local functions is well known and has been used quite a lot in the algebraic study of anomalies. The corresponding mathematical framework is the one of jet bundle theory (see e.g. [9, 10]). However, in order to keep the discussion simple, we shall not adopt here the jet bundle terminology. This is permissible because we shall assume that spacetime is R^n, so that there are no global subtleties.

Let V^0 be the space with coordinates (x, ϕ^i). More generally, let V^k be the space with coordinates $(x, \phi^i, \partial_\mu \phi^i, ..., \partial_{\mu_1...\mu_k} \phi^i)$. If f is a smooth local function, then there exists k such that $f \in C^\infty(V^k)$. For this reason, the V^k's are the natural spaces in which to analyze locality. These spaces arose first in the geometric study of differential equations, which can naturally be regarded as representing surfaces in the V^k's. In that context, the spaces V^k are called k-th jet bundles and are denoted by $J^k(E)$.

We stress that the jet bundle spaces are quite familiar not only in mathematics but also in physics since these are the spaces in which the Lagrangians of local field theories live. These spaces are finite dimensional for each k. For this reason, all the standard algebraic tools of the antifield formalism (contracting homotopy, counting operators, recursive introduction of the antifields of antifields by successive killing of unwanted cohomology, model for the exterior derivative along the gauge orbits, antibracket cohomology, role of zeroth order terms - see [5]) are available in the jet bundle spaces without functional complications.

In order to discuss local functionals, it is useful to consider the algebra $A_k \equiv C^\infty(V^k) \otimes \bigwedge[dx^\mu]$ of exterior forms on R^n with coefficients that are functions on V^k,

$$\omega \in A_k \Leftrightarrow \omega = \Sigma\, \omega_{\nu_1 \ldots \nu_j}(x, \phi^i, \partial_\mu \phi^i, \ldots, \partial_{\mu_1 \ldots \mu_k} \phi^i)\, dx^{\nu_1} \wedge \ldots \wedge dx^{\nu_j} \qquad (3)$$

One can define a differential $d : A_k \to A_{k+1}$ as follows,

$$d\omega = \Sigma\, d\omega_{\mu_1 \ldots \mu_j} \wedge dx^{\mu_1} \wedge \ldots \wedge dx^{\mu_j} \qquad (4)$$

where d acting on a function $f \in A_k$ is defined by

$$df = \frac{\partial^T f}{\partial x^\mu}\, dx^\mu, \qquad (5)$$

$$\frac{\partial^T f}{\partial x^\mu} \equiv \frac{\partial f}{\partial x^\mu} + \frac{\partial f}{\partial \phi^i} \partial_\mu \phi^i + \ldots + \frac{\partial f}{\partial(\partial_{\mu_1 \ldots \mu_k})\phi^i} \partial_{\mu_1 \ldots \mu_k \mu} \phi^i. \qquad (6)$$

One crucial property of d is that

$$\int d\omega = 0 \qquad (7)$$

(we assume here and throughout that the boundary conditions are such that the surface terms appearing in the equations vanish. If not, one must carefully keep track of the relevant surface integrals).

Conversely let ρ be a n-form such that $\int \rho = 0$ for all field configurations. Then $\rho = d\omega$ (see e.g. [5]). Accordingly, two local functions determine the same local functional if and only if they differ by a d-exact term. For that reason, one can, following Gel'fand and Dorfman [11], identify local functionals with the quotient space $H^n(d)$ of local n-forms (which are automatically closed) modulo exact ones.

The Lagrangian $\mathcal{L}(\phi^i, \partial_\mu \phi^i, \ldots, \partial_{\mu_1 \ldots \mu_s} \phi^i)$ of the theory is a smooth function on V^s. The equations of motion[1]

$$\frac{\delta \mathcal{L}}{\delta \phi^i} \equiv \frac{\partial \mathcal{L}}{\partial \phi^i} - \partial_\mu \frac{\partial \mathcal{L}}{\partial(\partial_\mu \phi^i)} + \ldots + (-1)^s \partial_{\mu_1 \ldots \mu_s} \frac{\partial \mathcal{L}}{\partial(\partial_{\mu_1 \ldots \mu_s} \phi^i)}, \qquad (8)$$

together with their derivatives $\partial_\mu(\delta \mathcal{L}/\delta \phi^i) = 0$, $\partial_{\mu_1 \mu_2}(\delta \mathcal{L}/\delta \phi^i) = 0$ ... determine surfaces Σ_k in V^k. For a fixed k, only a finite number of equations are relevant. The surfaces Σ_k are called "stationary surfaces".

---

[1]From now on, we shall drop the suffix T on ∂_μ^T: ∂_μ always stands for ∂_μ^T.

In the antifield formalism, the algebra $C^\infty(\Sigma_k)$ of smooth functions on Σ_k plays an important role because it is related to the observables [5]. The Koszul-Tate construction provides a resolution of $C^\infty(\Sigma_k)$ for each k. The idea is to view $C^\infty(\Sigma_k)$ as the quotient algebra $C^\infty(V^k)/\mathcal{N}_k$, where $\mathcal{N}_k$ is the ideal of functions of $C^\infty(V^k)$ that vanish on Σ_k. The Koszul-Tate differential is such that the elements of $\mathcal{N}_k$ are exact, i.e., are pure boundaries.

3   The Koszul-Tate differential for the massless scalar field

To illustrate the construction, we consider first the massless Klein-Gordon theory. One has a single scalar field ϕ with Lagrangian

$$\mathcal{L} = -\frac{1}{2}\partial_\mu\phi\partial^\mu\phi \tag{9}$$

The equations of motion are

$$\Delta\phi \equiv \partial_\mu\partial^\mu\phi = 0. \tag{10}$$

In V^0, the equations of motion imply no relation and Σ_0 is empty: two functions f and g in V^0 coincide "on-shell" (i.e., when the equations of motion hold) if and only if they are identical. Similarly, there is no relation in V^1. One has to go to V^2 to see the first effect of the equations of motion, which restrict the second derivatives of ϕ. The surface Σ_2 is defined by $\Delta\phi = 0$ in V^2. Then, in V^3, Σ_3 is the surface $\Delta\phi = 0$, $\partial_\mu\Delta\phi = 0$. More generally, the surface Σ_k in V^k is defined by the equations

$$\Sigma_k : \Delta\phi = 0, ..., \Delta\partial_{\mu_1}...\partial_{\mu_{k-2}}\phi = 0. \tag{11}$$

The equations of motion (11) are independent in V^k. This is most easily seen by introducing a new coordinate system in V^k, which has the left hand side of the equations (11) as independent coordinates. One such coordinate system is given by

$$\phi, \partial_\mu\phi, \partial_{m_1 m_2}\phi, \partial_{m_1 0}\phi, \Delta\phi, ..., \partial_{m_1...m_{k-3}m_k}\phi, \partial_{m_1...m_{k-1}0}\phi, \partial_{\mu_1...\mu_{k-2}}\Delta\phi. \tag{12}$$

One can easily verify that any function f on V^k that vanishes on Σ_k ($f \approx 0$) takes the form,

$$f \approx 0 \Leftrightarrow f = h\Delta\phi + h^\mu\partial_\mu\Delta\phi + ... + h^{\mu_1\cdots\mu_{k-2}}\Delta\partial_{\mu_1}...\partial_{\mu_{k-2}}\phi \tag{13}$$

where the h's are functions on V^k (see for instance [5], chapter 1 with $\phi_m = 0$ replaced by (11)).

In order to construct a resolution of $C^\infty(\Sigma_k)$, one introduces one independent odd generator for each (independent) equation (11). That is, one considers the differential algebra $C^\infty(V^k) \otimes \bigwedge[\phi^*, \partial_\mu\phi^*, ..., \partial_{\mu_1}...\partial_{\mu_{k-2}}\phi^*]$ with differential

$$\delta\phi = 0, \delta\phi^* = \Delta\phi, \tag{14}$$

extended to the derivatives of the field and "antifield" ϕ^* so as to commute with ∂_μ,

$$\delta\partial_{\mu_1...\mu_j}\phi = 0, \delta\partial_{\mu_1...\mu_j}\phi^* = \partial_{\mu_1...\mu_j}\Delta\phi. \tag{15}$$

One defines also the antighost number through

$$antigh(\phi) = 0, \; antigh(\phi^*) = 1. \tag{16}$$

By (14), (15), every equation of motion is δ-exact and so, is identified with zero when one passes to the δ-homology. More precisely, standard arguments from homological algebra show that

$$H_0(\delta) = C^\infty(\Sigma_k), \; H_j(\delta) = 0 \text{ for } j \neq 0. \tag{17}$$

This result may be derived by observing that the coordinates of $C^\infty(V^k) \otimes \bigwedge[\phi^*, \partial_\mu\phi^*, ..., \partial_{\mu_1}...\partial_{\mu_{k-2}}\phi^*]$ split into three groups $(x_i, z_\alpha, J\mathcal{P}_\alpha)$ such that δ takes the form

$$\delta x_i = 0, \; \delta\mathcal{P}_\alpha = z_\alpha, \; \delta z_\alpha = 0 \tag{18}$$

or equivalently

$$\delta = z_\alpha \frac{\partial}{\partial\mathcal{P}_\alpha}. \tag{19}$$

Explicitly, the coordinates x_i stand for the field ϕ and its derivatives with at most one ∂_0, the z_α stand for $\Delta\phi$ and its derivatives, while the $\mathcal{P}_\alpha$ stand for ϕ^* and its derivatives. A contracting homotopy may be defined through

$$\sigma x_i = 0, \; \sigma\mathcal{P}_\alpha = 0, \; \sigma z_\alpha = \mathcal{P}_\alpha \Leftrightarrow \sigma = \mathcal{P}_\alpha \frac{\partial}{\partial z_\alpha}, \tag{20}$$

i.e.,

$$\sigma = \phi^* \frac{\partial}{\partial(\Delta\phi)} + \partial_\mu\phi^* \frac{\partial}{\partial(\partial_\mu\Delta\phi)} + ... + \partial_{\mu_1}...\partial_{\mu_{k-2}}\phi^* \frac{\partial}{\partial(\partial_{\mu_1...\mu_{k-2}}\Delta\phi)} \tag{21}$$

where the derivatives with respect to $\partial_{\mu_1...\mu_j}\Delta\phi$ are computed in the coordinates (12) of V^k. One has

$$\sigma\delta + \delta\sigma = N \tag{22}$$

where N

$$N = \mathcal{P}_\alpha \frac{\partial}{\partial\mathcal{P}_\alpha} + z_\alpha \frac{\partial}{\partial z_\alpha} \tag{23}$$

is the operator counting the number of $\mathcal{P}_\alpha$ and z_α. The relation (22) crucially uses the derivation property of $\partial/\partial z_\alpha$. It follows from (22) and (23) that $\mathcal{P}_\alpha$ and z_α drop from the homology of δ ("they belong to the contractible part of the complex"), which is given by the functions of x_i ([5], sections 8.3.2 and 9.A.2. The G_a's there play the role of the equations of motion here). Since the functions of x_i are the functions on Σ_k and have antighost number equal to zero, formula (17) is established.

The argument is valid for any k, i.e. for any local function involving the derivatives of the field and antifield up to an arbitrarily high (but finite) order. One sometimes summarize (17) by saying that δ is acyclic in the space of local functions.

It should be noted that even though covariant-looking, the contracting homotopy (21) is not covariant. For instance, one finds

$$\sigma(\partial_\mu \partial_\nu \phi) = \delta_{\mu 0} \delta_{\nu 0} \phi^*. \tag{24}$$

Nevertherless, one can show that the homology of δ in the algebra of Lorentz invariant functions is trivial for positive k; that is, if $\delta f = 0$ and $antigh(f) = k \neq 0$, where f is Lorentz invariant, then $f = \delta g$ where g may also be taken to be Lorentz invariant. This can be proved either by redefining the homotopy, or equivalently, by following the methods of [12], theorem 2.

We close this section by a few remarks concerning incorrect statements that have been made in the literature.

1. First, it should be stressed that $f \approx 0$ does not imply $f = h\Delta\phi$ with h a local function. Rather, f may also involve the derivatives of $\Delta\phi$, i.e., one has the full expansion (13).

2. The homotopy σ given by (21) is well defined everywhere because the equations of motion are simple. For more general theories, however, a globally defined homotopy constructed along the above lines may just simply not exist. This is because obstructions for defining the derivation $\partial/\partial(\delta\mathcal{L}/\delta\phi^i)$ may be present (one needs to tell what is kept fixed when differentiating with respect to $\delta\mathcal{L}/\delta\phi^i$). Attempts for using a formula similar to (21) would then necessarily fail. This would show up in non convergence of power series, etc., which must be handled carefully. One way to handle correctly this problem is to introduce partitions of unity, as in [5], appendix 9A.

To make this point clear, consider the Lagrangian $L = L(q)$ where the function $h(q) \equiv dL/dq$ is such that (i) $h(q) = -1$ for $q \leq -1$; (ii) $h(q) = 1$ for $q \geq 1$; and (iii) $h(q)$ interpolates in a smooth way from -1 to $+1$ between -1 to $+1$ and vanishes only at the origin where $h'(0) = 1$. It is clear that it is impossible to define df/dh for all functions f's (with d/dh a derivation) since this would imply in particular that dq/dh is well-defined and such that $(dq/dh)(dh/dq) = 1$, in contradiction with $dh/dq = 0$ for $q \leq -1$ or $q \geq 1$. It turns out not to be necessary, however, to define df/dh in the open sets where $h \neq 0$. Indeed, in those sets ("of type V" according to [5]), any δ-closed function f is trivially δ-exact, $f = \delta(q^* f/h)$. The proof of acyclicity of δ proceeds by patching the V-sets with an open set covering the origin by means of a partition of unity.

One may also construct polynomial counterexamples. For instance, the Lagrangian

$$L(q) = \frac{1}{4}q^4 + \frac{5}{3}q^3 + \frac{1}{2}q^2 + 5q \tag{25}$$

for a real variable q leads to the equation of motion $h(q) \equiv dL/dq = (q + 5)(q^2 + 1) = 0$, whose sole solution is $q = -5$. The equation of motion is regular ($h'(q) \neq 0$ on-shell), but yet, one cannot define dq/dh everywhere since dh/dq

has two real roots. One may build other counteramples based on a non trivial topology of the stationary surface.

4 The Koszul-Tate differential for the electromagnetic field

We now turn to the electromagnetic case. The equations of motion are

$$\mathcal{L}^\rho \equiv \frac{\delta \mathcal{L}}{\delta A_\rho} = \partial_\mu F^{\mu\rho} = 0 \tag{26}$$

and define a surface in V^2. The new feature compared with the previous situation is that the derived equations

$$\partial_\mu \mathcal{L}^\rho = 0, \partial_{\mu_1 \mu_2} \mathcal{L}^\rho = 0, ... \tag{27}$$

in V^3, V^4, ... are no longer independent. Because of the gauge invariance of the electromagnetic field Lagrangian, one has rather (identically)

$$\partial_\rho \mathcal{L}^\rho \equiv 0, \partial_{\mu_1}(\partial_\rho \mathcal{L}^\rho) \equiv 0... \tag{28}$$

(for any field configuration). For that reason, one needs "antifields of antifields" J[3, 5].

We start with V^2. There are clearly no relations among the equations $\mathcal{L}^\rho = 0$ in V^2 since one can solve these equations for n of the coordinates in V^2 (we work in n dimensions). Namely, one can solve $\mathcal{L}^k = 0$ for $\partial_{00} A_k$ and $\mathcal{L}^0 = 0$ for $\partial_{11} A_0$ (say). Hence, if one defines in $C^\infty(V^2) \otimes \bigwedge(A^{*\mu})$ the differential

$$\delta A_\mu = 0, \delta \partial_\rho A_\mu = 0, \delta \partial_{\rho\sigma} A_\mu = 0, \delta A^{*\mu} = \partial_\nu F^{\nu\mu} \tag{29}$$

one gets that $H_k(\delta) = 0$ for $k \neq 0$ and $H_0(\delta) = C^\infty(\Sigma_2)$. To verify this statement, one repeats the argument of the previous section and splits the variables of the complex in three groups. The coordinates A_μ, $\partial_\rho A_\mu$, $\partial_{\rho\sigma} A^k$ $((\rho,\sigma) \neq (0,0))$ and $\partial_{\rho\sigma} A^0$ $((\rho,\sigma) \neq (1,1))$ are of the x_i-type, the coordinates $\mathcal{L}^\rho$ are of the z_α-type, while the $A^{*\mu}$ are of the $\mathcal{P}_\alpha$-type. The appropriate contracting homotopy in $C^\infty(V^2) \otimes \bigwedge(A^{*\mu})$ reads

$$\sigma = A^{*\mu} \frac{\partial}{\partial \mathcal{L}^\mu}. \tag{30}$$

Thus, only the variables not constrained by the equations of motion, namely, A_μ, $\partial_\rho A_\mu$, $\partial_{\rho\sigma} A^k$ $((\rho,\sigma) \neq (0,0))$ and $\partial_{\rho\sigma} A^0$ $((\rho,\sigma) \neq (1,1))$ remain in homology. The other variables drop out.

Turn now to $C^\infty(V^3) \otimes \bigwedge(A^{*\mu}, \partial_\rho A^{*\mu})$, with differential δ (29) extended to the derivatives so that

$$\delta \partial_\mu = \partial_\mu \delta \tag{31}$$

i.e.,

$$\delta \partial_{\rho\sigma\alpha} A_\mu = 0, \quad \delta \partial_\rho A^{*J\mu} = \partial_\rho(\partial_\nu F^{\nu\mu}) \tag{32}$$

The equations $\partial_\nu F^{\nu\mu} = 0$ and $\partial_\sigma \partial_\nu F^{\nu\mu} = 0$ are *not* independent in V^3 since they are subject to the (single) condition $\partial_\rho \mathcal{L}^\rho = 0$. There are no other identity in V^3 because one can solve $n^2 + n - 1$ of the $n^2 + n$ equations $\mathcal{L}^\rho = 0$, $\partial_\mu \mathcal{L}^\rho = 0$ for $n^2 + n - 1$ independent variables, namely $\partial_{00} A_k$ (from $\mathcal{L}^k = 0$), $\partial_{11} A_0$ (from $\mathcal{L}^0 = 0$), $\partial_{\rho 00} A_k$ (from $\partial_\rho \mathcal{L}^k = 0$) and $\partial_{s11} A_0$ (from $\partial_s \mathcal{L}^0 = 0$). The derivative $\partial_{011} A^0$ cannot be determined from $\partial_0 \mathcal{L}^0 = 0$, which is not an independent equation ($\partial_0 \mathcal{L}^0 = -\partial_k \mathcal{L}^k$). Hence, in V^3, there are $n^2 + n - 1$ independent equations and 1 dependent one.

Because the equations of motion in V^3 are not independent, there is one non trivial cycle at antighost number 1, namely $\partial_\rho A^{*\rho}$. Thus, $H_1(\delta) \neq 0$ in $C^\infty(V^3) \otimes \bigwedge (A^{*\mu}, \partial_\rho A^{*\mu})$. In order to achieve acyclicity of the Koszul-Tate differential, one needs to introduce one further even variable, denoted by C^* and called "antifield of antifield" [5], with grading

$$antigh C^* = 2. \tag{33}$$

This new variable must kill the non trivial cycle $\partial_\rho A^{*\rho}$ in homology, so that one defines

$$\delta C^* = \partial_\rho A^{*\rho}. \tag{34}$$

Once C^* is introduced, one can redefine the variables of the differential complex $C^\infty(V^3) \otimes C[A^{*\mu}, \partial_\rho A^{*\mu}, C^*]$ in such a way that δ takes again the characteristic form[2]

$$\delta x_i = 0, \quad \delta \mathcal{P}_\alpha = z_\alpha, \quad \delta z_\alpha = 0, \tag{35}$$

which makes manifest that $H_*(\delta) = C^\infty(x_i)$. The variables x_i have antighost number zero and parametrize Σ_3. They are explicitly given by A_μ, $\partial_\rho A_\mu$, $\partial_{\rho\sigma} A_k$ $((\rho, \sigma) \neq (0,0))$, $\partial_{\rho\sigma} A_0$ $((\rho, \sigma) \neq (1,1))$, $\partial_{\rho\sigma\nu} A_k$ (with at most one time derivative) and $\partial_{\rho\sigma\nu} A_0$ (with $(\rho, \sigma, \nu) \neq (k, 1, 1)$ even up to a permutation). The variables $\mathcal{P}_\alpha$ are $A^{*\mu}$, $\partial_\alpha A^{*k}$, $\partial_k A^{*0}$ and C^*. The variables z_α are the left hand sides of the equations of motion $\mathcal{L}^\rho$, $\partial_\alpha \mathcal{L}^k$, $\partial_k \mathcal{L}^0$ and $\partial_\rho A^{*\rho}$.

The same pattern goes on with the higher order derivatives. In $C^\infty(V^k) \otimes C[A^{*\mu}, \partial_\rho A^{*\mu}, ..., \partial_{\rho_1...\rho_{k-2}} A^{*\mu}, C^*, ..., \partial_{\rho_1...\rho_{k-3}} C^*]$, one may introduce new coordinates as follows:
(i) Coordinates of x_i-type : A_k and its derivatives with at most one ∂_0; A_0 and its derivatives except $\partial_{s_1 s_2...s_m} A_0$ with at least two ∂_1. These variables parametrize Σ_k.
(ii) Coordinates of z_α-type: $\mathcal{L}^k$ and its derivatives; $\mathcal{L}^0$ and its spatial derivatives; $\partial_\rho A^{*\rho}$ and its derivatives.
(iii) Coordinates of $\mathcal{P}_\alpha$-type : A^{*k} and its derivatives; A^{*0} and its spatial derivatives; C^* and its derivatives.

Thus, again, $H_0(\delta) = C^\infty(V^k)$ and $H_m(\delta) = 0$, $m \neq 0$. The contracting homotopy has the standard form

$$\sigma = \mathcal{P}_\alpha \frac{\partial}{\partial z_\alpha}, \tag{36}$$

---

[2]From now on, we shall use the notation $C[A^{*\mu}, \partial_\rho A^{*\mu}, C^*]$ for the algebra $\bigwedge (A^{*\mu}, \partial_\rho A^{*\mu}) \otimes R[C^*]$. The symmetry properties are taken care of by the gradings of $A^{*\mu}$ (odd) and C^* (even).

where the sum runs over all the z_α's. At each stage, one can separate the equations $\mathcal{L}^\rho = 0$ and their derivatives into independent ones and dependent ones *without going out of the spaces V^k, i.e., in a manner compatible with spacetime locality*. Statements to the contrary are thus wrong.

It is true that the dependent equations at order $k+1$ are not just the derivatives of the dependent equations at order k. One cannot separate the n equations $\mathcal{L}^\rho = 0$ into two groups, so that the independent (respectively, dependent) equations would simply be all the derivatives of the equations of the first (respectively second) group. To achieve this property, one would have to make a non local split. But a split with this property is not necessary once one formulates the problem in terms of the standard spaces V^k of jet bundle theory, as appropriate for dealing with locality.

Similarly, although we have not done it, one could define a Lorentz-invariant homotopy by decomposing the derivatives of the fields along the irreducible representations of the Lorentz group. Hence, acyclicity of the Koszul-Tate differential also holds in the algebra of Lorentz-invariant local functions. This same result can equivalently be established along the lines of [12], by using the facts that δ commutes with the representation and that the Lorentz group is semi-simple.

5 The Koszul-Tate differential for the Yang-Mills field

The Yang-Mills case can be treated in the same manner. This is because the terms with the highest (second) order derivatives of the gauge potential in the Yang-Mills equations of motion are exactly the same as in the Abelian case. Hence, the change of variables such that the left hand sides of the equations of motion and their derivatives are new coordinates is still permissible, and one can proceed as above.

For instance, in V^3, one would take as new variables $A_\mu^a, \partial_\rho A_\mu^a, \partial_{\rho J\sigma} A_k^a$ $((\rho,\sigma) \neq (0,0))$, $\partial_{\rho J\sigma} A_0^a$, $((\rho,\sigma) \neq (1,1))$ and $\mathcal{L}_a^\mu$. The expression of $\partial_{00} A_k^a$ in terms of $\mathcal{L}_a^k$ is the same as in the abelian case up to terms containing lower order derivatives (which are independent coordinates in the previous space V^2). A similar analysis holds for higher order derivatives.

We leave it to the reader to check also that an analogous derivation can be performed for p-form gauge fields. The only difference is that one needs this time more antifields for antifields because the reducibility equations are not independent.

6 Acyclicity of Koszul-Tate differential and local functionals

The above sections establish the acyclicity of δ in the space of local functions. Does this property also hold in the space of local functionals? That is, if f is a n-form such that

$$\delta \int f = 0, \ antigh f \geq 1 \tag{37}$$

does one have

$$\int f = \delta \int g \tag{38}$$

for some n-form g? [f and g are n-forms with coefficients that are local functions]. Equivalently, in terms of the integrands, does

$$\delta f = dj, \ antigh f \geq 1 \tag{39}$$

imply

$$f = \delta g + dk \tag{40}$$

for some n-form g and $n - 1$-form k? The presence of the d-exact terms in (39), (40) follows from (7) and *must be taken into account*. Failure to do so would be incorrect. The extra d-terms in (39) and (40) show that the relevant cohomology when dealing with local functionals is the cohomology of δ modulo d in the space of local n-forms. The corresponding cohomological spaces are denoted $H_k(\delta/d)$.

As pointed out in [8], the answer to this question is in general negative. Constants of the motion define non trivial solutions of $H_1(\delta/d)$. Indeed, the equation $\delta f + dj$ with $antigh f = -1$ and $antigh j = 0$ defines a conserved current j. If f is trivial (of the form (40)), then j is a trivial conserved current ($j = -\delta k + dm$). Since there exist in general non trivial conserved currents, $H_1(\delta/d)$ is not empty.

However, if f involves the ghosts[3] - which is the case encountered in homological perturbation theory -, then (39) does imply (40). To see this, consider first the case where f is linear in the ghosts. By making integrations by parts if necessary, one can assume that f does not involve the derivatives of the C^α,

$$f = \lambda_\alpha C^\alpha, \ antigh \lambda_\alpha = 0. \tag{41}$$

Then, $\delta f = \delta(\lambda_\alpha) C^\alpha$. If $\delta f = dj$, then δf *and* dj must separately vanish because dj would otherwise necessarily involve derivatives of the ghosts. Thus $\delta \lambda_\alpha = 0$, which implies $\lambda_\alpha = \delta \mu_\alpha$ since $H_k(\delta) = 0$ in the space of local functions. Consequently, $f = (\delta \mu_\alpha) C^\alpha = \delta(\mu_\alpha C^\alpha)$, which is the sought-for result. How to formalize the argument so that it applies also to forms f that are non linear in the ghosts is done in [8]. Thus, acyclicity of δ holds in the space of local functionals involving both the antifields and the ghosts.

---

[3]How the ghosts are introduced may be found for example in [5]. The ghosts will be denoted by C^α and are annihilated by the differential δ. Once the ghosts are introduced, the cohomology of δ is given by $C^\infty(\Sigma_k) \otimes \bigwedge(C^\alpha, \partial_\rho C^\alpha ...)$.

146

7 Conclusion

We have illustrated in this paper how to handle locality in the case of the antifield-antibracket formalism for gauge field theories. The tools involve both standard homological algebraic techniques applied to finitely generated algebras and ideas from jet bundle theory. We have shown in particular how the equations of motion for electromagnetism and Yang-Mills theory can split into independent and dependent ones in the "jet bundle" spaces V^k. The tools illustrated here have been used recently to prove a long-standing conjecture on the renormalization of Yang-Mills models [13].

We close this letter with two observations :

(i) The method of homological perturbation theory is quite general and does not depend on the precise form of the differential algebra on which the derivations act, provided these derivations fulfill the properties explained in [5] (chapter 8). Thus, one may modify the algebra of local functions by imposing restrictions if one wishes to do so. For instance, the well-known theorem that a BRST cohomological class is determined by its component of order zero in the antifields is quite standard and follows from the general principles of homological perturbation theory (see again [5], chapter 8, proof of main theorem and section 8.4.4).

(ii) Similarly, one may consider field theories for which the equations of motion are not "regular", in the sense that their gradients would vanish on the stationary surface. A theory with equation of motion $\delta\mathcal{L}'/\delta\phi = \phi^2 = 0$ (rather than the equivalent equation $\delta\mathcal{L}/\delta\phi = \phi = 0$) would provide such an example. This case does not arise in usual gauge theories, as we have just seen, but does occur in, say, Siegel formulation of chiral bosons [14]. Again, a lot of work already exists on this subject, especially in the Hamiltonian context. The algebraic framework is well developed. The real question is, however, what is the physical meaning of the BRST construction in those cases. The relation between the BRST cohomology and the cohomology of the geometrical longitudinal derivative on the stationary surface may no longer hold (this is why the BRST analysis performed in chapters 9 and 10 of [5] excludes these somewhat pathological cases). To the author, the question has not been fully resolved.

8 Acknowledgements

The author is grateful to G. Barnich, P. Grégoire, T. Kimura and J. Stasheff for discussions. This work has been supported in part by a research grant from F.N.R.S. and a research contract with the Commission of the European Community.

References

[1] I.A. Batalin and G.A. Vilkovisky, Phys. Lett. **102B** (1981) 27; Phys. Rev. **D28** (1983) 2567.

[2] J. Fisch, M. Henneaux, J.D. Stasheff and C. Teitelboim, Commun. Math. Phys. **120** (1989) 379.

[3] J.M.L. Fisch and M. Henneaux, Commun. Math. Phys. **128** (1990) 627.

[4] M. Henneaux, Nucl. Phys. Proc. Suppl. **18A** (1990) 47.

[5] M. Henneaux and C. Teitelboim, *Quantization of Gauge Systems*, Princeton University Press (Princeton : 1992).

[6] J.L. Koszul, Bull. Soc. Math. France **78** (1950) 5.

[7] J. Tate, Ill. J. Math. **1** (1957) 14.

[8] M. Henneaux, Commun. Math. Phys. **140** (1991) 1.

[9] I.M. Anderson, *The variational bicomplex*, Academic Press (Boston : 1994).

[10] R.L. Bryant, S.S. Chern, R.B. Gardner, H.L. Goldschmidt and P.A. Griffiths, *Exterior Differential Systems*, Springer Verlag (New York : 1991).

[11] I.M. Gel'fand and I. Ya. Dorfman, Funct. Anal. Appl. **13** (1979) 174.

[12] M. Henneaux, Phys. Lett. **B313** (1993) 35.

[13] G. Barnich and M. Henneaux, preprint hepth 9312206, to appear in Phys. Rev. Lett.

[14] W. Siegel, Nucl. Phys. **B238** (1984) 307.

On Supersymmetric and Topological Quantum Mechanical Models

Laurent Baulieu

LPTHE, Universités Pierre et Marie Curie and Denis Diderot
4 place Jussieu
75005 Paris, France

Abstract. We explain the details of two supersymmetric quantum mechanical models. Their simplicity make them solvable although they share the characteristics of more sophisticated models based on the gauge fixing of topological invariants. The first model is a supersymmetric quantum mechanical system defined on a punctured plane and leads to topological observables which we compute. The absences of a ground state and of a mass gap are special features of this system. The second model is the supersymmetric description of spin-one particles moving in D-dimensional space-time. We show that it is a topological model in a space with two more dimensions.

1 Introduction

During the last years, Topological Quantum Field Theories have emerged as possible realizations of general coordinates invariant symmetries [1][2].

One of the special features of these theories is their ability to produce space-time metric independent correlations functions, although they are defined from a local action.

In Topological Quantum Field Theories, an important symmetry operator which is at disposal is the BRST operator Q, such that the Hamiltonian is $H = \frac{1}{2}[Q, \overline{Q}]$. Q and $\overline{Q}$ can be often understood as 'twisted" deformations of $N = 2$ supersymmetry generators.

An attractive scheme is to introduce Topological Quantum Field Theories by the path integral quantization of topological terms. The techniques relies on the BRST formalism. More precisely, one can often start from a topological term, expressed as the integral over a manifold of a Lagrangian locally equal to a pure divergency which is a function of a set of given fields. Such a "classical" action is for instance a characteristic number, or any given invariant depending only of the topology of field configurations and/ or the space over which the fields are defined. No classical dynamics is generated. However, the existence of a gauge symmetry of the Lagrangian, namely the group of arbitrary infinitesimal deformations of fields, permits the quantization of the theory through the general formalism of BRST invariant gauge fixing. Our present knowledge makes this construction quite generic, provided one gets the intuition of (i) which manifold should be studied, and (ii) which fields should be introduced for this purpose. Actually, it is interesting to speculate that the symmetries of nature could be fundamentally of the topological type, and that the observed gauge symmetries would be obtained by gauge-fixing the huge topological symmetry in a BRST invariant way, leaving therefore an $N = 2$ supersymmetric theory of particles.

Not surprisingly, the problem of computing observables in Topological Quantum Field Theories is often technically complicated. The basic idea is the introduction of fields with positive and negative degrees of freedom (classical and ghost fields) which permit the exploration of topological properties through the computation of Green functions whose coefficient turn out to be topological invariants. Once the theory has been defined, dimensional reductions may appear as the only possible technichal way to perform realistic computations. The technicity of these computations may hide the beautiful simplicity of the idea! As an example, to compute the knot polynomials associated to the Chern-simons theory, one reduces the $3 - D$ theory into $2 - D$ conformal theories [3].

One usually defines the physical Hilbert space of Topological Quantum Field Theories as the cohomology of Q (states which are annihilated by Q without being the Q transformation of other states). This definition of the physical Hilbert space is perfectly suited for ordinary gauge theories. For Topological Field Theories there are doubts on the general validity of this definition. Due to properties of the vaccuum, other relevant observables than those defined by the BRST cohomology could exist. In particular, Q−exact observables with non vanishing mean values can exist. This is for instance the case in topological models of the type of those introduced by Witten in [1]. In other topological ones, based on first order actions like the Chern-Simons action [3], formal arguments show that the situation is similar. In all these models, one sees furthermore that a local version of of the topological BRST symmetry seems to single out the form of the supersymmetric potential [5][6].

The simplest examples of Topological Quantum Field Theories are zero-dimensional and turn out to be are $N = 2$ supersymmetric quantum mechanics models. Interestingly enough, two models exist which illustrate both extrem cases: (i) the Hilbert space is made of pure topological observables and (ii) the Hilbert space is made of particle degrees of freedom. In these notes we find it interesting to detail them. Indeed they provide elementary examples showing the basic rules of the BRST invariant topological gauge fixing procedure. In particular, they address the questions of the selection of gauge functions and of the calculability of observables (which can be completely worked out in the case (i)). The example (ii) is intriguing since it might be generalized to other particle or string models with N=2 supersymmetry.

These notes are the result of joint works with R. Attal and E. Rabinovicci.

2 Model (i): Supersymmetric quantum mechanics on a punctured plane

We wish to work with a simple topological classical Lagrangian that is a candidate to generate a topological quantum mechanics. We consider as a target space a plane from which we exclude the origin, so that one has a non trivial, although very simple topological structure defined by the winding number around the origin of the trajectories of a particle. We denote the time by the real variable t and the Euclidian time by τ, with $t = i\tau$ and τ real. The cartesian coordinates on the plane are q_i, with $i = 1, 2$. We select trajectories with periodic conditions, namely such that between the initial and final times $t = 0$ and $t = T$ the particle ends up at its starting point so an integer value of the winding number can be assigned to its trajectory.

From our understanding of the nature of a topological field theory [4], we start from a topological classical action $\mathcal{I}_{cl}[\vec{q}]$. $\mathcal{I}_{cl}[\vec{q}]$ must not depend on the time metric. This condition is satisfied if it is the integral of a locally closed form. The natural candidate is

$$\mathcal{I}_{cl}[\vec{q}] = \int f d\tau = \int_0^T d\tau f \dot{\tau}(\tau)$$
$$= \int_0^T d\tau \, f \frac{\epsilon^{ij} \dot{q}_i q_j}{\vec{q}^2} \tag{2.1}$$

where f is a real number. This action measures the winding number of the particle times $f/2\pi$. It shares analogy with the second Chern class $\int d^4x \, \mathrm{tr} \, F \wedge F$ where F is the curvature of a Yang-Mills field. Here and in what follows the symbol $\dot{X}$ denotes $\frac{dX}{d\tau}$.

To obtain the Topological Quantum Theory associated to our space, we need to give sense to the Euclidian path integral

$$\int \mathcal{D}[\vec{q}] \exp -\mathcal{I}_{cl}[\vec{q}] \tag{2.2}$$

as well as to compute topological quantities from Green functions

$$\text{Topological information} = \int \mathcal{D}[\vec{q}] \, O \exp -\mathcal{I}_{cl}[\vec{q}] \tag{2.3}$$

where O is a well chosen composite operator.

The difficulty for realizing this objective is that our action is different from that of conventional quantum mechanics where classical degrees of freeedom exist at the classical level and quantum fluctuations occur around the solutions of equations of motion. Here the Lagrangian is locally a pure derivative, the Hamiltonian vanishes and one has no equation of motion. On the other hand, one observes that the action $\mathcal{I}_{cl}[\vec{q}]$ is invariant under the gauge symmetry

$$\vec{q}(t) \rightarrow \vec{q}(t) + \vec{\epsilon}(t) \tag{2.4}$$

where $\epsilon(t)$ is any given local shift of the particle position $q(t)$ which does not change the winding number of the trajectory. Using the BRST technique it is then possible to define the path integrals (2.2) and (2.3) by a conventional gauge fixing of the action $\mathcal{I}_{cl}[\vec{q}]$.

The BRST transformation laws associated to the symmetry (2.4) are of the simple form

$$s\vec{q} = \vec{\Psi} \quad s\vec{\Psi} = 0 \quad s\vec{\bar{\Psi}} = \vec{\tau} \quad s\vec{\tau} = 0 \tag{2.5}$$

The anticommuting fields $\vec{\Psi}(t)$ and $\vec{\bar{\Psi}}(t)$ are the topological ghosts and antighosts associated to the particle position $\vec{q}(t)$. $\vec{\tau}(t)$ is a Lagrange multiplier. s acts on field functions as a differential operator graded by the ghost number.

To get a gauge fixed action with a quadratic dependence on the velocity $\dot{\vec{q}}$, one choses a gauge function of the type $\dot{q}_i + \frac{\delta V}{\delta q_i}$, where the prepotential V is an arbitrary given function of $\vec{q}$. This yields the following gauge fixed BRST invariant action $\mathcal{I}_{gf}$ which is supersymmetric

$$
\begin{aligned}
\mathcal{I}_{gf}[\vec{q}, \vec{\Psi}, \vec{\bar{\Psi}}, \vec{\tau}] &= \int_0^T d\tau \left(f\dot{\tau} - s\bar{\Psi}_i(\frac{1}{2}\tau_i - i\dot{q}_i + \frac{\delta V}{\delta q_i}) \right) \\
&= \int_0^T d\tau \left(f\dot{\tau} - \frac{1}{2}\tau_i^2 + i\tau_i \left(\dot{q}_i + \frac{\delta V}{\delta q_i} \right) - \right. \\
&\qquad\qquad \left. -i\bar{\Psi}_i \left(\dot{\Psi}_i + \frac{\delta^2 V}{\delta q_i q_j}\Psi_j \right) \right)
\end{aligned}
\tag{2.6}
$$

The BRST symmetry $s\mathcal{I}_{gf}[\vec{q}, \vec{\Psi}, \vec{\bar{\Psi}}, \vec{\tau}] = 0$ holds true independently of the choice of the function $V(\vec{q})$ and the partition function and the mean values of BRST invariant observables

$$Z = \int \mathcal{D}[\vec{q}]\mathcal{D}[\vec{\Psi}]\mathcal{D}[\vec{\bar{\Psi}}]\mathcal{D}[\vec{\tau}] \exp -\mathcal{I}_{gf} \tag{2.7}$$

$$<O> = \int \mathcal{D}[\vec{q}]\mathcal{D}[\vec{\Psi}]\mathcal{D}[\vec{\bar{\Psi}}]\mathcal{D}[\vec{\tau}] O \exp -\mathcal{I}_{gf} \tag{2.8}$$

are now well defined Euclidian path integrals. To understand $\dot{q}_i + \frac{\delta V}{\delta q_i}$ as a gauge function for the quantum variable $\vec{q}$, one may interpret the result of the integration over the ghosts as a determinant. The BRST invariance of the field polynomial O allows one to prove, at least formally, the topological properties of $< O >$. On the other hand our knowledge of supersymmetric quantum mechanics tells us that this mean value may depend on the class of the function V. What happens is that in the case of topological field theories, the Euclidian path integral explores the moduli space of the equation $\dot{q}_i + \frac{\delta V}{\delta q_i} = 0$, as a result of the gauge fixing.

The question of finding a symmetry principle which would select the prepotential $V(\vec{q})$ leading to interesting topological information was investigated in

[5]. The idea is to ask for the invariance of the action under a symmetry which is more restrictive than the topologocal BRST symmetry, namely a local version of it, for which the parameter becomes an affine function of the time, with arbitrary infinitesimal coefficients. One requires

$$\delta_l \mathcal{I}_{gf}[\vec{q}, \vec{\Psi}, \vec{\bar{\Psi}}, \vec{\tau}] = 0 \qquad (2.9)$$

where the "local" BRST transformations δ_l are

$$\delta l\vec{q} = \eta(t)\vec{\Psi} \quad \delta_l \vec{\Psi} = 0 \quad \delta_l \vec{\bar{\Psi}} = \eta(t)\vec{\tau} - \dot{\eta}(t)\vec{q} \quad \delta_l \vec{\tau} = \dot{\eta}(t)\vec{\Psi} \qquad (2.10)$$

and $\eta(t) = a+bt$ where a and b are constant anticommuting parameters. The idea of local BRST symmetry was considered in [9] for the sake of interpreting higher order cocycles which occurs when solving the anomaly consistency conditions, and has been shown to play a role in topological field theories in [6].

Imposing this local symmetry implies that V satisfies the constraint [5]

$$\frac{\delta V}{\delta q_i} + q_j \frac{\delta^2 V}{\delta q_i \, \delta q_j} = 0 \qquad (2.11)$$

This constraint is solved for $V(\vec{q}) = f\tau$ where τ is the angle such that $q_1 + iq_2 = |\vec{q}| \exp i\tau$ and f is a number [5][1]. By putting this value of $\dot{q}_i + \frac{\delta V}{\delta q_i}$ in (2.6) and eliminating the Lagrange multiplier τ by its equation of motion we obtain

$$\mathcal{I}_{gf}[\vec{q}, \vec{\Psi}, \vec{\bar{\Psi}}] = \int_0^T d\tau \left(\frac{1}{2}\dot{q}_i^2 + \frac{f^2}{2\vec{q}^2} - \vec{\Psi}_i \left(\dot{\Psi}_i \pm f\frac{\delta^2 \tau}{\delta q_i \delta q_j}\Psi_j \right) \right) \qquad (2.12)$$

Notice that

$$\begin{aligned}
\frac{\delta^2 \tau}{\delta q_i \delta q_j} &= \frac{1}{\vec{q}^2} \begin{pmatrix} -\sin 2\tau & \cos 2\tau \\ \cos 2\tau & \sin 2\tau \end{pmatrix}_{ij} \\
&= \frac{1}{\vec{q}^2} \begin{pmatrix} \cos\tau & -\sin\tau \\ \sin\tau & \cos\tau \end{pmatrix} \begin{pmatrix} 0 & -1 \\ -1 & 0 \end{pmatrix} \begin{pmatrix} \cos\tau & \sin\tau \\ -\sin\tau & \cos\tau \end{pmatrix}_{ij}
\end{aligned} \qquad (2.13)$$

The superconformal potential $1/\vec{q}^2$ has been already studied in [7][10]. We shall shortly compute the observables which seems interesting to us from the topological point of view in the canonical quantization formalism. We will show that a very specific supersymmetry breaking mechanism occurs and implies the existence of non vanishing Q exact observables which are metric independent as well as of a fractional Witten indexD.

We believe that the signal that the theory truly carries some topological information is the existence of an interesting instanton structure. Let us remember that, from our gauge fixing in the Euclidian time region, we have obtained an

---

[1]For the case of one variable x we would obtain $V = \log x$, with quite similar properties of the supersymmetric system, but the geometrical interpretation would be less clear and no meaningful observable exists

action whose bosonic part is the square of the gauge function. It follows that the solutions to the Euclidien equations of motion can be written as

$$\dot{q}_i + \frac{\epsilon^{ij} q_j}{\vec{q}^2} = 0 \tag{2.14}$$

$$\dot{\Psi}_i \pm f \frac{\delta^2 \tau}{\delta q_i \delta q_j} \Psi_j = 0 \tag{2.15}$$

If we introduce $z = q_1 + iq_2$ and $\Psi_z = \Psi_1 + i\Psi_2$, with $sz = \Psi_z$, we can write these equations as

$$\dot{z} + \frac{i}{z^*} = 0 \tag{2.16}$$

$$\dot{\Psi}_z - \frac{i}{(z^*)^2} \Psi_z^* = 0 \tag{2.17}$$

Assuming periodic boundary conditions, the solutions for $\vec{q}$ are circles described at constant velocities and indexed by an integer n

$$z^{(n)} = \sqrt{\frac{T}{2n\pi}} \exp -i \frac{2nt}{T} \qquad n \in Z \tag{2.18}$$

while for the ghost

$$\Psi_z^{(n)} = \eta \exp i \frac{2nt}{T} \tag{2.19}$$

where η is a constant fermion. The Euclidian energy and angular momentum of the action evalueted for these field configurations vanish for all valuese of n.

Due to the existence of these degenerate zero modes of the action we expect that BRST invariant observables should exist and that their mean values should be non zero as well as energy and time reparametrization independent. The corresponding numbers should to be expressable as a series over an integer related to the one which label the instanton solutions. This is the conjecture that we shall now verify.

To compute observables in the canonical, we will formalism. We do a Wick rotation to recover the real Minkowski time t by setting $\tau = it$, and change the quantum mechanichal variables into operators. The Hamiltonian associated to the action $\mathcal{I}_{gf}$ is

$$H = \frac{1}{2}\varphi^2 + \frac{f^2}{2\vec{q}^2} - f\overline{\Psi}_i \frac{\delta^2 \tau}{\delta q_i \delta q_j} \Psi_j \tag{2.20}$$

where the quantization rules are (remember that $q_i = (x, y)$ stands for the cartesian coordinates on the plane)

$$[p_i, q_j] = -i\delta_{ij} \qquad [\overline{\Psi}_i, \Psi_j]_+ = \delta_{ij}$$

$$[\Psi_i, \Psi_j]_+ = [\overline{\Psi}_i, \overline{\Psi}_j]_+ = [\Psi_i, p_j] = [\overline{\Psi}_i, p_j] = [\Psi_i, q_j] = [\overline{\Psi}_i, q_j] = 0 \qquad (2.21)$$

By construction H can be written as

$$H = \frac{1}{2}\{Q, \overline{Q}\} \qquad (2.22)$$

with

$$Q = \Psi_i(p_i + if\frac{\delta\tau}{\delta q_i}) \qquad \overline{Q} = \Psi_i(p_i - if\frac{\delta\tau}{\delta q_i}) \qquad (2.23)$$

Following [10], we use the following matricial representation for the ghost and antighost operator

$$\Psi_1 = \begin{pmatrix} 0 & 1 & 0 & 0 \\ 0 & 0 & 0 & 0 \\ 0 & 0 & 0 & 1 \\ 0 & 0 & 0 & 0 \end{pmatrix} \qquad \Psi_2 = \begin{pmatrix} 0 & 0 & -1 & 0 \\ 0 & 0 & 0 & 1 \\ 0 & 0 & 0 & 0 \\ 0 & 0 & 0 & 0 \end{pmatrix} \qquad (2.24)$$

One has $\overline{\Psi} = \Psi^\dagger$ and $p_i = -i\delta/\delta q_i$. In this representation

$$H = \begin{pmatrix} H_0 & 0 & 0 & 0 \\ 0 & H_{11} & H_{12} & 0 \\ 0 & H_{21} & H_{22} & 0 \\ 0 & 0 & 0 & H_2 \end{pmatrix} \qquad (2.25)$$

where

$$H_0 = H_2 = -\frac{1}{2r}\frac{\delta}{\delta r}r\frac{\delta}{\delta r} - \frac{1}{2r^2}\frac{\delta^2}{\delta\tau^2} + \frac{f^2}{2r^2} \qquad (2.26)$$

and

$$\begin{aligned}
\begin{pmatrix} H_{11} & H_{12} \\ H_{21} & H_{22} \end{pmatrix} &= H_0\delta_{ij} + f\frac{\delta^2\tau}{\delta q_i\delta q_j} \\
&= R_{-\tau}\left(-\frac{1}{2r}\frac{\delta}{\delta r}r\frac{\delta}{\delta r} + \frac{f^2 + 1 - \frac{\delta^2}{\delta\tau^2}}{2r^2} - \right. \\
&\quad \left. -\frac{1}{r^2}\begin{pmatrix} 0 & f + i\frac{\delta}{\delta\tau} \\ f - i\frac{\delta}{\delta\tau}\tau & 0 \end{pmatrix} \right)R_\tau
\end{aligned} \qquad (2.27)$$

where r and τ are the polar coordinates on the plane and

$$R_\tau = \begin{pmatrix} \cos\tau & \sin\tau \\ -\sin\tau & \cos\tau \end{pmatrix} \qquad (2.28)$$

The spectrum of H is straightforward to derive in this representation. One uses the usual strategy based on the fact that if an eigenstate of H has energy E, its Q and $\overline{Q}$ transforms are either zero or an eigenstate of H with the same energy.

States are labelled by their non negative energy E, angular momentum n and fermion number α, that is ghost number. We denote them as $|E, n, \alpha >$. For each value E and n, one has four states labelled by $\alpha = 1, 2, 3, 4$. The states with $\alpha = 1$ and $\alpha = 4$ are respectively anihilated by Q and $\overline{Q}$. This is due to the fact that states $|\phi >$ which are BRST invariant, $Q|\phi >= 0$, are such that

$$|\phi >= \begin{pmatrix} |E, n > \\ 0 \\ 0 \\ 0 \end{pmatrix} \qquad \overline{Q}|\phi >= \begin{pmatrix} 0 \\ (p_1 - if\frac{\delta \tau}{\delta q_1})|E, n > \\ -(p_2 - if\frac{\delta \tau}{\delta q_2})|E, n > \\ 0 \end{pmatrix}$$

$$H|\phi >= \begin{pmatrix} H_0|E, n > \\ 0 \\ 0 \\ 0 \end{pmatrix} \qquad (2.29)$$

One has similar relations for states $|\overline{\phi} >$ satisfying $\overline{Q}|\overline{\phi} >= 0$.

Let us define $g_{E,n} =< r, \tau|E, n >$. This function is the solution of the equation

$$< r, \tau|H_0|E, n >= \left(-\frac{1}{2r}\frac{\delta}{\delta r}r\frac{\delta}{\delta r} - \frac{1}{2r^2}\frac{\delta^2}{\delta \tau^2} + \frac{f^2}{2r^2} \right) g_{E,n} = Eg_{E,n} \qquad (2.30)$$

$g_{E,n}$ is also the solution of the ghost number 2 equation $< r, \tau|H_2|E, n >= E|E, n >$. Its knowledge is sufficient to get the full spectrum for $E \neq 0$. One has indeed

$$|E, n, 1 >= \begin{pmatrix} |E, n > \\ 0 \\ 0 \\ 0 \end{pmatrix} \qquad |E, n, 2 >= \frac{1}{\sqrt{E}}\overline{Q}|E, n, 1 >$$

$$|E, n, 4 >= \begin{pmatrix} 0 \\ 0 \\ 0 \\ |E, n > \end{pmatrix} \qquad |E, n, 3 >= \frac{1}{\sqrt{E}}\overline{Q}|E, n, 4 > \qquad (2.31)$$

The diagonalization of the part with ghost number one of the Hamiltonian (2.27) amounts to solve the equations

$$\left(-\frac{1}{2r}\frac{\delta}{\delta r}r\frac{\delta}{\delta r} + \frac{f^2 + 1 - \frac{\delta^2}{\delta \tau^2} \pm 2\sqrt{f^2 - \frac{\delta^2}{\delta \tau^2}}}{2r^2} \right) g_{E,n,\pm} = Eg_{E,n,\pm} \qquad (2.32)$$

which are of the same type as (2.30).

To solve (2.30) and (2.32) we set

$$g_{E,n} = \frac{1}{\sqrt{2\pi}} \exp in\tau \, f_{E,n}(r) \qquad n \in Z$$

$$g_{E,n,\pm} = \frac{1}{\sqrt{2\pi}} \exp in\tau \, f_{E,n,\pm}(r) \qquad n \in Z \qquad (2.33)$$

For $E \neq 0$, $f_{E,n}(r)$ and $f_{E,n,\pm}$ are expressable as a Bessel function $J_\nu(\sqrt{2}Er)$ of order ν, with

$$f_{E,n}(r) = \frac{1}{\sqrt{2}} J_{\sqrt{n^2+f^2}}(\sqrt{2}Er) \tag{2.34}$$

and

$$f_{E,n,\pm}(r) = \frac{1}{\sqrt{2}} \exp in\tau J_{\sqrt{f^2+1+n^2\pm2\sqrt{f^2+n^2}}}(\sqrt{2}Er) \tag{2.35}$$

These states are normalizable as plane waves in one dimension. This is a consequence of the continuity of the spectrum in the radial direction. They build an appropriate basis of stationary solutions since, with the normalization factor which is explicit in (2.34), one has $\sum_n \int_{E>0} dE |E, n> <E, n| = 1$. On the other hand, for $E = 0$, the Schrödinger equations (2.30) and (2.32) have no admissible normalizable solution. Thus we have a continuum spectrum, bounded from below, with a spin degeneracy equal to 4 and an infinite degeneracy in the angular momentum quantum number n. The peculiarity of this spectrum is that there is no ground state, since we have states with energy as little as we want, but we cannot have $E = 0$. This is a consequence of the conformal property of the potential $\frac{1}{|\vec{q}|^2}$.

Since we cannot reach the energy zero which woud be the only Q and $\overline{Q}$ invariant state, we conclude that supersymmetry is broken.

It is useful for what follows to redefine the ghost and antighost operators into

$$\begin{pmatrix} \Psi_r \\ \Psi_\tau \end{pmatrix} = \begin{pmatrix} \cos\tau & \sin\tau \\ -\sin\tau & \cos\tau \end{pmatrix} \begin{pmatrix} \Psi_1 \\ \Psi_2 \end{pmatrix} \quad \begin{pmatrix} \overline{\Psi}_r \\ \overline{\Psi}_\tau \end{pmatrix} = \begin{pmatrix} \cos\tau & \sin\tau \\ -\sin\tau & \cos\tau \end{pmatrix} \begin{pmatrix} \overline{\Psi}_1 \\ \overline{\Psi}_2 \end{pmatrix} \tag{2.36}$$

These rotated ghost operators satisfy similar anticommutation rotations as the Ψ_i and $\overline{\Psi}_i$. On the other hand, notice that

$$[\frac{\delta}{\delta\tau}, \begin{pmatrix} \overline{\Psi}_r \\ \overline{\Psi}_\tau \end{pmatrix}]_+ = \begin{pmatrix} \overline{\Psi}_\tau \\ -\overline{\Psi}_r \end{pmatrix} \qquad [\frac{\delta}{\delta\tau}, \begin{pmatrix} \Psi_r \\ \Psi_\tau \end{pmatrix}] = \begin{pmatrix} \Psi_\tau \\ -\Psi_r \end{pmatrix} \tag{2.37}$$

$$[\frac{\delta}{\delta\tau}, \begin{pmatrix} \overline{\Psi}_r \\ \overline{\Psi}_\tau \end{pmatrix}] = [\frac{\delta}{\delta\tau}, \begin{pmatrix} \Psi_r \\ \Psi_\tau \end{pmatrix}]_+ = 0 \tag{2.38}$$

One has the following expression of Q and $\overline{Q}$ which will be used shortly

$$Q = -i\Psi_r\frac{\delta}{\delta r} - i\frac{1}{r}\Psi_\tau(\frac{\delta}{\delta\tau} - f) \qquad \overline{Q} = -i\overline{\Psi}_r\frac{\delta}{\delta r} - i\frac{1}{r}\overline{\Psi}_\tau(\frac{\delta}{\delta\tau} + f) \tag{2.39}$$

These expressions in curved coordinates could be obtained from the general formalism of [11]].

We now turn to tomputation of BRST Invariant Observables. We have just seen that supersymmetry is broken in a very special way. This opens the possibility of having non vanishing BRST-exact Green functions which are topological in the sense that they are scale independent, that is independent of time, or energy, rescalings.

From dimensional arguments the candidates for such commutators are

$$O_r = [Q, r\overline{\Psi}_r]_+ = [\overline{Q}, r\Psi_r]_+^\dagger \qquad O_r = [Q, r\overline{\Psi}_r]_+ = [\overline{Q}, r\Psi_r]_+^\dagger \qquad (2.40)$$

The mean values of these operators between normalized states are

$$\frac{< E, n|[Q, r\overline{\Psi}_r]_+|E, n >}{< E, n|E, n >} = n + if \qquad (2.41)$$

and

$$\frac{< E, n|[Q, r\overline{\Psi}_r]_+|E, n >}{< E, n|E, n >} = \lim_{L \to \infty} \frac{L^2 J^2_{\sqrt{n^2+f^2}}(L)}{\int_0^L dr J_{\sqrt{n^2+f^2}}(r)} \qquad (2.42)$$

The last quantity is bounded but ill-defined, so we reject it. We get therefore that for any normalized state $|\phi_n> = \int dE\rho(E)|E, n >$ with a given angular momentum n, the expectation value of $[Q, r\overline{\Psi}_r]_+$ is

$$< \phi_n|[Q, r\overline{\Psi}_r]_+|\phi_n >= n + if \qquad (2.43)$$

indepently of the weighting function ρ.

If we now sum over all values of n, what remains is the topological number

$$< O_r >= \sum_n < \phi_n|[Q, r\overline{\Psi}_r]_+|\phi_n >= \sum_n n + if \sum_n 1 \qquad (2.44)$$

¿From a topological point of view, our result mean that there are two observables, organized in a complex form, in the cohomology of the punctured plane. The summation over the index n, that is the angular momentum, could have expected from the formal argument that in the path integral one gets a single finite contribution from each instanton solution to the mean value of a topological observable, so that

$$\text{Topological information} = \int \mathcal{D}[\bar{q}] O_f \exp -\mathcal{I}_{cl}[\bar{q}] \sim \sum_n f(n) \qquad (2.45)$$

Our computation shows the existence of a BRST invariant observable with non zero mean value which is Q-closed. The supersymmetry breaking mechanism made possible by our potential choice (on the basis of local BRST symmetry) is responsible of this situation. With other potentials than the one that we have chosen , either supersymmetry would be unbroken, or a mass gap would occur. In the previous case all Q-exact observable would vanish; in the latter case they could be nonzero but they would be scale dependent.

As another topological observable of the theory, we may consider the Witten index [12] [13]. The idea is that although there is no normalizable vacuum in the theory, we can consider the trace

$$\Delta = \text{Tr}(-)^F \exp -\beta H \qquad (2.46)$$

where the trace means a sum over angular momentum as well as over all energy including energy zero, and $(-)^F$ is the ghost or fermion number operator. The result should be finite because, although the state with energy zero is not normalizable, it contributes only over a domain of integration with zero measure. Indeed, since supersymmetric compensations occur for $E \neq 0$ and provided one uses a BRST symmetry preserving regularization, the full contribution to Δ should come from the domain of integration concentrated at $E \sim 0$, while the topological nature of the theory should warranty that Δ is non zero and independent on β.

By using the suitably normalized eigenfunctions of the Hamiltonian, eqs.(2.34) and (2.35), one can write the index Δ as follows

$$\Delta = \sum_n \int_0^\infty dE \exp{-\beta E} \int r dr \frac{1}{2} \left(2 J^2_{\sqrt{n^2+f^2}}(\sqrt{2Er}) - \right.$$
$$\left. -J^2_{\sqrt{f^2+1+n^2+\sqrt{f^2+n^2}}}(\sqrt{2Er}) - J^2_{\sqrt{f^2+1+n^2-\sqrt{f^2+n^2}}}(\sqrt{2Er}) \right)$$
(2.47)

To compute this double integral one needs a regularisation. Following for instance [13], we can use a dimensional regularization . Thus we change dr into $r^\epsilon dr$. Then, the analytic comtinuation of the result when $\epsilon \to 0$ is

$$\Delta = \sum_n \frac{1}{2} \left(2\sqrt{f^2+n^2} - \sqrt{f^2+1+n^2+\sqrt{f^2+n^2}} - \right.$$
$$\left. - \sqrt{f^2+1+n^2-\sqrt{f^2+n^2}} \right)$$
(2.48)

As announced this result is independent on β. As a series, it diverges logarimically as $\sum 1/n$ which is presumably the consequence of the conformal invariance of the potential. We see that the contribution of each topological sector is n dependant.

Let us now summarize what we understood from this model. We have shown an example for which the requirement of local BRST symmetry for topological quantum mechanics results in selecting a superconformal quantum mechanichal system. As a result, the spectrum of the theory has no ground state and a supersymmetry breaking mechanism occurs, without the the presence of a dimensionful parameter. Our goal was to understand the mechanism which provide topological observables. We observed that the special properties of the potential allows the computation of energy independant quantities although they are of mean values of BRST exact observables between non zero energy states. These quantities deserve to be called topological and they get a contribution from the whole spectrum of the theory. We have also singled out the Witten index, in a computation which includes a contribution from the non normalizable state of zero energy. The generalization of these observations to quantum field theory is an interesting open question.

3 Model (ii): The supersymmetric Lagrangian for spin-one particles

Supersymmetric quantum mechanics can be used to describe the dynamics of spinning point particles. The use of anticommuting variables to describe spinning particles was introduced in [14]. Then, it was found that local supersymmetry of rank $2S$ on the worldline is necessary to describe consistently a particle of spin S. The resulting constrained system [17] [18] requires a careful gauge-fixing of the einbein and the gravitini. One obtains eventually a tractable Lagrangian formulation [19], [20]. (There are many references on the subject, of which we quote very few) as well as to compute a certain number of topological invariants of the target space [12].

Using these facts, we will now point out an example showing that topological quantum theories may exhibit a phase with a Hilbert space made of particle degrees of freedom. We will interpret local supersymmetry on the worldline as a residue of a more fundamental topological symmetry, defined in a target-space with two extra dimensions. One of the coordinates is eventually identified as the einbein on the worldline. Other fields must be introduced to enforce the topological BRST invariance. They can be eliminated by their equations of motion and decouple from the physical sector. To obtain in a natural way a nowhere vanishing einbein, we use a disconnected higher dimensional target-space where the hyperplane $\{e = 0\}$ is a priori extracted. Thus, one introduces some topology before any gauge-fixing. Two disconnected topological sectors exist, $\{e > 0\}$ and $\{e < 0\}$, which correspond to the prescription $\pm i\epsilon$ for the propagators. It is fundamental that the gauge functions be compatible with the topology of space: they must induce a potential which rejects the trajectories from the hyperplane $\{e = 0\}$.

We will first review the supersymmetric description of a relativistic spinning particle in a Riemannian space-time. Then we will consider the case of $N = 2$ supersymmetry and show a link between the supersymmetric description of scalar or spin-one particles and topological quantum mechanics in a higher dimensional target-space. Finally, we will verify that the constraints of the theory identify its physical content and illustrate the result by computing the deviation of the trajectories from geodesics due to the interactions between geometry and spin.

Consider a spin-S particle in a D-dimensional space-time. Classically, it follows a worldline whose coordinates $X^\mu(\tau)$ are parametrized by a real number τ. If the particle is massive, a natural choice of this parameter is the proper-time. The idea originating from [14] is to describe the spin of the particle by assigning to each value of τ a vector with anticommuting coordinates $\Psi_i^\mu(\tau)$ where the vector index μ runs between 1 and D and i between 1 and $2S$. Indeed, in the case of a flat space-time and spin one-half, the Lagrangian density introduced in [14] is

$$\mathcal{L} = \frac{1}{2}(\dot{X}^2(\tau) - \Psi^\mu(\tau)\dot{\Psi}_\mu(\tau)) \tag{3.1}$$

where the dot $\dot{}$ means ∂_τ, τ being a parametrization of the worldline. Upon canonical quantization $\Psi^\mu(\tau)$ is replaced by a τ-independent operator $\hat{\Psi}^\mu$ which satisfies anticommutation relations

$$\{\hat{\Psi}^\mu, \hat{\Psi}_\nu\}_+ = 2\delta^\mu_\nu \qquad (3.2)$$

The Hamiltonian is

$$H = \frac{1}{2}p^2 = \frac{1}{2}Q^2 \qquad (3.3)$$

with $Q = p_\mu \hat{\Psi}^\mu$. Due to (3.2) the $\hat{\Psi}$'s can be represented by Dirac matrices and Q is the free Dirac operator. Q commutes with H and it makes sense to consider the restriction of the Hilbert space to the set of states $|\varphi >$ satisfying

$$Q|\varphi >= 0 \qquad (3.4)$$

By definition of Q, this equation means that the $|\varphi >$ are the states of a massless spin one-half particle. The extension to the case of a massive particle implies the introduction of an additional Grassmannian variable Ψ^{D+1} and the generalization of $\mathcal{L}$ to

$$\mathcal{L} = \frac{1}{2}(\dot{X}^2(\tau) - \Psi^\mu(\tau)\dot{\Psi}_\mu(\tau) - \Psi^{D+1}(\tau)\dot{\Psi}^{D+1}(\tau) + m^2) \qquad (3.5)$$

(Formally, $\dot{X}^{D+1} \to m$), so that

$$H = \frac{1}{2}(p^2 - m^2) = \frac{1}{2}Q^2 \qquad (3.6)$$

with

$$Q = p_\mu \hat{\Psi}^\mu + m\hat{\Psi}^{D+1} \qquad (3.7)$$

and one has in addition to (3.2)

$$\{\hat{\Psi}^{D+1}, \hat{\Psi}^{D+1}\}_+ = -2 \qquad \{\hat{\Psi}^\mu, \hat{\Psi}^{D+1}\}_+ = 0 \qquad (3.8)$$

The condition (3.4) is now the free Dirac equation for a spin one-half particle of mass m, multiplied by $\hat{\Psi}^{D+1}$. The generalization to the case of an arbitrary spin is obtained by duplicating $2S$ times the components of Ψ, $\Psi^\mu \to \Psi^\mu_i, 1 \le i \le 2S$, as can be seen by constructing the representations of $SO(D)$ by suitable tensor products of spin one-half representations [21] [22].

To understand the constraint (3.4), it is in fact necessary to promote the global supersymmetry of the action, corresponding to the commutation of H and Q, into a local supersymmetry. Indeed, when time flows, the state of the particle must evolve from a solution of the Dirac equation to another solution of this equation, without any possibility to collapse in an unphysical state (out of $Ker(Q)$). A natural way to reach such a unitarity requirement is to impose the supersymmetry independently for all values of τ, that is, to gauge the supersymmetry on the worldline. In this way, the condition (3.4) appears as the definition of physical states in a gauge theory with generator Q which ensures unitarity, like the transversality condition of gauge bosons in ordinary Yang-Mills theory. For consistency, the diffeomorphism invariance on the worldline must be also imposed since the commutator of two supersymmetry transformations contains a diffeomorphism. One thus introduces gauge fields for these symmetries, the einbein $e(\tau)$ and the (anticommuting) gravitino $\alpha(\tau)$. By minimal coupling on the worldline, (3.5) is thus generalized to the following Lagrangian which is locally supersymmetric and reparametrization invariant, up to a pure derivative with respect to τ :

$$\mathcal{L} = \frac{1}{2}\left(e^{-1}\dot{X}^2 - \Psi(\dot{\Psi} + \alpha e^{-1}\dot{X}) - \Psi^{D+1}(\dot{\Psi}^{D+1} + m\alpha) + em^2\right) \qquad (3.9)$$

(we will now omit the vector and spin indices). Formally, $\dot{X}^{D+1} \to me$. The transformation laws of e and α are those of one-dimensional supergravity of rank $2S$.

The gauge-fixing $e(\tau) = 1$ and $\alpha(\tau) = 0$ identifies (2.5) and (2.9), up to Faddeev-Popov ghost terms. These ghost terms have a supersymmetric form $b\dot{c}+ \beta\dot{\gamma}$. They decouple effectively, since their effect is to multiply all the amplitudes by a ratio of determinants, independent of the metric in space-time. This gauge-fixing is however inconsistent because it is too strong, since the Lagrangian is gauge invariant only up to boundary terms. Therefore, given a general gauge transformation, one must put restrictions on its parameters to get the invariance of the action, and there are not enough degrees of freedom in the symmetry to enforce the gauge $e(\tau) = 1$ and $\alpha(\tau) = 0$. One can at most set $e(\tau) = e_0$ and $\alpha(\tau) = \alpha_0$, letting the constants $e_0 > 0$ and α_0 free, that is, doing an ordinary integration over e_0 and α_0 in the path integral after the gauge-fixing [19]. This yields the following partition function for the theory

$$Z = \int_0^\infty de_0 \int d\alpha_0 \int [dX(\tau)][d\Psi(\tau)] \exp - \int_0^1 d\tau \mathcal{L}_0 \qquad (3.10)$$

with

$$\mathcal{L}_0 = \frac{1}{2}\left(e_0^{-1}\dot{X}^2 + e_0 m^2 - \Psi(\dot{\Psi} + \alpha_0 e_0^{-1}\dot{X}) - \Psi^{D+1}(\dot{\Psi}^{D+1} + m\alpha_0)\right) \qquad (3.11)$$

Using the Lagrangian (3.5) instead of (3.11) implies that one misses crucial spin-orbit interactions described by the Grassmannian integration over the constant α_0 which induces the fermionic constraint $\int d\tau(\Psi\dot{X} + me_0\Psi^{D+1}) = 0$. The use of (3.5) leads indeed to a spin-zero particle propagator while (2.11) leads to the expected spin one-half propagator. One gets the $\pm i\epsilon$ propagators depending on the choice of the integration domain $\{e_0 > 0\}$ or $\{e_0 < 0\}$. Notice that the e-dependence of the Lagrangian (3.5) gives a negligible weight in the path integral (3.10) to the trajectories with points near the hyperplane $\{e_0 = 0\}$. The integration over e_0 and α_0 has a simple interpretation in Hamiltonian formalism. The Hamiltonian associated to (3.11) is

$$H = \frac{e_0}{2}(p^2 - m^2) + \frac{\alpha_0}{2}(p_\mu\hat{\Psi}^\mu + m\hat{\Psi}^{D+1})$$

$$= \frac{e_0}{2}(p^2 - m^2) + \frac{\alpha_0}{2}Q \tag{3.12}$$

The constants e_0 and α_0 are thus Lagrange multipliers which force the particle to satisfy the Klein-Gordon equation and the Dirac equation (or its higher spin generalizations $Q_i|\varphi>=0$). Observe that in Lagrangian formalism, the Klein-Gordon equation is not a consequence of the Dirac equation, due to the anticommutativity of Grassmann variables, and the two constraints $Q|\varphi>=0$ and $H|\varphi>=0$ must be used separately. Therefore, we have a theory where the Hamiltonian is a sum of constraints, which leads to known technical difficulties [17][18]. In Lagrangian formalism, supergravity on the worldline and its correct gauge-fixing take care of all details [19].

The above description is valid for a flat space-time. It can be generalized to the case where the particle moves in a curved space-time and/or couples to an external electromagnetic field, by minimal coupling in the target-space. The compatibility between the worldline diffeomorphism invariance and local supersymmetry with reparametrization invariance in the target-space for a general metric $g_{\mu\nu}$ is however possible only for $N \leq 2$ [21]. This phenomenon is possibly related to the limited number of consistent supergravities [23].

We will now consider the case $N = 2$ and show n the link of the theory with a topological model.

The $N = 2$ supersymmetric Lagrangian with a general background metric $g_{\mu\nu}$ is

$$
\begin{aligned}
\mathcal{L}_{SUSY} &= \frac{1}{2e}g_{\mu\nu}\dot{X}^\mu \dot{X}^\nu - \overline{\Psi}^\mu(g_{\mu\nu}\dot{\Psi}^\nu + e\Gamma_{\mu\nu\rho}\dot{X}^\nu\Psi^\rho) + e^{-1}g_{\mu\nu}\dot{X}^\mu(\overline{\Psi}^\nu\alpha + \overline{\alpha}\Psi^\nu) \\
&+ \frac{em^2}{2} - \overline{\Psi}^{D+1}\dot{\Psi}^{D+1} + m(\overline{\Psi}^{D+1}\alpha + \overline{\alpha}\Psi^{D+1}) \\
&- e^{-1}\overline{\alpha}\alpha\overline{\Psi}\Psi + \frac{e}{2}R_{\mu\nu\rho\sigma}\overline{\Psi}^\mu\Psi^\nu\overline{\Psi}^\rho\Psi^\sigma
\end{aligned}
\tag{3.13}
$$

where Ψ and $\overline{\Psi}$ are independent Grassmannian coordinates. (Compare with [21]). The Lagrangian (3.13) has two local supersymmetries, with generators Q and $\overline{Q}$. An $O(2)$ symmetry between Ψ and $\overline{\Psi}$ can be enforced by introducing a single gauge field $f(\tau)$ and adding a term $f\overline{\Psi}\Psi$. However, no new information is provided, since one increases the symmetry by one generator, which is compensated by the introduction of the additional degree of freedom carried by f. The latter can indeed be gauge-fixed to zero and one recovers (3.13). Moreover, in view of identifying Ψ and $\overline{\Psi}$ as ghosts and antighosts, one wishes to freeze the symmetry between these two fields. We thus ignore the possibility of gauging the $O(2)$ symmetry. We will check shortly that the Hilbert space associated to the Lagrangian (3.1) contains spin-one particles.

The Lagrangian (3.13) can be conveniently rewritten in first order formalism by introducing a Lagrange multiplier $b^\mu(\tau)$. One gets the equivalent form

$$
\begin{aligned}
\mathcal{L}_{SUSY} \sim &-\frac{e}{2}(g_{\mu\nu}b^\mu b^\nu - m^2) + g_{\mu\nu}b^\mu(\dot{X}^\nu + e\Gamma^\nu_{\rho\sigma}\overline{\Psi}^\rho\Psi^\sigma + \overline{\Psi}^\nu\alpha + \overline{\alpha}\Psi^\nu) \\
&-\overline{\Psi}^\mu(g_{\mu\nu}\dot{\Psi}^\nu + e\partial_\rho g_{\mu\nu}\dot{X}^\nu\Psi^\rho) - \Gamma_{\nu\rho\sigma}\overline{\Psi}^\rho\Psi^\sigma(\overline{\Psi}^\nu\alpha + \overline{\alpha}\Psi^\nu)
\end{aligned}
$$

$$-\overline{\Psi}^{D+1}\dot{\Psi}^{D+1} + m(\overline{\Psi}^{D+1}\alpha + \overline{\alpha}\Psi^{D+1}) - \frac{e}{2}\partial_\nu\Gamma_{\mu\rho\sigma}\overline{\Psi}^\mu\Psi^\nu\overline{\Psi}^\rho\Psi^\sigma \quad (3.14)$$

(The symbol $\sim$ means that the two Lagrangians differ by a term which can be eliminated using an algebraic equation of motion, and, consequently, define the same quantum theory). For $e = 1$, $\alpha = \overline{\alpha} = 0$ and $\Psi^{D+1} = \overline{\Psi}^{D+1} = 0$, the Lagrangian (3.2) can be interpreted as the gauge-fixing of zero or of a term invariant under isotopies of the curve X [4]. In this interpretation the Ψ are topological ghosts and the $\overline{\Psi}$ are antighosts. The BRST graded differential operator s of the topological symmetry is defined by

$$sX^\mu = \Psi^\mu$$
$$s\Psi^\mu = 0$$
$$s\overline{\Psi}^\mu = b^\mu$$
$$sb^\mu = 0 \quad\quad\quad (3.15)$$

and the gauge-fixing Lagrangian is s-exact modulo a pure derivative

$$\mathcal{L}_{GF} = s(\overline{\Psi}_\mu(-\frac{1}{2}b^\mu + \dot{X}^\mu + \frac{1}{2}\Gamma^\mu_{\rho\sigma}\overline{\Psi}^\rho\Psi^\sigma)) \quad (3.16)$$

(Since $s^2 = 0$, $\mathcal{L}_{GF}$ is s-invariant.) To identify (3.1) as a topological Lagrangian, we must introduce new ingredients. We will enlarge the target-space with two additional components, and add a ghost of ghost. We will eventually identify one of the extra coordinates with the einbein e and the other one will be forced to vary in a Gaussian way around an arbitrary scale, with an arbitrary width. The gravitini α and $\overline{\alpha}$ of the effective worldline supergravity will be interpreted as ghosts of the topological symmetry. The $O(2)$ invariance corresponds to the ghost number conservation.

We consider a $(D + 2)$-dimensional space-time with coordinates $X^A = (X^\mu, X^{D+1} = e, X^{D+2})$. We exclude from the space the hyperplane $\{X^{D+1} = 0\}$ which yields two separated half-spaces, characterized by the value of $sign(e)$. We wish to define a partition function through a path integration over the curves $X^A(\tau)$, with a topological action which is invariant under the BRST symmetry associated to isotopies of this curve in each half-space. In other words we wish to construct an action by consistently gauge-fixing the topological Lagrangian $sign(e)$. In a way which is analogous to the case of topological Yang-Mills symmetry, where one gauge-fixes the second Chern class $\int Tr\ F^2$ [4], we combine the pure topological symmetry, with topological ghosts $\Psi^A_{top}(\tau)$, to the diffeomorphism symmetry on the curve, with Faddeev-Popov ghost $c(\tau)$. The apparent redundancy in the number of ghost variables $\Psi^A_{top}(\tau)$ and $c(\tau)$, which exceeds the number of bosonic classical variables, is counterbalanced by the introduction of a ghost of ghosts $\Phi(\tau)$ with ghost number two. The action of the BRST differential s is defined by

$$sX^\mu = \Psi^\mu_{top} + c\dot{X}^\mu = \Psi^\mu$$
$$se = \Psi^e_{top} + c\dot{e} = 2\eta = \alpha + \dot{\Psi}^{D+1}$$

$$
\begin{aligned}
sX^{D+2} &= \Psi^{D+2}_{top} + c\dot{X}^{D+2} = \Psi^{D+2} \\
s\Psi^\mu &= 0 \\
s\Psi^{D+2} &= 0 \\
s\Psi^{D+1} &= \Phi \\
s\alpha &= -\dot{\Phi} \\
s\Phi &= 0
\end{aligned}
\tag{3.17}
$$

In agreement with the art of BRST invariant gauge-fixing, we introduce $D+2$ antighosts with ghost number (-1) and the associated Lagrange multipliers for the gauge conditions on the X^A's. We also introduce an antighost $\overline{\overline{\Phi}}$ with ghost number (-2) and its fermionic partner $\overline{\eta}$ with ghost number (-1) which we will use as a fermionic Lagrange multiplier for the gauge condition in the ghost sector. In this sector the action of s is

$$
\begin{aligned}
s\overline{\Psi}^A &= b^A \\
sb^A &= 0 \\
s\overline{\overline{\Phi}} &= \overline{\eta} \\
s\overline{\eta} &= 0
\end{aligned}
\tag{3.18}
$$

The gauge-fixing Lagrangian must be written as an s-exact term

$$
\mathcal{L}^X + \mathcal{L}^{D+1} + \mathcal{L}^{D+2} + \mathcal{L}^\Phi = s\left(\overline{\Psi}^A(\ldots)_A + \overline{\overline{\Phi}}(\ldots)\right)
\tag{3.19}
$$

For the gauge-fixing in the X-sector, we choose

$$
\begin{aligned}
\mathcal{L}^X &= s\left(-\frac{e}{2}g_{\mu\nu}\overline{\Psi}^\nu b^\mu + g_{\mu\nu}\overline{\Psi}^\mu(\dot{X}^\nu + \overline{\eta}\Psi^\nu + \frac{e}{2}\Gamma^\nu_{\rho\sigma}\overline{\Psi}^\rho\Psi^\sigma)\right) \\
&= -\frac{e}{2}g_{\mu\nu}b^\mu b^\nu + g_{\mu\nu}b^\mu(\dot{X}^\nu + e\Gamma^\nu_{\rho\sigma}\overline{\Psi}^\rho\Psi^\sigma + \overline{\Psi}^\nu\eta + \overline{\eta}\Psi^\nu) \\
&\quad - \overline{\Psi}^\nu(g_{\mu\nu}\dot{\Psi}^\mu + e\partial_\rho g_{\mu\nu}\dot{X}^\nu\Psi^\rho) - \frac{e}{2}\partial_\nu\Gamma_{\mu\rho\sigma}\overline{\Psi}^\mu\Psi^\nu\overline{\Psi}^\rho\Psi^\sigma - \Gamma_{\mu\rho\sigma}\overline{\Psi}^\mu\eta\overline{\Psi}^\rho\Psi^\sigma
\end{aligned}
\tag{3.20}
$$

For the gauge-fixing in the e-sector, we choose

$$
\begin{aligned}
\mathcal{L}^{D+1} &= -s\left(\overline{\Psi}^{D+1}e(m + \frac{b^{D+1}}{2})\right) \\
&= -e\frac{(b^{D+1})^2}{2} + b^{D+1}(-me + \overline{\Psi}^{D+1}\eta) + 2m\overline{\Psi}^{D+1}\eta
\end{aligned}
\tag{3.21}
$$

After elimination of the field b^{D+1}, we obtain

$$
\mathcal{L}^{D+1} \sim \frac{em^2}{2} + m\overline{\Psi}^{D+1}\eta
\tag{3.22}
$$

For the gauge-fixing in the X^{D+2}-sector, we choose

$$
\begin{aligned}
\mathcal{L}^{D+2} &= s\left(\overline{\Psi}^{D+2}\left(-\frac{a}{2}b^{D+2} + X^{D+2} - C - \frac{1}{a}\frac{\overline{\Psi}^{D+1}\dot{\Psi}^{D+1}}{X^{D+2}-C}\right)\right) \\
&= -\frac{a}{2}(b^{D+2})^2 + b^{D+2}\left(X^{D+2} - C - a\frac{\overline{\Psi}^{D+1}\dot{\Psi}^{D+1}}{X^{D+2}-C}\right) \\
&\quad -\overline{\Psi}^{D+2}\left(\Psi^{D+2} - as\left(\frac{\overline{\Psi}^{D+1}\dot{\Psi}^{D+1}}{X^{D+2}-C}\right)\right)
\end{aligned}
\tag{3.23}
$$

a and C are arbitrarily chosen real numbers. After elimination of the field b^{D+2}, we find

$$
\begin{aligned}
\mathcal{L}^{D+2} \sim &-\overline{\Psi}^{D+1}\dot{\Psi}^{D+1} + \frac{1}{2a}(X^{D+2}-C)^2 - \\
&-\overline{\Psi}^{D+2}\left(\Psi^{D+2} - as\left(\frac{\overline{\Psi}^{D+1}\dot{\Psi}^{D+1}}{X^{D+2}-C}\right)\right)
\end{aligned}
\tag{3.24}
$$

The variable X^{D+2} can be eliminated by its algebraic equation of motion as well as the corresponding ghosts Ψ^{D+2} and $\overline{\Psi}^{D+2}$, after some field redefinitions. X^{D+2} is concentrated in a Gaussian way around the arbitrary scale C, with an arbitrary width a. We are thus left with the propagating term for Ψ^{D+1} and $\overline{\Psi}^{D+1}$ which was missing in $\mathcal{L}^X$ and $\mathcal{L}^{D+1}$

$$
\mathcal{L}^{D+2} \sim -\overline{\Psi}^{D+1}\dot{\Psi}^{D+1}
\tag{3.25}
$$

We finally choose the gauge-fixing in the ghost sector. To recover the full Lagrangian (3.14) and eventually identify the coordinate e as the einbein of the projection of the particle trajectory in the D-dimensional physical space-time, we need a term linear in $\overline{\eta}$ as well as another term to get rid of unwanted higher order fermionic terms. We define

$$
\begin{aligned}
\mathcal{L}^{\Phi} &= s(\overline{\Phi}(m\Psi^{D+1} - \Gamma_{\nu\rho\sigma}\overline{\Psi}^{\rho}\Psi^{\sigma}\Psi^{\nu})) \\
&= \overline{\eta}(m\Psi^{D+1} - \Gamma_{\nu\rho\sigma}\overline{\Psi}^{\rho}\Psi^{\sigma}\Psi^{\nu}) + \overline{\Phi}(m\Phi - s(\Gamma_{\nu\rho\sigma}\overline{\Psi}^{\rho}\Psi^{\sigma}\Psi^{\nu}))
\end{aligned}
\tag{3.26}
$$

The dependence on the ghosts of ghosts Φ and $\overline{\Phi}$ is trivial: these fields decouple after a Gaussian integration. One has thus

$$
\mathcal{L}^{\Phi} \sim m\overline{\eta}\Psi^{D+1} - \Gamma_{\nu\rho\sigma}\overline{\Psi}^{\rho}\Psi^{\sigma}\overline{\eta}\Psi^{\nu}
\tag{3.27}
$$

Adding all terms (3.20), (3.22), (3.25) and (3.27), we finally recognize that $\mathcal{L}^X + \mathcal{L}^{D+1} + \mathcal{L}^{D+2} + \mathcal{L}^{\Phi}$ is equivalent to the Lagrangian (3.2), modulo the elimination of auxiliary fields and the change of notation $(\eta, \overline{\eta}) \to (\alpha, \overline{\alpha})$. We have therefore shown the announced result: the $N=2$ local supersymmetry of the Lagrangian describing spin-one particles is a residual symmetry coming from a topological model after a suitable gauge-fixing.

To verify the physical content of the model presented just above, we consider a flat space-time, and choose the gauge where the einbein and gravitini are constants over which we integrate. The Hamiltonian is

$$H = \frac{e_0}{2}(p^2 - m^2) + \overline{\alpha}_0 Q + \alpha_0 \overline{Q} \qquad (3.28)$$

with

$$Q = p_\mu \Psi^\mu + m\Psi^{D+1}$$
$$\overline{Q} = p_\mu \overline{\Psi}^\mu + m\overline{\Psi}^{D+1} \qquad (3.29)$$

The matrices Ψ and $\overline{\Psi}$ satisfy the Clifford algebra

$$\{\Psi^A, \overline{\Psi}^B\}_+ = \eta^{AB} \quad , \quad \{\Psi^A, \Psi^B\}_+ = \{\overline{\Psi}^A, \overline{\Psi}^B\}_+ = 0 \qquad (3.30)$$

for $A, B = 1, ..., D + 1$. Since the underlying gauge symmetry has Q and $\overline{Q}$ as generators, the physical states satisfy

$$Q|\phi> = 0 \qquad \overline{Q}|\phi> = 0 \qquad (3.31)$$

in addition to

$$(p^2 - m^2)|\phi> = 0 \qquad (3.32)$$

The Ψ and $\overline{\Psi}$ are generalizations of the Pauli matrices, and it is convenient to use a Schwinger type construction, in order to exploit directly their Clifford algebra structure. One introduces a spin vacuum $|0>$ annihilated by the Ψ's. Then, the $\overline{\Psi}$'s can be identified as their adjoints and act as creation operators. In the X representation, we can write a general state as

$$|\phi> = \left(\varphi_0 + \varphi_\mu \overline{\Psi}^\mu + \varphi_{\mu_1 \mu_2}\overline{\Psi}^{\mu_1}\overline{\Psi}^{\mu_2} + ... + \varphi_{\mu_1...\mu_D}\overline{\Psi}^{\mu_1}...\overline{\Psi}^{\mu_D}\right)|0>$$
$$+\overline{\Psi}^{D+1}\left(\overline{\varphi}_0 + \overline{\varphi}_\mu \overline{\Psi}^\mu + \overline{\varphi}_{\mu_1 \mu_2}\overline{\Psi}^{\mu_1}\overline{\Psi}^{\mu_2} + ... + \overline{\varphi}_{\mu_1...\mu_D}...\overline{\Psi}^{\mu_D}\right)|0> \quad (3.33)$$

The wave functions $\varphi_{\mu_1...\mu_p}(X)$ and $\overline{\varphi}_{\mu_1...\mu_p}(X)$ are antisymmetric and it is useful to consider the differential forms

$$\varphi_p = \frac{1}{p!}dX^{\mu_1}...dX^{\mu_p}\varphi_{\mu_1...\mu_p}(X)$$
$$\overline{\varphi}_p = \frac{1}{p!}dX^{\mu_1}...dX^{\mu_p}\overline{\varphi}_{\mu_1...\mu_p}(X) \qquad (3.34)$$

for $0 \leq p \leq D$. The constraints (3.31) can be conveniently written as

$$d\varphi_p + im\overline{\varphi}_{p+1} = 0 \qquad (3.35)$$

$$d^*\overline{\varphi}_p + im\varphi_{p-1} = 0 \qquad (3.36)$$

$$d\overline{\varphi}_p = 0 \qquad (3.37)$$

$$d^*\varphi_p = 0 \qquad (3.38)$$

Where $d = dx^\mu \partial_\mu$ and d^* is its Hodge dual. One has also

$$(d^*d + dd^*)\varphi = -m^2\varphi \qquad (d^*d + dd^*)\overline{\varphi} = -m^2\overline{\varphi} \qquad (3.39)$$

These equations determine the independent degrees of freedom. When $m \neq 0$, they couple the two sectors of opposite chiralities. Moreover, when D is even, the first one contains $\frac{D}{2}$ forms, namely one scalar (φ_0), one vector (φ_1), ..., and one ($\frac{D}{2} - 1$)-form ($\varphi_{\frac{D}{2}-1}$). The other one has a dual structure ($\overline{\varphi}_{\frac{D}{2}+1},...,\overline{\varphi}_D$). For ($\varphi_1$), the constraints (4.4) can be rewritten:

$$\partial_\mu \overline{\varphi}^{\mu\nu} + im\varphi^\nu = 0$$
$$\partial_\mu \varphi^\mu = 0$$
$$\partial_{[\mu}\varphi_{\nu]} + im\overline{\varphi}_{\mu\nu} = 0 \qquad (3.40)$$

Thus the vector wave function φ_1 satisfies Proca's equations, and descibes a spin-one particle with mass m. It follows that the field equations of φ_1 and $\overline{\varphi}_1$ can be derived by minimizing Proca's Lagrangian

$$\mathcal{L}_{Proca} = \frac{m}{2}\overline{\varphi}_{\mu\nu}\varphi^{\mu\nu} - \frac{i}{2}\varphi^{\mu\nu}(\partial_\mu\varphi_\nu - \partial_\nu\varphi_\mu)$$
$$- \frac{i}{2}\overline{\varphi}^{\mu\nu}(\partial_\mu\overline{\varphi}_\nu - \partial_\nu\overline{\varphi}_\mu) + m\overline{\varphi}_\mu\varphi^\mu \qquad (3.41)$$

When $m = 0$, the two sectors of opposite chiralities decouple. In each sector, the independent degrees of freedom are now one 0-form A_0 (with $\varphi_1 = dA_0$), one 1-form A_1 (with $\varphi_2 = dA_1$),..., one (D-2)-form A_{D-2} (with $\varphi_{D-1} = dA_{D-2}$). The φ_p's are closed and co-closed, i.e. the A_p's satisfy Maxwell's equations and are defined up to gauge transformations. Consequently, φ_2 can be identified with the field strength of a photon. If we consider the case $D = 4$ and $m \neq 0$, the spectrum reduces to two scalars and two massive spin-one particles, and contains 8=2(1+3) degrees of freedom. For $m = 0$, we have two massless scalars and two massless vectors, so that we still have 8=2(1+1+2) independent degrees of freedom.

As an application of this formalism, we study the classical behavior of spinning particles in a curved space-time. We are interested in the approximation where the trajectory of the particle is classical, while the spin effects are visible as it would be the case in a Stern-Gerlach experiment. This situation occurs if the order of magnitude of the interaction energy between the spin and the curvature, which is essentially proportional to the space-time curvature times $\hbar$ (analogously to the interaction between the the magnetic field and a magnetic moment due to the spin), is comparable to the kinematical energy of the particle. One must also measure the position of the particle on a domain much larger than its Compton wavelength. In this limit the position X^μ and momentum P_μ are ordinary numbers and the quantum Hamiltonian becomes simply a matrix built from the Ψ's and $\overline{\Psi}$'s acting in the spin-space with coefficients depending on the classical position X and momentum P. The τ-dependence of the classical dynamics of the particle can be expressed by applying Hamilton-Jacobi's method with this matricial Hamiltonian. The only quantum effects are due to the spin interaction with the space-time curvature. (In a fully classical approximation, $\hbar = 0$, and the spin effects disappear, since all the fermionic operators are proportional to $\sqrt{\hbar}$.) One can always find a basis for the spin states, which depends on the space-time position and such that the Hamiltonian is diagonal. In this basis the spin value is conserved through evolution, i.e. the spin observables are paralelly transported along the trajectory. By diagonalization in spin space, H determines independent Hamilton-Jacobi's equations for each spin degree of freedom of the particle. For the spin-one case, we expect three different trajectories corresponding to the values 1, 0 and -1 for the projection of the spin on a spatial axis in the rest frame of the particle.

We consider the case of a Schwarzschild gravitational field in four dimensional space-time $(ds^2 = (1 - \frac{r_0}{r})dt^2 - (1 - \frac{r_0}{r})^{-1}dr^2 - r^2(d\theta^2 + sin^2\theta\ d\phi^2)$ with $r_0 = 2GM/c^2$.) We will compute the correction, due to the spin, to Einstein's formula predicting the shift of the perihelion of a spinless point particle. For the other classical test of general relativity, i.e. the bending of light rays in a gravitational field, we will find that the wave vector of a polarized photon deviates from geodesic motions by a relative shift proportional to $\hbar$. These results are in agreement with the fact that a particle with an angular momentum interacts with the space-time curvature, as first pointed out by Papapetrou for a rotating body [25]. The advantage of a supersymmetric Hamiltonian is that it defines unambiguously the spin effects. Since we work to first non-trivial order in $\hbar$, we restore from now on the $\hbar$ dependence in the formulae. The matricial Hamilton-Jacobi's equation is obtained by replacing in the supersymmetric Hamiltonian the classical momentum p_μ by $\frac{\delta S}{\delta X^\mu}$ where $S[X^\mu, \tau]$ is the action of the classical trajectory of the particle in a given spin state, with arbitrarily chosen initial and final boundary conditions. Notice that keeping the lowest order in $\hbar$ means that we only retain the covariant derivative of the fermionic variables and not the curvature term. This yields

$$g^{\mu\nu}\frac{\delta S}{\delta X^\mu}\frac{\delta S}{\delta X^\nu} - m^2 + 2\hbar\frac{\delta S}{\delta X^\mu}\omega^\mu_{ab}\Sigma^{ab} + O(\hbar^2) = 0 \qquad (3.42)$$

The space-time spin-connection ω is related to the space-time vierbein E and to Christoffel's symbol Γ

$$\omega_{\mu ab} = E_a^\alpha E_b^\beta \Gamma_{\alpha\mu\beta} \qquad (3.43)$$

$$E_\alpha^a E_\beta^b \eta_{ab} = g_{\alpha\beta} \qquad (3.44)$$

$$\Gamma_{\alpha\mu\beta} = \frac{1}{2}(\partial_\mu g_{\alpha\beta} + \partial_\beta g_{\alpha\mu} - \partial_\alpha g_{\beta\mu}) \qquad (3.45)$$

The $\Sigma^{ab} = \frac{i}{2}\left(\overline{\Psi}^a\Psi^b - \overline{\Psi}^b\Psi^a\right)$ are the generators of the (reducible) 32-dimensional representation of the Lorentz group defined by the algebra (4.3) and acting on the states solving (4.6). If the matrix form of $\overline{\Psi}^5$ is chosen diagonal, the spin operators Σ^{ab} become block-diagonal with two independent sectors of opposite chiralities, corresponding to the eigenvalues 0 and 1 of $\overline{\Psi}^5 \Psi^5$, so the 32-dimensional representation splits into two independent 16-dimensional representation, each one containing five sectors of dimensions 1,4,6,4,1 corresponding respectively to 0-forms, 1-forms, 2-forms, 3-forms, and 4-forms. As explained above, the constraints imply that only two block-sectors made of one 0-form and one 1-form sectors are independent wave-functions. The one-form sector, and the corresponding 4×4 Hamiltonian matrix, determine the dynamics of spin-one particles. Moreover, in a Schwarzschild metric with characteristic radius r_0,the motion is planar, so one can separate the variables and write

$$S = -Et + L\varphi + S_r(r) \tag{3.46}$$

The spin-dependent part of Hamilton-Jacobi's equation is obtained by the substitution

$$p_\mu \omega^\mu_{ab} \Sigma^{ab} = \frac{r_0 E}{r^2} \Sigma^{01} - \frac{2L}{r^2} \left(1 - \frac{r_0}{r}\right)^{1/2} \Sigma^{13} \tag{3.47}$$

where

$$\Sigma^{01} = \frac{1}{2} \begin{pmatrix} 0 & 1 & 0 & 0 \\ -1 & 0 & 0 & 0 \\ 0 & 0 & 0 & 0 \\ 0 & 0 & 0 & 0 \end{pmatrix}, \quad \Sigma^{13} = \frac{1}{2} \begin{pmatrix} 0 & 0 & 0 & 0 \\ 0 & 0 & 0 & i \\ 0 & 0 & 0 & 0 \\ 0 & -i & 0 & 0 \end{pmatrix} \tag{3.48}$$

By inserting (3.46) and (3.47) into Hamilton-Jacobi's equation (3.42), one obtains a matricial equation for $\frac{\partial S}{\partial r}$. The diagonalization can be done easily, and one gets three possibilities S_ϵ for the classical action, indexed by $\epsilon = 0, \pm 1$

$$\frac{E^2}{c^2} \left(1 - \frac{r_0}{r}\right)^{-1} - \left(\frac{L^2}{r^2} + m^2\right) - \left(1 - \frac{r_0}{r}\right) \left(\frac{\partial S_\epsilon}{\partial r}\right)$$

$$+2\epsilon \frac{\hbar}{r^2} \left(L^2 \left(1 - \frac{r_0}{r}\right) - \left(\frac{r_0 E}{2c}\right)^2\right)^{1/2} = 0 \tag{3.49}$$

(We have restored the dependence in the speed of light c.) The energy E and the angular momentum L are constants of motion of the particle. The values $\epsilon = 0, \pm 1$ correspond to the three possible projections of the spin along a given spatial axis in the rest frame of the particle. The case $\epsilon = 0$ corresponds to the geodesic trajectory followed by the scalar particle. Far from the Schwarzschild horizon, we can use the standard techniques of integration of Hamilton-Jacobi's equation to determine the three possibilities for the shift of the perihelion over a quasi-periodic trajectory. This amounts to replace L in the classical formulas [26] by an effective angular momentum L_ϵ defined by

$$L_\epsilon^2 = L^2 + 2\epsilon\hbar L \sqrt{1 - \left(\frac{r_0 E}{2Lc}\right)^2} \tag{3.50}$$

(Notice that near the horizon, unitarity breaks down). In the case of a massive particle, the shift of the perihelion is thus given by:

$$\delta\phi_\epsilon = \frac{3\pi}{2}\left(\frac{mcr_0}{L_\epsilon}\right)^2 \sim \delta\phi_0 \left(1 - \epsilon\frac{\hbar}{L}\sqrt{1 - \left(\frac{r_0 E}{2Lc}\right)^2}\right) \tag{3.51}$$

For a non-relativistic Z^0 orbiting quasi-tangentially to the sun at a speed of $10^5 m/s$, which is approximately the circular velocity around the sun , we find $|\delta\phi_+ - \delta\phi_0|/\delta\phi_0 \sim \hbar/L \sim 10^{-21}$, which is much to small to be detected.

The solutions of Hamilton-Jacobi's equation are continuous when $m \to 0$. However, in this limit the interpretation of its solution S is different. The particle is a photon following the laws of the geometrical optics, S is the eikonal of the light ray, and $\frac{\delta S}{\delta X^\mu}$ is its wave-vector. The solution $\epsilon = 0$ must then be rejected. In this massless case, one finds for the deflections of the two helicities $\epsilon = \pm 1$ the following formula

$$\delta\phi_\epsilon = \frac{2r_0\omega}{cL_\epsilon} \sim \delta\phi_0 \left(1 - \epsilon\frac{\hbar}{2L}\sqrt{1 - \left(\frac{r_0 E}{2Lc}\right)^2}\right) \tag{3.52}$$

where $\omega = \frac{E}{\hbar}$. For an optical photon of wavelength $\lambda = 7 \times 10^{-7}m$ (red) skimming past the sun, we find $|\delta\phi_+ - \delta\phi_0|/\delta\phi_0 \sim \frac{\hbar}{2L} = \frac{\lambda}{2R_{sun}} \sim 10^{-15}$. (Note that this ratio does not depend on $\hbar$: the gravitational field interacts classically with the two polarizations of the electromagnetic field.) However, this doubling of Einstein's rings is to small to be detected.

References

[1] E. Witten, *Comm. of Math. Phys.* **117**, (1988), 353; *Comm. of Math. Phys.* **118**, (1988) 601; *Phys. Lett.* **B206** , 1988.

[2] For a review see O. Birmingham, M. Blau, M. Rakowski and G. Thomson, *Physics Reports* **209**, (1991), 129 and references therein.

[3] E. Witten, *Comm. of Math. Phys.* **121**, (1989), 351.

[4] L. Baulieu and I.M. Singer, *Nucl. Phys. Proc. Suppl.* **5B**, (1988), 12; *Comm. of Math. Phys.* **125**, (1989), 227; *Comm. of Math. Phys.* **135**, (1991), 253.

[5] L. Baulieu and C. Aragao de Carvalho *Phys. Lett.* **B275** (1991)323; *Phys. Lett.* **B275** (1991)335 .

[6] D. Birmingham, M. Rakowski and G. Thompson, *Nucl. Phys.* **B329** (1990) 83; D. Birmingham and M. Rakowski *Mod. Phys. Lett.* **A4** (1989) 1753; F. Delduc, F. Gieres and S.P. Sorella *Phys. Lett.* **B225** (1989) 367.

[7] S. Fubini and E. Rabinovici *Nucl. Phys.* **B245** (1984) 17; V. de Alfaro, S. Fubini and G. Furlan *Nuovo Cimento* **34 A**, (1976), 569.

[8] E. Witten, *Nucl. Phys.* **B323** (1989) 113.

[9] L. Baulieu, B. Grossman and R. Stora, *Phys. Lett.* **B180** (1986) 95.

[10] A. Forge and E. Rabinovici *Phys. Rev.* **D32**, (1985), 927.

[11] A. C. Davis, A. J. Macfarlane, P. C. Popat, and J. W. Van Holten *J. Phys. A Math. Gen.* **17**, (1984), 2945.

[12] E. Witten, *Nucl. Phys.* **B202**, (1982), 253. L. Alvarez Gaumé, *Comm. of Math. Phys.* **90**, (1983), 161; D. Friedan and P. Windey *Nucl. Phys.* **B235**, (1984), 395.

[13] N.A. Alvez, H. Aratyn and A.H. Zimmerman *Phys. Rev.* **D31**, (1985), 3298; R. Akhoury and A. Comtet *Nucl. Phys.* **B246**, (1984), 253.

[14] F.A. Berezin and M.S. Marinov, *JETP Lett.* **21** (1975) 320 and *Ann. Phys. NY* **104** (1977) 336; R. Casalbuoni, *Nuovo Cimento* **33A** (1976) 389 and *Phys. Lett.* **62B** (1976) 49; A. Barducci, R. Casalbuoni and L. Lusanna, *Nuovo Cimento* **35A** (1976) 377; L. Brink, S. Deser, B. Zumino, P. di Vecchia and P.S. Howe *Phys. Lett.* **64B** (1976) 43.

[15] L. Brink and J.H. Schwarz, *Nucl. Phys.* **B121** (1977) 285; L. Brink, P. di Vecchia and P.S. Howe, *Nucl. Phys.* **B118** (1977) 76 and *Phys. Lett.* **65B** (1976) 471; S. Deser and B. Zumino *Phys. Lett.* **65B** (1976) 369.

[16] A.M. Polyakov, *Phys. Lett.* **103B** (1981) 211.

[17] P.A.M. Dirac. Lectures on Quantum Mechanics, (Belfer Graduate School of Science, Yeshiva University; New York; 1964); M. Henneaux and C. Teitelboim, Quantization of Gauge Systems (Princeton University Press; 1992).

[18] W. Siegel, Introduction to String Field Theory. (World Scientific; 1988).

[19] A.M. Polyakov, Gauge Fields and Strings (Harwood Academic Publishers; 1987); Vl. S. Dotsenko, *Nucl. Phys.* **B285** (1987) 45.

[20] R.H. Rietdijk and J.W. van Holten, *Class. Quantum Grav.* **7** (1990) 247.

[21] P. Howe, S. Penati, M. Pernici and P. Townsend, *Phys. Lett.* **215B** (1988) 555.

[22] R. Marnelius and U. Martensson, *Nucl. Phys.* **B335** (1990) 395; U. Martensson, Preprint Goteborg-92-3 (Jan. 92).

[23] P. Van Nieuwenhuyzen, *Phys. Rep.* **68C** (1981) 189.

[24] R.H. Rietdijk and J.W. van Holten, *Class. Quantum Grav.* **10** (1993) 575.

[25] A. Papapetrou, *Proc. Roy. Soc. London* **A209** (1951) 248.

[26] L.D. Landau, The Classical Theory of Fields. (Pergamon Press; London; 1971).

Structures of K.Saito Theory of Primitive Form in Topological Theories Coupled to Topological Gravity

A. Losev*

Institute of Theoretical and Experimental Physics
B.Cheremushkinskaya 25
117259 Moscow, Russia

Abstract. Structure of topological theory coupled to topological gravity is studied on a typical example - Landau-Ginzburg theory. It is shown that all main ingredients of K.Saito theory of primitive form are implied in such gravity theory. Filtration, that he considers turns out to be filtration by degrees of Morita-Mamford classes, and can be considered as a filtration of equivariant cohomologies (equivariance with respect to rotation of local coordinate). Higher residue pairing are nothing by pairing in equivariant cohomologies induced by integration over C^d. Section is the kernel of a contact term map. Axioms on goods sections follow from the symmetry of n-point correlation functions on genus zero.

1 Introduction

Topological theories coupled to topological gravity[1, 3] seem to be interesting because:
1. They look like the simplest string-like theories, and by studying them we can get some information about what string theory is. Moreover, some of them are equivalent[1, 2, 9] to noncritical strings for $c < 1$
2. These theories naturally connect geometry of the space of complex structures[1, 6], geometry of algebraic manifolds[1, 7] and theory of integrable systems[5, 8, 17]

In this paper we try to study general formalism of topological theory coupled to gravity on example of Landau-Ginzburg model. We show that the topological matter is descibed by the ring and a linear functional on it. In the case of Landau-Ginsburg theory ring is a Jacobian ring of singularity and linear functional is

---
*This work was partially supported by Grant 93-02-14365 of the Russian foundation of fundamental research.

given by zero-th higher residue pairing of K.Saito [11]. This linear functional involves additional data - holomorphic top form due to fermionic anomaly. We discuss coupling theory to topological gravity and stress the problems in performing such coupling for massive theory: these problems come from the boundary of the moduli space. Then we show how in conformal theory descendents[15, 16, 10] could appear in matter theory from fields with logarithmic anomalous conformal dimension, and illustrate it on example of Landau-Ginzburg theory. We find that observables in topological thery coupled to gravity theory form filtration in the degree of gravitational descendents, and identify this filtration with the space of equivariant cohomologies with respect to rotations of local coordinates. This is the same filtration that appeared in K.Saito's theory [11]. His higher residue pairing is identified with the pairing in equivariant cohomologies induced by integration over target space. Integration over position of marked points leads to flow on the space of theories equipped with the connection in the bundle of observables (fiber of this bundle is a filtration of equivariant cohomologies). This connection arises from the contact term that we calculated up to the kernel of this map(that is a subspace in the fitration with dimension of the ring). Such a kernel is called section in K.Saito theory. Integration over positions of the marked points could be interchanged, moreover, 3-point function could be obtained in different ways. Physical requirement of consistency leads exactly to K.Saito axioms of good section, that involve higher residue pairing. K.Saito proved the integrability of connection arised from good section. To fix invariance over diffeomorphisms we choose some versal deformation of singularity. Then flows due to descendents lead to change of holomorphic top form. Finally we describe generating function for correlators in genus zero generalising algebraic solution to dispersionless KP found in this context by I.Krichever [8].

2   Topological matter

By topological matter[18, 14, 19] we mean a 2-d theory with a scalar fermionic symmetry Q, such that its square equals to zero, and with an action S_m, that on the Riemann surface with the metric g has the following form:

$$S_m(\phi, g) = S_{top}(\phi) + Q(R(\phi, g)) \tag{1}$$

Here we assume that action is invariant under diffeomorphisms that act on both matter fields ϕ and metric g; the first term in (1) is metric independent and thus we call it topological term. The second term in (1) depends on metric but is Q-trivial. Function R here and below we will call regulator.

From the form of action (1) we easily get that the energy-momentum tensor T is Q-exact.Really, if we define

$$G(x) = \frac{\delta R}{\delta g} \tag{2}$$

then

$$T(x) = \frac{\delta S_m}{\delta g(x)} = Q(G) \tag{3}$$

i.e. G is a superpatner of the energy-momentum tensor.

Let us denote by $<>^M$ correlator of local fields in the topological matter theory(here superscript M means matter , not to be confused with correlator in topological matter coupled to topological gravity, that would appear later):

$$< \Phi_1(z_1),\ldots,\Phi_n(z_n) >^M = \int D\phi_m \Phi_1(z_1),\ldots,\Phi_n(z_n)\exp(S_m) \qquad (4)$$

Here ϕ_m stands for fields in the matter theory, z_i denotes different points on the Riemann surface.

Since Q is a nilpotent symmetry of the theory, the exactness of T leads to independence of correlator of Q-closed local operators on metric on the Riemann surface and thus on their positions. It means that correlators of such operators are topological, i.e. depend only on the type of operators and genus of the worldsheet.

Definition. Local fields that are Q-closed are called local observables in the topological matter theory. Correlator of local observables in topological theory is called topological correlator.

It is obvious that topological correlator of a Q-exact local observable is zero, thus topological correlators are non-zero on the Q-cohomologies of the space of local operators.

These cohomologies form a ring, that we will denote J. Multiplication can be defined by putting two local observables at two different points and moving them to each other. Since energy-momentum tensor is exact, one can show that different limits of such a motion would differ on a Q-exact local field. Topological correlator could be computed by putting all observables together, i.e. multiplying them in cohomology ring J, described above. Thus, any correlator on a Riemann surface could be reduced to one-point correlator, and for n-point correlator in genus q in topological matter theory we get :

$$< \Phi_1(z_1),\ldots,\Phi_n(z_n) >_q^M = < (\Phi_1 * \Phi_2 * \ldots * \Phi_n) >_q^M \qquad (5)$$

where the star denotes multiplication in cohomology ring J, and $<>_q^M$ is just a linear functional on cohomologies[1].

3 Partition function in topological gravity coupled to topological matter

Following [3] we define topological gravity as a theory with the following fields: metrics g on Riemann surfaces and its superpartner Ψ, that is an element of the tangent bundle to the space of all metrics; this theory has fermionic Q_G-symmetry that acts as an external derivative:

$$Q_G(g(x)) = \Psi(x) \qquad (6)$$

---

[1] Using "handle gluing" operation one can show that linear functionals for all genera could be expressed just through structure constants of the ring and linear functional for genus zero

and topological part of the action is zero. Thus, topological gravity itself should be some cohomological theory on the space of all metrics on a Riemann surface. However, such a theory is ill-defined, since this space is infinitely-dimensional and non-compact. The way out is to consider equivariant cohomologies with respect to the action of some Lie group, i.e. to fix the supergroup, associated with this group[2]. We will take this group to be the product of group of conformal transformations of metrics and group of diffeomorphisms. It would reduce integration to an integral over the moduli space of conformal structures.

Still there is a problem of constructing a form to be integrated over moduli space. Originally this form was constructed from ghosts, that appear in the procedure of gauge fixing[3]. It is possible to treat ghosts, conformal factor, and their superpartners as some specific conformal topological theory coupled to "gravity" (the only thing that left from gravity is the space of moduli of complex structures). Obvious generalization is to change such a special topological theory on a general one.

We suppose that after coupling topological matter to topological gravity the total fermonic symmetry is a sum of a matter fermionic symmetry and gravitational one:

$$Q_t = Q_G + Q \tag{7}$$

thus, the action is a sum of S_{top} and Q_t-exact term,i.e.

$$S_{M+G} = S_{top} + Q_t(R) = S_m + Q_G(R) \tag{8}$$

Space of metrics forms a bundle over the moduli space of conformal structures. Fixing supergroup means integration over the section of this bundle[3].

Conformal regulator

If we take such a regulator R, that trace of G is zero, then $\exp(S_{M+G})$ as a form on the space of metrics is invariant under conformal transformations and horizontal(its contraction with the vector, tangent to the fiber is zero). Let $\sigma(m)$ be a section from the moduli space to the space of all metrics, perhaps with vertical jumps. Let σ' be a continuous hypersurface that is obtained by gluing to σ hypersurfaces W_i, that are vertical, i.e. tangent space to any point of W_i contains vertical vector. Then we define

$$Z = \int_{\sigma'} \int D\phi_m \exp(S_{M+G}) = \int_{\sigma'} < (Q_G(R))^{6q-6} >^M \tag{9}$$

Since trace of G is zero, an integral over vertical hypersurfaces is zero. Using this, one can show that for conformal regulator partition function Z does not depend on the section. In this way it could be continued on the compactified moduli space that we will consider below in applications.

---

[2]Consider supersymmetric system as a space of differential forms on some space; then the action of the element of the Lie algebra (associated with the Lie group) on forms is represented by taking Lie derivatives along the vector. Superpartner to this element of the Lie algebra acts on forms by internal contraction of them with this vector field

4 Anomalies

This section was done in collaboration with M.Bershadsky

Here we will make preliminary attempt to study anomalies in topological gravity. In this section we will consider non-compactified moduli space.

Naively, action (8) looks like an action in theory with fixed ordinary gauge symmetry, where we expect that nothing depends on the regulator R. Let us check what happens with expression (9) over noncompactified moduli space even for a smooth section when we a)change the section, b) change the regulator. We will find that anomaly(dependence on section and regulator) comes from the boundary of that moduli space.

Section anomaly for massive regulator

Let us define the form on the space of metrics:

$$F = \int D\phi_m \exp(S_{M+G}) \tag{10}$$

One can explicitly check that this form is closed, so its integral over the boundary of some manifold in the space of metrics is zero. If two sections from moduli to metrics coincide near the boundary of the moduli space, then they form a boundary and integral of F over these sections coincide. If two sections do not coincide, one can form a cylinder-like (vertical) hypersurface, that is a boundary: its upper and lower bases are sections ; its lateral surface L is formed by vertical segments, connecting points on two sections, that are projecting on the same point on the boundary of the moduli space.

Thus, we have the following anomaly in the massive theory

$$\int_{\sigma_1} F - \int_{\sigma_2} F = \int_L F \tag{11}$$

Anomaly in changing a regulator

Anomalous dependence of Z on R could be found as follows: one can show, that under the change of R

$$\delta F = d_G(\int D\phi_m \delta R \exp(S_{M+G})), \tag{12}$$

Let ∂M denote the boundary of the moduli space, then

$$\delta Z = \int_{\partial M} \int D\phi_m \delta R \exp(S_{M+G}) \tag{13}$$

This dependence on R is quite similar to holomorphic anomaly, found in [7].

That is why below we will consider only conformal regulators. If for some theories it is impossible to find such a regulator (it happens, for example, in topological sigma models of type A on a manifolds with non-zero first Chern class) it seems that this theory is should be considered in the asymptotically conformal limit, i.e. when trace of G is very small and we can neglect it.

5 Local observables in theories coupled to topological gravity

Like in the string theory, we will place local observables at marked points on the Riemann surface. In 2-d gravity it means that such Riemann surfaces are obtained by cutting n nonintersecting discs from the compact surface, and gluing semiinfinite flat cylinders to n boundaries. Placing local observables at marked points means that first we take cylinders to be of the finite length l, then we glue the boundary of the cylinders by a disc with local observable inserted in the middle of it, and finally we take the limit l goes to infinity. Typical example of such metrics are Penner metrics, constructed for fat graphs. Under conformal projection on a compact Riemann surface discs are projected to points (called marked points); these points never coincide.

Along with the obvious conditions on local observables, namely, local observable should be Q-closed and be primary conformal field of zero dimension, we will impose one additional condition, namely equivariance(horizontality) condition:

$$G_{L,0}\Phi = 0; G_{R,0}\Phi = 0, \tag{14}$$

here subscripts mean components of G ,acting on the left and right movers respectively. Obvious conditions are needed to preserve Q_t invariance. To make correlator in conformal theory covariant under coordinate transformations one has to multiply field of dimension Δ by the $(\det g)^{\Delta}$, so condition of being a conformal field with zero dimension is needed for Q_G invariance of the observable.

Appearance of the equivariance condition was explained in string theory[20, 21] using operator formalism. In that formalism a form on the space $Z(M)$ of complex structures of Riemann surfaces with local coordinates [3] at marked points was constructed. In the case of genus zero with n marked points, for example, the pairing of this form with the $2(n-3)$ vectors, tangent to $Z(M)$, is given by the functional integral:

$$\int D\phi_m \prod_{j=n-2}^{n} \Phi_j(P_j) \prod_{i=1}^{n-3} \int_{\gamma_i} v_i(z)G_L \int_{\gamma_i} \bar{v}_i G_R \Phi_i(P_i) \exp(S_m) \tag{15}$$

here γ_i is a contour around marked point P_i, and v_i is a holomorphic vector fields inside this contour. If this vector field equals zero at P_i, it corresponds to change of local coordinates leaving position of the marked point P_i fixed, such vectors fields are tangent to the fiber of the bundle $Z(M)$. If the vector field is nonzero at marked point, its value at P_i is identified with the tangent vector to M, and corresponds to the motion of the marked point P_i along $v_i(P_i)$ on the Riemann surface (accompanied with local coordinate).

To reduce this form to the base M one has to impose horizontality conditions, that for zero dimensional primary conformal fields corresponds to (14).

---

[3]By local coordinate Z_i at a marked point P_i we mean holomorphic function in a neighborhood of P_i that has a simple zero at P_i

Similarly, observable is trivial if it is of the form $Q(\Phi)$, where Φ is a horizontal field of zero conformal dimension.

Gravitational descendents

Sometimes it happens[16] that theory contains a local field A, that has conformal dimension zero, and such that:

$$(G_{L,0} - G_{R,0})A = 1 \tag{16}$$

Consider A inserted at i-th marked point as a form on the principal $U(1)$ bundle $E_i(M)$ of phases of local coordinate at i-th point over space of complex structures. Condition of zero conformal dimension means that A is invariant along the fiber. Condition (16) means that restriction of A to the fiber produces an invariant volume form on it. Thus, A is a connection form on this bundle, and its Q derivative is a curvature of this bundle. Bundle $E_i(M)$ is canonically associated with the bundle $T_i(M)$ of tangent planes at marked point, and holomorphic line bundle $L_i(M)$ of holomorphic one forms at marked point. For example, the map from bundle of local coordinates to $L_i(M)$ is given by

$$Z_i \to dZ_i(P_i)$$

So, $Q(A)$ is a first Chern class of L_i or the Euler class of T_i.

Such strange fields as connections appear rather naturally in topological theories by the following mechanism. In conformal theories the fermionic symmetry Q could be decomposed as a sum of two symmetries, Q_L and Q_R, that act on the left and the right movers respectively:

$$\{Q_L, G_L\} = T_L; \{Q_R, G_R\} = T_R \tag{17}$$

This decompositions corresponds to decomposition of the external derivative on the moduli space on the holomorphic and antiholomorphic one. Suppose we have a local field H that has logarithmic conformal dimension, namely:

$$T_{L,0}(H) = T_{R,0}(H) = 1 \tag{18}$$

Such a field could be interpreted as a logarithm of the hermitean metric on the bundle L_i. Then natural connection could be constructed as

$$A = (Q_L - Q_R)(H) \tag{19}$$

So first Chern class of the line bundle L_i (by tradition denoted as $\sigma_1(1)$) takes the form

$$\sigma_1(1) = Q_L Q_R(H) \tag{20}$$

Fields with logarithmic conformal dimension could be introduced into the theory from the very beginning, like conformal factor of the metric[3] (or by coupling a scalar field to the curvature of the background metric). Such fields have normal(or classical) logarithmic dimensions.

It is interesting that there is a phenomena of getting anomalous logarithmic conformal dimension due to normal ordering in Landau-Ginsburg theory, that we will consider below.

6 Structure of the Landau-Ginzburg model

Landau-Ginzburg model[14, 4, 13] is a type B twisted sigma model on C^d (non-compact flat space) with holomorphic superpotential $W(X)$ (that prevents fields from flying away). To describe its action we need also a Kahler potential that we will take to be flat . Moreover, we start with potential, that gives nonsingular metric on the full C^d, i.e. with

$$K(X, \bar{X}) = \sum_{A=1}^{d} X^A \bar{X}^A \qquad (21)$$

Action in this theory could be found from the twisted version of the superfield formalizm, and is a sum of kinetic D-term, arizing from the Kahler potential, F-term, described by superpotential $W(X)$, and $\bar{F}$ term , described by superpotential $\bar{W}(\bar{X})$. Note, that we do not insist on $\bar{W}$ being complex conjugate to W. Let $X, \psi_z, \psi_{\bar{z}}, F_{z\bar{z}}$ be components of the chiral B type twisted superfield $\hat{X}$, and $\bar{X}, \rho, \rho_*, \bar{F}$ are components of twisted antichiral superfield $\hat{\bar{X}}$; here all fields of antichiral supermultiplet are worldsheet scalars, while two ψ fields form a worldsheet one form, and auxillary field F is a worldsheet two-form. Then the Lagrangian of the F-term is a topological one:

$$L_{top} = \frac{\partial^2 W}{\partial X^A \partial X^B} \psi^A \wedge \psi^B + \frac{\partial W}{\partial X^A} F^A \qquad (22)$$

Lagrangian of D-term is conformal and equals to

$$L_{conf} = X^A \square \bar{X}^A + \psi^A \bar{\partial} \rho^A + \psi^A \partial \rho_*^A + F_{z\bar{z}}^A \bar{F}^A \qquad (23)$$

and the $\bar{F}$ term depends on the conformal factor of the metric:

$$L_{massive} = \sqrt{\det g} \left(\frac{\partial^2 \bar{W}}{\partial \bar{X}^A \partial \bar{X}^B} \rho^A \wedge \rho^B + \frac{\partial \bar{W}}{\partial \bar{X}^A} \bar{F}^A \right) \qquad (24)$$

From the superfield formalizm we get that the theory for a given complex structure of the worldsheet has two scalar fermionic symmetries that we will call Q_L and Q_R. The offshell symmetry Q_L acts as follows:

$$Q_L(X) = 0, \ Q_L(\psi_z) = \partial_z X, \ Q_L(\psi_{\bar{z}}) = 0, \ Q_L(F) = \partial_z \psi_{\bar{z}}$$
$$Q_L(\bar{X}) = \rho, \ Q_L(\rho) = 0, \ Q_L(\rho_*) = -\bar{F}, \ Q_L(\bar{F}) = 0 \qquad (25)$$

while Q_R acts as

$$Q_R(X) = 0, \ Q_R(\psi_z) = 0, \ Q_R(\psi_{\bar{z}}) = \partial_{\bar{z}} X, \ Q_R(F) = \partial_{\bar{z}} \psi_z$$
$$Q_R(\bar{X}) = \rho_*, \ Q_R(\rho_*) = 0, \ Q_R(\rho) = \bar{F}, \ Q_R(\bar{F}) = 0 \qquad (26)$$

From these formula we see that the sum $Q = Q_L + Q_R$ acts as a scalar independently on complex structure of the worldsheet and can be taken as a Q-symmetry of the topological matter theory.

One can check that the massive part of Lagrangian (that depends on $\bar{W}$) is exact, conformal is exact up to the total derivative, and topological term is closed up to the total derivative. Thus, the second and third terms could be considered as regulators, and the first as a density of the topological action.

Above we stressed that there are problems in defining massive topological theories coupled to topological gravity, that is why below we will take $\bar{W}$ equal to zero, or ,equivalently , consider sections of the almost zero area.

Then theory turns out to be conformal. Really, after exclusion of auxillary fields, we find that

$$F = 0; \quad \bar{F}^A = \frac{\partial W(X)}{\partial X^A} \tag{27}$$

and after substitution of (27) we get a theory with the Lagrangian:

$$X^A \Box \bar{X}^A + \psi_z^A \bar{\partial}\rho^A + \psi_{\bar{z}}^A \partial\rho_*^A + \frac{\partial^2 W}{\partial X^A \partial X^B}\psi^A \wedge \psi^B \tag{28}$$

This theory is really conformal even at the quantum level, since the only non-quadratic term could not be contracted with itself by propagator, and thus, it only changes the classical equations of motion in the theory. For fields from chiral supermultiplet they look exactly like in the free theory

$$\Box X = 0; \quad \partial_z \psi_{\bar{z}} = 0; \quad \partial_{\bar{z}} \psi_z = 0, \tag{29}$$

while for fields from antichiral supermultiplet they are a little bit different:

$$\begin{aligned}
\Box \bar{X}^C &= \frac{\partial^3 W}{\partial X^A \partial X^B \partial X^C}\psi^A \wedge \psi^B \\
\partial_z \rho_*^B &= \frac{\partial^2 W}{\partial X^A \partial X^B}\psi_z^A \\
\partial_{\bar{z}} \rho^B &= \frac{\partial^2 W}{\partial X^A \partial X^B}\psi_{\bar{z}}^A
\end{aligned} \tag{30}$$

Using these equations of motion one can prove that theory contains two conserved fermionic currents:

$$J_L = \rho^A \partial_z X^A + \frac{\partial W}{\partial X^A}\psi_{\bar{z}}^A \tag{31}$$

$$J_R = \rho_*^A \partial_{\bar{z}} X^A + \frac{\partial W}{\partial X^A}\psi_z^A \tag{32}$$

Note, that each of these currents contains both holomorphic and antiholomorphic components along the worldsheet, and thus is conserved as a current of a global symmetry, i.e. $dJ = 0$. These currents generate symmetries Q_L and Q_R respectively. Being conformal, this theory also contains holomorphic and antiholomorphic energy-momentum tensors T_L and T_R, and their superpartners G_L and G_R, that have the standard form of the free theory, for example for the left-movers:

$$T_L = \partial_z X^A \partial_z \bar{X}^A + \psi_z^A \partial_z \rho^A \tag{33}$$

$$G_L = \psi_z^A \partial_z \bar{X}^A \qquad (34)$$

These fields are holomorphic:

$$\partial_{\bar{z}} T_L = \partial_{\bar{z}} G_L = 0 \qquad (35)$$

Similar expressions are valid for the right-movers; moreover, action of symmetries $Q_{L,R}$ on $G_{L,R}$ are standart:

$$Q_L(G_L) = T_L, \ Q_R(G_L) = 0 \qquad (36)$$

Summing up all properties of the theory, we may say that topological LG theory with *nonhomogenious* superpotential is nonunitary conformal, has traceless superpartner of the energy-momentum tensor, and two fermionic symmetries(scalars for a given complex structure). These fermionic symmetries and superpartners of energy-momentum tensor commute like BRST-symmetries and b fields in ordinary string theories, so it is possible to construct a measure on the moduli space exactly like in string theory. At the same time, fermionic symmetries are global, not holomorphic, that is why there are no holomorphic conserved currents, apriory no notion of anomaly in the current and this theory in general could not be obtained by twisting of $N = 2$ superconformal theory(sometimes such theories could be obtained from twisted superconformal $N = 2$ theory by perturbing the action with the perturbation, that violates twisted $N = 2$ superconformal algebra).

7 Calculation of the ring and the linear functonal in LG topological matter theory

The ring of observables

Let us first qualitevly estimate the structure of the ring and the linear functional on it. From offshell supersymmetry transformations it is clear that all polynomials in fields X are Q-closed fields. Since $\bar{F}$ is Q-exact in the offshell algebra, and in the process of eliminating the auxillary fields we found (27), that $\bar{F}$ equals to the gradient of W, we can expect, that ring of observables is the factor ring over the gradient ideal. Really, one can show that there are no other Q-cohomologies, thus

$$J = \frac{C[X]}{\{\frac{\partial W}{\partial X^A}\}} \qquad (37)$$

Homotopy to deal with zero modes

We will calculate k-point correlator in conformal ($\bar{W} = 0$) version of the theory. Before we start calculation, we should notice that after substituting polynomials of X into the fuctional integral, the integral is ill defined: namely an integral over bosonic zero modes implies that the integral is infinite, while an integral over the zero modes of ρ fermions turns integral to be zero; so we have typical "zero times infinity" problem. To solve this problem we will use instead one of polynomials an element in the same Q-cohomology class, namely, for some function $\bar{V}(\bar{X})$, we construct homotopy:

$$P(X) \rightarrow \exp(\tau Q(\frac{\partial \bar{V}}{\partial \bar{X}^A} \rho^A))P(X) \tag{38}$$

We are mostly interesed in the case when τ tends to zero. At the point $\tau = 0$ we have the initial problem. Consider τ very small, so we could ignore interaction of nonzero modes, that is proportional to τ, and pick up only contribution from zero modes, that should appear(as we expect) in the zero-th order in τ. The only restriction on function $\bar{V}(\bar{X})$ is that the integral over bosonic zero modes should converge. For example we can take for $\bar{V}$ complex conjugate to $W(X)$.

Anomaly in the measure and holomorphic top form

Before we worked with the LG theory on the level of the action. To compute functional integral we need the measure. In ordinary nontwisted sigma models measure is defined by the metric $G_{A\bar{B}}$ on the target space and the metric on the worldsheet. After twisting situation is different. Really, it is impossible to construct measure on the fermions with the help of metric on the target space since fermions, that are holomorphic tangent vectors to the target space (ψ fermions) are one-forms on the worldsheet, while fermions that are antiholomorphic tangent vectors(ρ fermions), are scalars on the worldsheet. That is why it seems impossible to construct a non-degenerate scalar product (something like an integral of $\psi^A \rho^{\bar{B}} G_{A\bar{B}}$) not only because holomorphic square root of worldsheet metric is needed, but also because on a surface different from the torus there are different number of ψ and ρ fields and the difference is $2(g-1)$, where g is the genus of the surface.

Moreover, one would expect that in topological theories nothing depends on the worldsheet metric, since energy-momentum tensor is Q-exact. Typically this is a result of existance of equal number of bosons and fermions with the same target-space properties, so that formal measure $DbosonDfermion$ written in some coordinates is well defined as a prescription to integrate over the space of all bosons and consider fermions as differentials of corresponding bosons(in other words, to define such a measure one can choose some coordinates on the space of bosons, this induces coordinates on the space of fermions, and under changes of such coordinates measure is invariant, since Jacobians cancel each other).

In the case of type B twist, this takes place in the space of antichiral fields: one can really define a formal measure on the space of $\bar{X}, \rho, \rho_*, \bar{F}$ since all of them are worldsheet scalars. But it is not the case for the chiral sector: X, ψ, F - the difference between numbers of bosons and fermions (in any reasonable regularization) is $2(1-g)$. That is why to define a measure in the functional integral we need an additional data[14, 12, 19] *holomorphic top form* Ω. We expect, the answer on genus g to depend on this form like $\Omega^{2(1-g)}$. If we still want to define a measure on bosons in terms of metric $G_{A\bar{B}}$, we have to say that measure is defined on the ψ-fermions themself by Ω, on ρ fermions themself by $\bar{\Omega}$, and measures on bosons and fermions correspond to each other if

$$\Omega\bar{\Omega} = \det(G_{A\bar{B}}) \tag{39}$$

Moreover, holomorphic top form needs to be holomorphic to preserve supersymmetry of the measure (we will see how this condition appears in calculations of the correlators). Note, that on the compact target spaces compatibility between holomorphic top form and metric(together with nondegeneracy of metric) means that the target space is a Calaby-Yau manifold.

Calculations

We will compute in genus zero $< P_1, \ldots, P_n >^M$ on C^d with the metric

$$G_{A\bar{B}} = \delta_{A\bar{B}}$$

and holomorphic top form

$$\Omega = \prod_{A=1}^{d} dX^A$$

One can easily see that determinants on nonzero modes cancel each other and we are left with the finite dimensional integral over zero modes:

$$< P_1, \ldots, P_n >^M = \int d\rho d\rho_* dX d\bar{X} P_1(X) \ldots P_n(X)$$

$$\exp(\tau \rho^{\bar{A}} \rho_*^{\bar{B}} \frac{\partial \bar{V}}{\partial \bar{X}^A \partial \bar{X}^B} - \tau \frac{\partial \bar{V}}{\partial \bar{X}^A} \frac{\partial W}{\partial X^A}) \tag{40}$$

If $d = 1$, then an integral could be calculated as follows. Consider a manifold M obtained from the full C by cutting out small discs around the points, where derivative of $W(X)$ equals to zero, and substitute integral over C on integral over M. On M the integral takes the following form

$$< P_1, \ldots, P_n >^M = \int_M \bar{\delta}(P_1(X) \ldots P_n(X) \frac{\exp(-\tau \partial W \bar{\partial} \bar{V})}{\partial W}) \tag{41}$$

Thus, integral over M reduces to boundaries of M, and we get

$$< P_1, \ldots, P_n >^M = \int_\Gamma \frac{P_1(X) \ldots P_n(X)(dX)^2}{dW} \tag{42}$$

Here Γ is a contour that goes around zeroes of dW. Note, that dX^2 in the numerator stands for holomorphic top form to the second power, as we expected before.

Note, that result we obtained is independent of τ. If we choose nonholomorphic top form, it would not be the case because Q symmetry would be breaken by measure.

In the general case calculations become a little bit more tricky. Nevertheless they could be done using the following observation of Vafa [13]. One can show that not only functional integral but also finite dimensional integral is independent of τ and $\bar{V}$. Then we can take $\bar{V}$ equal to complex conjugated superpotential, put τ to infinity. We observe that the integral gets localized on the critical points of W, and we get the following result:

$$< P_1, \ldots, P_n >^M = \int_\Gamma \frac{P_1(X) \ldots P_n(X)(\Omega)^2}{\prod_{A=1}^{d} \partial_{X^A} W dX^A} \tag{43}$$

Here Γ is defined by equations

$$|\partial_{X^A} W| = c$$

for some small positive number c.

Moreover, here we restored dependence on the holomorphic top form, so correlator here is put in a coordinate invariant form.

Thus, we justified that ring of observable is a factor over the gradient ideal of superpotential and linear functional is determined by a holomorphic top form.

8   Descendents and equivariant cohomology

Descendents constructed from matter fields

In order to find descendents, consider the following Q-exact field

$$\Phi = P_A(X)\frac{\partial W(X)}{\partial X^A} \tag{44}$$

where P_A are some polynomials. One can check that this field is a local observable in a theory coupled to topological gravity(apriori it could be zero obzervable).Note the following fact:

$$\Phi = Q_L Q_R(P_A(X)\bar{X}^{\bar{A}}) \tag{45}$$

while field in the round brackets has nontrivial logarithmic conformal dimension due to contraction of fields X and $\bar{X}$. Namely:

$$T_{L,0}(P(X)_A\bar{X}^{\bar{A}}) = T_{R,0}(P_A(X)\bar{X}^{\bar{A}}) = \frac{\partial P_A(X)}{\partial X^A} \tag{46}$$

For example, for $P_A(X) = X^B\delta_{AB}$ observable Φ corresponds to the first Chern class of the line bundle L_i.

More physically, since correlator between X and $\bar{X}$ is singular, operation of placing these two fields at the same marked point is ill-defined. It is natural to regularize it by point splitting. Making a point splitting means to choose a nonzero vector from a tangent plane, and smooth point splitting everywhere on the moduli space means picking up a nonzero vector field from the bundle $T_i(M)$(or smooth section of $E_i(M)$). The obstacle to do it is just the Euler class of this bundle.

Higher descendents

Now let us consider the general polynomial observable $P(X)$. In order to show that such an observable could represent powers of the first Chern classes we use the following trick. Consider the tensor product of the initial LG system with the trivial one (described in terms of N fields Y_i, the number N of fields Y can be as big as we wish) with the total superpotential

$$W_{coupled}(X,Y) = W(X) + \sum_{C=1}^{N} Y_C^2/2 \tag{47}$$

The matter ring in this theory is the same as in initial one, and if an observable does not contain fields Y, we can simply omit them. At the same time logarithmic divergency in the field

$$P_A(X)\bar{X}^{\bar{A}} - \bar{Y}_1 Y_1 \frac{\partial P_A(X)}{\partial X^A}$$

equals to zero, thus in the space of observables in topological theory coupled to topological gravity:

$$P_A(X)\frac{\partial W(X)}{\partial X^A} \sim (Y_1)^2 \frac{\partial P_A(X)}{\partial X^A} \tag{48}$$

where $\sim$ means that the left and right hand sides of above expression differ by the zero observable.

In other terms [16],

$$P_A(X)\frac{\partial W(X)}{\partial X^A} - (Y_1)^2 \frac{\partial P_A(X)}{\partial X^A} = Q(P_A(X)\rho_A - Y_1\rho_1 \frac{\partial P_A(X)}{\partial X^A}) \tag{49}$$

where ρ_1 is a fermion in the antichiral supermultiplet of $\bar{Y}_1$, and [4]

$$G_{0,-}(P_A(X)\rho_A - Y_1\rho_1 \frac{\partial P_A(X)}{\partial X^A}) = 0 \tag{50}$$

It may happen that polynomial in X in the right hand side of the equation (48) also belongs to gradient ideal of $W(X)$, then this procedure could be applied once again, but now we will compensate the logarithmic dimension with the help of Y_2 fields, e.t.c. Thus, in principle it could happen, that some polynomial $P(X)$ equals in the space of observables to:

$$P(X) \sim \prod_{C=1}^{k} (Y_C)^2 \tilde{P}(X) \tag{51}$$

Then, we represent $(Y_C)^2$ as $Q_L(Q_R(Y_C\bar{Y}_C)$ and split Y_C and $\bar{Y}_C$ along some section v_C of cotangent bundle.

Taking first functional integral over all Y we get the measure on the moduli space that has support on the common zero of all these sections. In the rest of the integral observable $\tilde{P}$ is inserted at the marked point we consider. In cohomologies of the moduli space the support of the measure before integration over X is the cycle dual to the k-th self intersection of the first Chern class. In such a situation we write $P = \sigma_k(\tilde{P})$.

The "state" representation and equivariant cohomology

The effect observed above has the following nice presentation. Let us represent local fields as differential forms of the type $(0, m)$, that take value in

---

[4]It is sufficient to consider $G_{0,-} = G_{0,L} - G_{0,R}$,since sum of G correspond to dilatations of local coordinate, i.e. to noncompact direction in the structure group of the bundle. It is wellknown that noncompact directions are always contractible, i.e. structure group of the bundle could be always reduced to its compact subgroup

antisymmetric powers of holomorphic vector bundle, namely, $\rho_+ = \rho + \rho_*$ corresponds to $d\bar{X}$, $\nu_A = G_{A\bar{B}}(\rho^{\bar{B}} - \rho_*^{\bar{B}})$ corresponds to ∂_A. Multiplication in the sence of field theory corresponds to wedge multiplication of forms.

In such representation operators Q and $G_{0,-}$ take the following form:

$$Q = \bar{\partial} + \partial_A W dX^A \qquad (52)$$

$$G_{0,-} = dX^A \partial_A \qquad (53)$$

Here dX^A acts by contraction with the polyvector of the bundle. It is easy to see,that $G_{0,-}$ is not a differentiation with respect to natural multiplication, and that is why observables in topological theory coupled to topological gravity do not form a ring under this product.

It is convenient to use another representation of the same space of local operators. Namely, introducing new fermionic variables θ^A,dual to ν_A, and making fermionic Laplace-like transformation with the measure determined by Ω we obtain (p, q) forms. For example, functions of X correspond to holomorphic top forms,specifically:

$$P_i(X) \to \Omega_i = P_i(X)\Omega \qquad (54)$$

Formulas (52,53) look the same but now dX means wedge product with the differential.

This transformation is not so unnatural as it looks like. It somehow represents states, corresponding to local fields, inserted in the middle of the small disk. Really, one can compute the fermionic number of the disc, that is a semi-sphere and find that it equals d - the dimension of the manifold(it is in agreement with the anomaly, that equals $2d$ that we observed on the sphere). It is nice that on "states" actions of Q and $G_{0,-}$ are independent of Ω and depend only on W. Moreover, on the "states" the operation of rotation of local coordinate is represented by a vector field, and $G_{0,-}$ has a geometrical meaning of contraction with such a vector. That is why "states" corresponding to observables are equivariant cohomologies, namely, they are cohomologies of the operator $D(z)$:

$$D(z) = D_1 - zD_2 = Q - zG_{0,-} = \bar{\partial} + \partial W - z\partial \qquad (55)$$

K.Saito higher residue pairing as pairing in equivariant cohomologies

In terms of equivariant cohomology it is possible to interpret K.Saito higher residue pairing, as a pairing on equivariant cohomologies (in the "state" picture) induced by integration of the ordinary wedge product of the forms over C^d. If we take representation of cohomologies as holomorphic top form we could not get the answer şince product of holomorphic top forms looks like zero but they are not in an integrable class. To cure the situation we can do homotopy like in (38), but now with $\{D(z), \bar{\partial}_A \bar{W} \nu^A\}$. We will obtain seria in z, the i-th term in that seria was called K.Saito as $K^{(i)}$.The zero-th term would be ordinary residue, and the first term would be the first higher residue pairing $K^{(1)}$, that has the following form:

$$K^{(1)}(\Omega_1, \Omega_2) = \int_\Gamma \frac{(P_1\partial_{X^B}P_2 - P_2\partial_{X^B}P_1)\Omega^2}{\partial_{X^B}W \prod_{A=1}^d \partial_{X^A}W dX^A} \tag{56}$$

Here P_i is connected with Ω_i like in (54).

Filtration on the space of equivariant cohomologies

Cohomology of $D(z)$ naturally form a $D[z]$ module. With this module we can assosiate filtration on the space of observables.

If in cohomologies of $D(z)$ $\Phi_1 = z^k\Phi_2$, then as we observed above Φ_1 is k-th gravitational descendent of Φ_2, this is denoted as $\Phi_1 = \sigma_k(\Phi_2)$. Let us define filtration in the space H of equivariant cohomologies due to degree of descendents: class of equivariant cohomologies belongs to the H_k subspace of the space H_0 it contains k-th descendent. Thus we have

$$H = H_0 \supset H_1 \supset H_2 \supset \ldots \tag{57}$$

In Landau-Ginsburg theory it is obvious, that

$$H_i/H_{i+1} = J \tag{58}$$

But apriori there is no natural splitting of this filtration.

It can happen that in some theory descendents could not be constructed from the matter fields. For example, this happens in the sigma model of type B on Calabi-Yau manifold. In that case

$$D(z) = \bar{\partial} + z\partial \tag{59}$$

and from $\partial\bar{\partial}$-Lemma it follows that there are no descendents constructed from matter fields in this model. In such case to reproduce gravitational descendents we have to couple theory to additional theory with trivial ring of observables, for example, to Landau-Ginsburg theory with quadratic superpotential.

9   Connection on the space of theories

Consider n-point correlator on a Riemann surface with n marked points. It turns out that this correlator could be reduced to a $n - 1$ point correlator in a nearby theory.

Really, moduli space of Riemann surfaces of genus g with n marked points (for $g = 0, n > 3$; for $g = 1, n > 1$) almost everywhere (except boundaries, where two marked points instead of collision decouple on a sphere) is fibered over the moduli space with $n - 1$ marked points. The fiber is a Riemann surface with $n - 1$ marked points. The idea is to integrate over the fiber, and this would give shift to a nearby theory. Including properly the boundary contribution we get a connection in the space of observables fibered over the space of theories.

188

Far from the boundary coordinate of n-th marked point could be considered as a moduli, this moduli corresponds to $G_{L,-1}G_{R,-1}P(X)(z_n)$ contribution to the functional integral (see (15)), the -1 means expansion in $z - z_n$), that gives

$$\partial_A \partial_B P(X)(z_n)\psi_z(z_n)^A \psi_{\bar{z}}(z_n)^B d^2 z_n$$

contribution, and thus integral over fiber symply corresponds to the shift in the action:

$$W \to W + tP$$

in the first order in t.

Calculation of boundary contribution(where n-th point decouples on a sphere together with i-th point) is more tricky, because we have to calculate the integral [3]:

$$|C(P_n, P_i) >= \int_{\tau_0}^{\infty} d\tau \int d\phi \exp(-\tau T_{0,+}) \exp(\phi T_{0,-})G_{0,+}G_{0,-}$$
$$P_n(X(1))|P_i(X(0)) > \tag{60}$$

Here $|P_i(X(0)) >$ is a state that represents a disc with polynomial $P_i(X)$ inserted at the middle, and polynomial field P_n is inserted at point 1. We will take τ_0 large enough, so that the disc covered by this integration has radius $\exp(-\tau_0)$ and is relatively small.

Commuting G operators with P_n for finite τ still could be considered as the same perturbation of the action on the small disc that we considered before, so the infinity in the upper limit of the integral is really important, since the effect we are trying to find could be considered as a delta-function contribution from infinity.

First note, that field P_n could be placed at zero, really , derivative of this field with respect to the worldsheet coordinate has conformal dimention 1 and is supressed like $\exp(\tau_0)$. Thus, $C(P_n, P_i)$ depends only on product of polynomials.

The problem of computation of the resulting expression comes from the fact, that G acting on the state give zero, but integral over τ up to infinity of the state with conformal dimension zero is divergent.

The only known way to calculate such an integral is to regularize it. This can be easily done when the product of polynomials belongs to the gradient ideal, i.e. for some field Ψ:

$$P_i P_n = Q(\Psi) \tag{61}$$

Then, we take Q from Ψ outside. Commutation of Q with $G_{0,-}$ is irrelevant, since it gives $T_{0,-}$, and integral over compact circle gives zero(this is true in all regularizations), so the only contribution comes from commutation of Q with $G_{0,+}$. (Q acting outside seems to be irrelevant, because it acts on the G_0^- exact state). So after taking integral over ϕ we get:

$$|C(P_n, P_i) >= \int_{\tau_0}^{\infty} d\tau \exp(-\tau T_{0,+})T_{0,+}G_{0,-}|\Psi > \tag{62}$$

This integral could be easily regularized(see also [10]) by substituting zero energy state $G_{0,-}|\Psi>$ by state with non-zero energy E. After computation of the integral we will put E to zero. We are left with the integral:

$$|C(P_n, P_i)>= \int_{\tau_0}^{\infty} d\tau \exp(-\tau E)E(G_{0,-}|\Psi>) = \exp(-\tau_0 E)(G_{0,-}|\Psi>) \quad (63)$$

Note, that main contribution to such an integral came from the region $\tau \sim 1/E$, that is infinity regularized by E. Now, putting E to zero we obtain the final answer:

$$|C(P_n, P_i)>= G_{0,-}|\Psi> \quad (64)$$

In other terms, if

$$|P_n P_i>= |\sigma_1(P_{n,i})>,$$

then

$$|C(P_n P_i)>= |P_{n,i}>.$$

We see, that contact term is completely defined as a linear operator C from the space of observables H to itself , such that it maps H^i on H^{i-1} for $i > 0$.

Definition of a section

The kernel V of the contact term map C in the space of observables H we call a section.

From derivation of the contact term we see that up to section it can be described in terms of filtration of decsendents, that in "state" representation (54) does not need the fixation of the holomorphic top form Ω. Below we will see from compatibility conditions that in "state" representation section is also independent of this form. That is why below we will mostly use this representation. Operator representation (i.e. representation in terms of polynomials), is needed when we want to assosiate to observable a shift in W.

10 Recursion relations and properties of the section

Above we constructed connection in the bundle of observables over the space of theories. It allows us to write the following recursion relations for correlators (if $g = 0, n > 3$, if $g = 1, n > 1$):

$$< \Omega_1, \ldots, \Omega_n >_{W,\Omega} = \frac{d}{dt} < \Omega_1, \ldots, \Omega_{n-1} >_{W+tP_n,\Omega} |_{t=0} +$$

$$+ \sum_{i=1}^{n-1} < \Omega_1, \ldots, C(P_n, \Omega_i), \ldots, \Omega_{n-1} >_{W,\Omega} \quad (65)$$

Here P_n is connected with Ω_n as in (54).

Connection given by contact term is really connection on equivariant cohomologies

To show this statement we have to prove that $D(z)$ trivial fields are mapped to themself by connection, induced by contact term. Note, that since all such fields are Q exact, we do not need to involve notion of section. Such a proof could be given at a formal level of equivariant cohomologies, we will show how it works in LG case. Really, contact term

$$C(P, z\partial\Omega_i - \partial W\Omega_i) = \partial P\Omega_i = \delta_P(z\partial\Omega_i - \partial W\Omega_i) \qquad (66)$$

thus

$$(z\partial\Omega_i - \partial W\Omega_i) + \delta_P(z\partial\Omega_i - \partial W\Omega_i) = z\partial\Omega_i - \partial(W + \delta_P W)\Omega_i \qquad (67)$$

So we can rewrite recursion relation as follows:

$$< \Omega_1, \ldots, \Omega_n >_{W,\Omega} = \frac{d}{dt} < \Omega_1(t), \ldots, \Omega_{n-1}(t) >_{W+tP_n,\Omega} |_{t=0} \qquad (68)$$

where $\Omega_i(t)$ is a result of trasport of a class of equivariant cohomologies Ω along the line $W + tP$ with connection, given by contact term.

Now we will analize the properties of this connection, that come from physical consistency conditions.

Symmetry of the four-point function and first K.Saito higher residue pairing

The four-point function on the sphere could be reduced to the three-point function(that we know how to calculate, since it is given by an answer in the topological matter theory) either by integration of the position of the first maked point or by integration over the second.

After some calculation one can show that the difference equals to

$$K^{(1)}(V(P_1 P_3), V(P_2 P_4)) - K^{(1)}(V(P_1 P_4), V(P_2 P_3)) \qquad (69)$$

that is why from the symmetry of the four-point function we get:

$$K^{(1)}(V_1, V_2) = 0 \qquad (70)$$

for any two elements of a section. Note, that this is the first axiom for a good section in the K.Saito theory of the primitive form.

Motion of punctures in different order and second K.Saito axiom of good section

The origin of additional requirements on section comes from the fact that we can integrate first over the position of the first point, then over the position of the second, but we can also change the order and start from integration over position of the second point. To see this effect we have to apply recursion relation twice, and we get:

$$< \Omega_1, \ldots, \Omega_n >_W = \frac{d}{dt_1}\frac{d}{dt_2} < \ldots, \Omega_i + t_1 C_W(P_1\Omega_i) + t_2 C_W(P_2\Omega_i) +$$
$$t_1 t_2 C_W(C_W(P_1 P_2)) + [t_1 t_2 C_W(P_2 C_W(P_1\Omega_i)) + t_2 C_{(W+t_1 P_1)}(P_2\Omega_i)]$$
$$>_{(W+t_1 P_1 + t_2 P_2 + t_1 t_2 C_W(P_1 P_2))} |_{t=0} \qquad (71)$$

The only asymmetric in 1 and 2 term stands in the square brackets, consistency demands that antysymmetric part of this term should equal to zero.

Suppose that $\Omega_i = V_i$, i.e. is an element of a good section. Suppose that $P_1 = 1$, then term in square brackets equals to zero. Changing 1 and 2 we get:

$$C_{(W+tP_2)}(V_i + tC(P_2V_i)) = 0 \tag{72}$$

in the first order in t. This is the second K.Saito axiom on the good section, that could be reformulated as follows:

Good section is invariant under connection determined by itself

Note, for general P_1 and P_2 the symmetry of the twise applied recursion relation means that connection, determined by good section, is flat. It was really shown in [11].

Invariance under diffeomorfisms and change of Ω

In our previous considerations we implied that after integration over position of a marked point the observable P simply shifts superpotential. Since observables could be of any degree it makes picture not so nice, namely we add terms to the superpotential, that change its behaviour at infinity, and we have to treat them perturbatively. It is possible to get rid of almost all of these terms because observables from H_1 that have form RdW in "state" representation look like $R^A\partial_A W$ as polynomials (index is raised with the help of the form Ω); they shift superpotential on a term that could be killed by diffeomorphism:

$$X^A \to X^A + R^A(X)$$

This would change holomorphic top form: $\Omega \to \Omega - dR$. One can check [22] that the total effect of contact term and diffeomorphism leave observables in the "state" representation invariant. It justifies our assumption that section in "state" representation is independent of Ω.

All this means, that we can restrict ourself to some particular versal deformation of W, we will call it $W(u)$, and represent any observable as a sum of two terms: change of W along versal deformation and diffeomorphism. Explecitely, in the state representation for any Ω_i we have:

$$\Omega_i = \delta_i(W(u)) + R_i dW \tag{73}$$

and recursion relation take the following form:

$$< \Omega_1, \ldots, \Omega_n >_{W,\Omega} = \frac{d}{dt} < \Omega_1(t), \ldots, \Omega_{n-1}(t) >_{W+t\delta_n W, \Omega - t dR_i} |_{t=0} \tag{74}$$

with $\Omega_i(t)$ understood as a parallel transport along the base of versal deformation in direction of $\delta_n W$.

Generating function for correlators

To write down the formula, containing all correlators, let us introduce the notion of formal exponent:

$$<; \exp \Phi >=<> + < \Phi > +1/2 < \Phi, \Phi > +1/6 < \Phi, \Phi, \Phi > + \ldots \tag{75}$$

We are going to consider correlators in presence of formal exponent. If recursion relations are self consistent, they imply, that:

$$< \Omega_1, \ldots, \Omega_n; \exp(t_i\Omega_i) >_{W,\Omega} = < \Omega_1(t), \ldots, \Omega_n(t) >_{W(u(t)),\Omega(t)} \qquad (76)$$

naimely, recursion equations could be considered as vector fields on the space of theories equipped with the connection in the bundle of observables, and formal exponent just integrates them. Parameters t become coordinates on the space of theories.

Relation (76) is self-similar,taking derivative with respect to t we get:

$$\Omega_i(t) = \frac{\partial W}{\partial t_i}\Omega(t) - z\frac{\partial \Omega(t)}{\partial t_i} \qquad (77)$$

and

$$\frac{\partial \Omega_j(t)}{\partial t_i} = C_W(t)(\frac{\partial W}{\partial t_i}, \Omega_j(t)) \qquad (78)$$

The second equation implies that contact terms form a flat connection in the bundle of observables over base of versal deformation. Such a connection as we understood is nothing else that K.Saito connection. Suppose we know its flat sections $\Omega_i(u)$. Then it turns out that it is possible to solve first equation, namely:

$$(t_i + t_i^0)\Omega_i(u) = z\Omega(t) \qquad (79)$$

where numbers t_i^0 are choosen to satisfy initial condition: $t_i^0\Omega_i = z\Omega$. Algebraic solution of this equation is generalization of the algebraic solution of dispersionless KP [8].

After solving this equation we know generating function for correlators in genus 0:

$$< \Omega_1, \Omega_2, \Omega_3; \exp(t_i\Omega_i) >_{W,\Omega} = \int_\Gamma \frac{\Omega_1(t)\frac{\partial W}{\partial t_2}\Omega_2(t)}{\prod_{A=1}^d \partial_{X^A}W dX^A} \qquad (80)$$

The author is deeply indebted to R.Dijkgraaf,A.Gerasimov, V.Fock, I.Krichever, S.Kharchev, A.Marshakov, A.Mironov, A.Morozov, N.Nekrasov, I.Polyubin, A.Rosly, K.Saito, A.Varchenko for fruitful discussions.

References

[1]   E.Witten, *Nucl.Phys*B340 (1990) 281

[2]   R.Dijkgraaf and E.Witten, *Nucl.Phys* B342 486

[3]   E.Verlinde and H.Verlinde,*Nucl.Phys.* B348 (1991) 457

[4]   R.Dijkgraaf, E.Verlinde, H.Verlinde, Proc. of the Trieste Spring School 1990, M.Green et al. eds., (World Scientific, 1991)

[5]   S.Kharchev et.al.,*Nucl.Phys.*, 380B (1992) 181

[6] M.Kontsevich,*Comm. Math. Phys.***147** (1992) 1

[7] M.Bershadsky et al , *Holomorphic Anomalies in Topological Field Theories*preprint HUTP-93/A008

[8] I.M.Krichever,*Comm. Math. Phys.***152**(1993) 539

[9] K.Li,*Nucl.Phys.* **354B**,(1991) 711

[10] R. Dijkgraaf, talk on RIMS conference "Infinite dimensional analisis",July,1993,

[11] K.Saito, Publ.RIMS, Kyoto Univ. 19 (1983) 1231

[12] S.Checotti and C.Vafa,*Nucl.Phys***367B**(1991) 359

[13] C. Vafa, *Mod.Phys.Lett.***A6(1990)** 337

[14] E.Witten,*Mirror Manifolds And Topological Field Theory*,in S.-T.Yau,ed., *Essays On Mirror Manifolds* (International Press, 1992)

[15] A.Losev,*Descendants Constructed From Matter Fields And K. Saito Higher Residue Pairing In Topological L-G Theories Coupled To Topological Gravity*, preprint TPI-MINN-40T (May,1992) ; A.Losev, *TMP(Teoreticheskaya i matematicheskaya Fizika,in English)*vol.**95**,n.**2**(1993) 307

[16] T.Eguchi et al,*Topological Strings, Flat Coordinates And Gravitational Descendants*, University of Tokyo preprint UT 630, Febr.1993

[17] B.Dubrovin, *Nucl. Phys.* **379B** (1992) 627

[18] E.Witten, Comm.Math.Phys.117(1988)353,

[19] C.Vafa, in *Essays on Mirror Manifolds*,ed. by S.T.Yau,International Press, 1992

[20] L.Alvarez-Gaume et al.,*Nucl.Phys.***B303**(1988) 455

[21] J.Distler and P.Nelson, Comm.Math.Phys.138(1991) 273

[22] A.Losev and I.Polyubin *On connection between Landau Ginzburg gravity and integrable theories*, hepth 9305079

String Theory and Classical Integrable Systems

A. Marshakov[*]

Theory Department
P.N.Lebedev Physics Institute
Leninsky prospect, 53, Moscow, 117 924, Russia,
and
Institute of Theoretical Physics, Uppsala University, Uppsala, Sweden

Abstract. We discuss different formulations and approaches to string theory and $2d$ quantum gravity. The generic idea to get a unique description of *many* different string vacua altogether is demonstrated on the examples in $2d$ conformal, topological and matrix formulations. The last one naturally brings us to the appearance of classical integrable systems in string theory. Physical meaning of the appearing structures is discussed and some attempts to find directions of generalizations to "higher-dimensional" models are made. We also speculate on the possible appearence of quantum integrable structures in string theory.

1 Introduction

String theory or theory of $2d$ gravity continues to be one of the main directions of investigation in mathematical physics. Recent years have brought us to some progress in understanding its relation to a much older field of interest for many mathematical physicists - integrable systems. At the moment we can already advocate that partially "integrable science" is directly related to string theory – that part connected with *classical* integrable equations and their hierarchies. The situation with *quantum* integrable systems is not yet as clear [1] so I will almost skip this question below (except for some minor speculations at the end). In contrast, the appearance of hierarchies of classical integrable equations in description of non-perturbative string amplitudes is already a well-known fact at least for low-dimensional string models [1, 2, 3, 4, 5].

---

[*]E-mail: mars@td.fian.free.net marshakov@nbivax.nbi.dk andrei@rhea.teorfys.uu.se

[1]though there are already many arguments making us believe that quantum integrable systems should play an essential role in formulation of string theory.

Below I will try to stress the most essential points of this formalism when the hierarchies of integrable equations appear in string theory and discuss the parallels with other languages. I will also try to clarify the target-space picture for these models and demonstrating possible "generalizations" to the case of "higher-dimensional" string theories.

Let me remind you now some questions that can appear in the modern string theory from a physical point of view [2]. If we beleive that string theory could have something to do with our reality then the idea is to find a convenient language suitable for computation of physical quantities - physical amplitudes. It would be marvelous if they could be computed exactly and in *any* background. Unfortunately nobody knows how to do this for most of string theories with the exception of those where amazing structure of integrable equations has appeared. However, there is no complete effective target space theory even for these models - if exists such a theory will be a good candidate for the role of string field theory [6, 7, 8].

The simplest example of string theory (mostly well-known) is topological pure gravity (which has lots of different equivalent formulations and is mostly well-studied)[9, 10, 11, 12]. Naively such theory should not have target space at all. Then the natural question is what is the origin of the nice structure appearing in the form of the Virasoro constraints acting to a partition function [13, 14, 15], the set of which is actually equivalent to the concept of integrability in these models?

Below we are going to pay attention to several points concerning integrability arising in description of low-dimensional string theory and try to show how the structure of hierarchies of classical integrable equations can be generalized to higher dimensional theories. The main idea is that the phase space of appearing classical integrable models may be considered as a phase space for effective string field theory, and its quantization can lead to the formulation of the second-quantized string theory in terms of (quantum?) integrable systems.

First, we review a little the $2d$ conformal field theory language – the basic Polyakov definition of string theory. We will concentrate on Liouville (physical) gravity coupled to most known non-critical string – so called (p, q) models, and try to discuss its target-space structure when taking the quasiclassical limit. As in any covariant description this one requires a lot of "extra" information of the theory, this leads to the situation when the original formulation of string theory is not effective for answering questions about its target-space structure etc (like it occurs in the theory of particles). We are going to consider an example of motion in the space of (p, q) theories - a step towards their effective description and then turn to the other ways to formulate the same theories.

At least part of these models $((p, 1)$ and/or $(1, p))$ allow to consider them by topological theory language [16, 17, 18] [3] . Coupling to reparameterization ghosts one gets total $c = 0$ central charge which allows an interpretation in terms of the twisted $N = 2$ theory [19]. Physical gravity is now included in

---

[2]with a hope that the word "physics" is right in this context
[3]see also A.Losev's contribution to this volume

"topological matter" – while topological gravity appears roughly speaking in the integration over module space - thus, generalization of the notion of *critical* string. In principle even 26-dimensional bosonic string can be considered as topological theory, but low-dimensional examples allow one to demonstrate better the explicit $N = 2$ cancellation of bosonic and fermionic two-dimensional determinants and target-space co-ordinates appear as corresponding zero modes. The result is very close to *localization* formulas appeared to be a useful tool when studying topological, quantum-mechanical and integrable models [20].

It appears that at least for topological models there exists a very effective exact formulation on the language of matrix models. Indeed, the matrix integral is a sort of counting of possible two-dimensional diagrams thus being a natural object in the theory of pure gravity - or string theory with an empty target space. These integrals naturally come from an intersection theory on module space [12], what would be more interesting is if one can successfully realize the same idea for the moduli of target-spaces (see recent papers [21, 22]).

Matrix models brought us to the understanding of possible role of integrable systems in string theory. Till the moment only for these "empty models" but there exists a way to compute the exact amplitudes in these string theories by prooving that their generation function is a τ-function of KP (or in general Toda-lattice) hierarchy satisfying some natural additional constraint. This constraint is an analog of unitarity condition and can be also interpreted in terms of some flow in the space of different low-dimensional models.

However, there are some puzzles, arising along this way. They are directly connected with the question of interpretation of $p - q$ duality, which is in order repated to the problem of string field theory background independence [23]. We are going to discuss them below (see also [24]).

2 2d conformal theories

Let us start with reminding that by canonical string theory one usually has in mind the induced two dimensional gravity, having the following form in the Polyakov's path integral approach:

$$\int DgD\phi e^{-S[\phi]} \sim \int Dg e^{\gamma \int R \frac{1}{\Delta} R + \mu^2 \sqrt{g}} \tag{1}$$

where ϕ stands for the integration over some $2d$ conformal field theory in the background world-sheet metric g.

It seems to be true that at the moment there is not still an existing consistent method of quantization the appearing in (1) Liouville theory (see however [25, 26] etc). The common beleif is that the following ideology is right [27, 28, 29]. Consider the integral in (1) as integral over *conformal field theory* consisting of (*i*) conformal matter; (*ii*) reparameterization ghosts b and c; (*iii*) *conformal* Liouville theory. The latter one should be *defined* as a conformal theory with the central charge $26 - c_{matter}$ in order to cancell the anomaly.

For the simplest example of so-called (p, q) models coupled to $2d$ gravity ϕ can be simply considered as a "deformed" scalar field. Choosing the gauge for metric $g_{ab} = e^{\varphi} \hat{g}_{ab}$ we reduce the problem to a conformal theory of two fields ϕ and φ with the stress-tensors

$$T_m = -\frac{1}{2}(\partial\phi)^2 + i\alpha_0 \partial^2 \phi$$

$$T_L = -\frac{1}{2}(\partial\varphi)^2 + \beta_0 \partial^2 \varphi \tag{2}$$

where

$$\partial\phi(z)\partial\phi(0) = -\frac{1}{z^2} + \dots$$

$$\partial\varphi(z)\partial\varphi(0) = -\frac{1}{z^2} + \dots \tag{3}$$

thus giving the Virasoro central charges

$$c_m = 1 - 12\alpha_0^2$$

$$c_L = 1 + 12\beta_0^2 \tag{4}$$

satisfying $c_m + c_L - 26 = 0$ with -26 coming from the reparameterization ghosts contribution. For the (p, q) theories

$$\alpha_0 = \sqrt{\frac{p}{2q}} - \sqrt{\frac{q}{2p}}$$

$$\beta_0 = \sqrt{\frac{p}{2q}} + \sqrt{\frac{q}{2p}} \tag{5}$$

or

$$\beta_0 = \sqrt{2}\cosh\theta$$

$$\alpha_0 = \sqrt{2}\sinh\theta \tag{6}$$

with

$$\theta = \frac{1}{2}\log\frac{p}{q} \tag{7}$$

For such system one has the following matter Kac spectrum

$$\alpha_+ = \sqrt{\frac{2p}{q}}$$

$$\alpha_- = -\sqrt{\frac{2q}{p}}$$

$$\alpha_{n,m} = \frac{1-n}{2}\alpha_+ + \frac{1-m}{2}\alpha_- = \frac{(1-n)p - (1-m)q}{\sqrt{2pq}}$$

$$\Delta_{n,m} = \frac{(np - mq)^2 - (p - q)^2}{4pq}$$

$$\Delta_{min} = \frac{1 - (p - q)^2}{4pq} \tag{8}$$

where in matter sector there exists a "periodicity" allowing one to restrict to [4]

$$n = 1, ..., q - 1$$
$$m = 1, ..., p - 1 \tag{9}$$

while the gravity sector is given by

$$\beta_\pm = \pm\alpha_\pm$$

$$\beta_{n,m} = \frac{p + q \pm (np - mq)}{\sqrt{2pq}} \rightarrow \frac{(1 - n)p + (1 + m)q}{\sqrt{2pq}}$$

$$\beta_{min} = \frac{p + q \pm 1}{\sqrt{2pq}} \tag{10}$$

where we have chosen a sign in order to make correspondence to the proper quasiclassical limit.

Indeed, we see that the conformal (p, q) model is totally symmetric under exchange of p and q. However, the difference between p and q becomes crucial when coupling to $2d$ gravity, or better to say when considering string theory. In fact, the asymetry appears if one takes the quasiclassical limit [5] then only half of the screening operators have well-defined limit (α_+ and β_+ for $q \rightarrow \infty$ and α_- and β_- for $p \rightarrow \infty$). The simplest way to see it is to consider (p, q) model as a Hamiltonian reduction of the WZNW theory [37] and for the WZNW theory it is known [38] that only one screening operator (having smooth limit for $k \rightarrow \infty$) appears naturally from classical action as a constraint on free fields. Physically it means that one has to choose the operator coupling to unity (or lowest dimensional one) - i.e. what is called the puncture operator in a theory.

2.1 Rotations in the space of free fields: the way to move in the space of theories

Now, let us turn to the question how one can describe the whole set of different (p, q) models. In fact this is rather hard to do using the technique of present section - more or less complete descriprion exists only based on the methods presented below. However, here we will try to use as much as possible of conformal methods in order to get understanding of possible flows in the space of theories.

One can consider the following rotation in the space of (p, q) theories

$$\tilde{\beta}_0 = \alpha_0 \sinh \vartheta + \beta_0 \cosh \vartheta$$

---

[4]and actually this "minimality" is broken by interacting with gravity (see for example [30])

[5]this quasiclassical limit is important if we want to discuss target-space properties of the theory

$$\tilde{\alpha}_0 = \alpha_0 \cosh \vartheta + \beta_0 \sinh \vartheta \tag{11}$$

The same rule one has for primary operators $e^{i\alpha\phi + \beta\varphi}$, labeled by (8), (10)

$$\tilde{\beta} = \alpha \sinh \vartheta + \beta \cosh \vartheta$$
$$\tilde{\alpha} = \alpha \cosh \vartheta + \beta \sinh \vartheta \tag{12}$$

with parameter of the transformation

$$\vartheta = \frac{1}{2} \log \frac{\tilde{p}}{\tilde{q}} \frac{q}{p} \tag{13}$$

Now it is easy to rewrite it in the space of fields [6]

$$\Phi_{n,m} = \exp\left(i\alpha_{n,m}\phi + \beta_{n,m}\varphi\right) \tag{14}$$

One finds that for $\Phi_{n,m}^{(p,q)} \rightarrow \Phi_{\tilde{n},\tilde{m}}^{(\tilde{p},\tilde{q})}$

$$\tilde{n} = \frac{p}{q}n$$
$$\tilde{m} = m \tag{15}$$

Now, let us consider an explicit example how the transformation (15) works. First, let us take (10) and make there a substitution

$$kq + r = np - mq \tag{16}$$

for (q, p) theory and

$$kp + r = np - mq \tag{17}$$

for (p, q) theory. Then (we restrict ourselves to the second choice (17))

$$k = n - \left[\frac{qm}{p}\right]$$
$$r = mq - p\left[\frac{qm}{p}\right] \tag{18}$$

where $[x]$ means integer part of x, and this give the correspondence [39]

$$\sigma_k(O_r) \sim \int \exp\left(i\alpha_{n,m}\phi + \beta_{n,m}\varphi\right) \tag{19}$$

i.e. r enumerates "topological primary fields" and we have defined (16) and (17) in order to have exactly $q - 1$ or $p - 1$ of them. In such terminology k counts their "gravitational descendants".

Now the transformation (15) works as follows: take for example $(p, 1)$ theory and consider $\sigma_1(1)$. 1 is given by zero-dimensional matter operator with

---

[6]Note that in particular such rotation makes from *real* Liouville field for $c = 1$ *complex-valued* for $c < 1$.

$$r = m = p - 1$$

$$k = n = 0$$

$$\Delta_1 = \Delta_{0,p-1} = 0 \qquad (20)$$

while for $\sigma_1(1)$ itself one has

$$r = m = p - 1$$

$$k = n = 1 \qquad (21)$$

Then, making transformation (15) we get

$$\tilde{m} = m = p - 1$$

$$\tilde{n} = p \qquad (22)$$

It means that the rotated field becomes *primary* one in the $(\tilde{p}, \tilde{q})$ model with $\tilde{p} = p$ and

$$\tilde{k} = \tilde{n} - \left[\frac{\tilde{q}m}{p} \right] = 0 \qquad (23)$$

which gives

$$\tilde{q} = p + \left[\frac{\tilde{q}}{p} \right] \qquad (24)$$

or just $\tilde{q} = p + 1$. For $p = 2$ such an operator drops us from pure topological gravity $(p, q) = (2, 1)$ to the pure physical gravity point $(\tilde{p}, \tilde{q}) = (2, 3)$.

The example considered above as just an illustration of the flow in the space of simplest string theories. We have seen that in the original conformal formulation they strongly depend on the basis one has to choose in the space of fields and/or observables. That is one of the reasons why more convenient target space description for string theory is necessary. On the language of matrix models these relations can be rewritten in the form of the Virasoro-W constraints and generalized KdV flows. We will see below, that the effective target-space description based on integrable systems gives much stronger possibilities to investigate this phenomenon.

3 Topological language

Now, let us make a sort of an intermideate step – to reformulate the above picture in the following way. Forget about the difference between conformal matter and conformal gravity since metric contrubution – Liouville and ghost sectors are also represented by certain conformal theories. Then one can generalize the above consideration restricting to the only requirement that the total central charge of a theory is equal to zero. The presence of gravity remains in the only fact that the result after all should be integrated over *module* space. Such object is usually meant by what is called *topological gravity* [9, 18]. From such point of view critical string is a good example of a topological theory interacting with topological gravity except for the only case that integral over module space might diverge.

Now consider this (conformal matter plus Liouville gravity plus reparameterization ghosts)

$$T_{gh} = -2b\partial c + c\partial b \tag{25}$$

($T = -jb\partial c - (1 - j)c\partial b$ for $j = 2$) as a twisted $N = 2$ superconformal theory [19]. In such case for (q, p) and (p, q) "untwisted" models two values of the central charge are

$$c_{(q,p)} = 3(1 - \frac{2p}{q})$$

$$c_{(p,q)} = 3(1 - \frac{2q}{p}) \tag{26}$$

This can be demonstrated for example as follows. First let us consider the $SU(2)_k$ WZNW model. Such theory posseses conformal symmetry with the Virasoro central charge

$$c_{SU(2)_k} = \frac{3k}{k+2} \tag{27}$$

One of the possible ways to get matter (p, q) model is via the Drinfeld-Sokolov reduction. Then, one easily finds that

$$k_{(q,p)} \equiv \tilde{k} = \frac{q}{p} - 2$$

$$k_{(p,q)} \equiv k = \frac{p}{q} - 2 \tag{28}$$

where asymmetry between p and q appeared exactly as we mentioned above when one has to distinguish the *classical* screening operator. The relation

$$k + 2 = \frac{1}{\tilde{k} + 2} \tag{29}$$

in particular demonstrates the duality between two "classical" limits when $k \to \infty$ corresponds to $\tilde{k} \to -2$ and vice versa. It can also clarify what is the meaning of the Wess-Zumino model with a *rational* central charge – considering it as a dual to that one with integer k in the above sense.

The most easy way to check the relations (26), (28) is using bosonization technique when performing the reduction [37]. Indeed, "twisting"

$$T_{WZNW} \to \tilde{T}_{WZNW} = T_{WZNW} - \partial H \tag{30}$$

where (see [38] for more detailed description of free field technique)

$$T_{WZNW} = w\partial\chi - \frac{1}{2}(\partial\phi)^2 - \frac{i}{\sqrt{2(k+2)}}\partial^2\phi$$

$$H = w\chi - \frac{i}{\sqrt{2}}\sqrt{k+2}\partial\phi \tag{31}$$

one gets

$$\tilde{T}_{WZNW} = -\partial w\chi - \frac{1}{2}(\partial\phi)^2 - \frac{i}{\sqrt{2}}\left(\frac{1}{\sqrt{k+2}} - \sqrt{k+2}\right)\partial^2\phi \tag{32}$$

and the field ϕ stands now for minimal model, i.e. the corresponding screening operators become the screening charges of the (p,q) model.

Another valuable relation exists between the $SU(2)_k$ WZNW model and $N=2$ minimal model A_k. Namely the $SU(2)_k$ Kac-Moody currents can be represented as

$$J_\pm = e^{\pm i\sqrt{\frac{2}{k}}h \mp i\sqrt{1+\frac{2}{k}}\Phi}G_\pm \tag{33}$$

with

$$H = i\sqrt{\frac{k}{2}}\partial h \tag{34}$$

– the Cartan current of the $SU(2)_k$ while

$$J = i\sqrt{\frac{k}{k+2}}\partial\Phi \tag{35}$$

being the $U(1)$ current of the $N=2$ minimal model and $G_\pm$ denote the corresponding superconformal symmmetry generators. The twisting of $N=2$ gives

$$T_{N=2} \to T_{N=2}^{tw} = T_{N=2} - \frac{i}{2}\sqrt{\frac{k}{k+2}}\partial^2\Phi \tag{36}$$

where [7]

$$T_{N=2} = T_{WZNW} - T_h + T_\Phi \tag{37}$$

Eqs. (32), (36) and (37) altogether give

$$T_{N=2}^{tw} = T_{WZNW}^{tw} + \hat{T}_\Phi - \hat{T}_h \tag{38}$$

where from the first term in the r.h.s. one can single out the (p,q) matter model, while the rest can be transformed by similiar technique into the Liuoville and ghost contributions.

---

[7] the equality should be understood schematically, i.e. in the sense of bosonization.

3.1 Landau-Ginzburg models

The particular class of topological theories which includes $N = 2$ superconformal minimal models is given by the Landau-Ginzburg models. The action can be written in the form:

$$\int \partial X \bar{\partial} X^* + \psi \bar{\partial} \psi^* + \bar{\psi} \partial \bar{\psi}^* + FF^* + W'(X)F + \psi \bar{\psi} W''(X) + W'(X^*)F^* +$$

$$+ \psi^* \bar{\psi}^* W''(X^*) \tag{39}$$

which is invariant under the $N = 2$ supersymmetry transformations, generated by

$$G = \psi \frac{\delta}{\delta X} + F \frac{\delta}{\delta \bar{\psi}} - \partial X^* \frac{\delta}{\delta \psi^*} - \partial \bar{\psi}^* \frac{\delta}{\delta F^*}$$

$$\bar{G} = \bar{\psi} \frac{\delta}{\delta X} - F \frac{\delta}{\delta \psi} - \bar{\partial} X^* \frac{\delta}{\delta \bar{\psi}^*} + \bar{\partial} \psi^* \frac{\delta}{\delta F^*}$$

$$G^* = \psi^* \frac{\delta}{\delta X^*} + F^* \frac{\delta}{\delta \bar{\psi}^*} - \partial X \frac{\delta}{\delta \psi} - \partial \bar{\psi} \frac{\delta}{\delta F}$$

$$\bar{G}^* = \bar{\psi}^* \frac{\delta}{\delta X^*} - F^* \frac{\delta}{\delta \psi^*} - \bar{\partial} X \frac{\delta}{\delta \bar{\psi}} + \bar{\partial} \psi \frac{\delta}{\delta F}$$

$$\{G, G^*\} = -2\partial$$

$$\{\bar{G}, \bar{G}^*\} = -2\bar{\partial} \tag{40}$$

After twisting, the lagrangian takes the form

$$\int \partial X \bar{\partial} X^* + \psi \bar{\partial} \psi^* + \bar{\psi} \partial \bar{\psi}^* + FF^* + \psi \bar{\psi} W''(X) + FW'(X) +$$

$$+ \sqrt{g} \left[F^* W'(X^*) + \psi^* \bar{\psi}^* W''(X^*) \right] =$$

$$= \int \frac{1}{2} \psi \bar{\partial} (\psi^* - \bar{\psi}^*) + \frac{1}{2} \bar{\psi} \partial (\bar{\psi}^* - \psi^*) + \psi \bar{\psi} W''(X) + FW'(X) + \{Q, V\} \tag{41}$$

with

$$Q = G^* + \bar{G}^* = \theta \frac{\delta}{\delta X^*} - F^* \frac{\delta}{\delta \eta} - \partial X \frac{\delta}{\delta \psi} - \bar{\partial} X \frac{\delta}{\delta \bar{\psi}} - (\partial \bar{\psi} + \bar{\partial} \psi) \frac{\delta}{\delta F} \tag{42}$$

ψdz, $\bar{\psi} d\bar{z}$ and $F dz d\bar{z}$ are forms and ψ^*, $\bar{\psi}^*$ and F^* are scalar functions and where

$$V = - \int \frac{1}{2} (\psi \bar{\partial} X^* + \bar{\psi} \partial X^*) + \sqrt{g} \eta W'(X^*) \tag{43}$$

where we have introduced

$$\theta = \psi^* + \bar{\psi}^*$$

$$\eta = \frac{1}{2}(\psi^* - \bar{\psi}^*) \tag{44}$$

The integral with the action (41) can be computed by localization technique [20]. It localizes on $Q = 0$, i.e.

$$\theta = 0$$

$$F^* = 0 \left(= \frac{\partial W}{\partial X} \right)$$

$$\partial X = 0$$

$$\bar{\partial} X = 0$$

$$\partial \bar{\psi} + \bar{\partial}\psi = 0 \tag{45}$$

Computation of the path integral for (41) gives zero for the trivial potential $W(X) = X$. This is the statement we will use below for $(1, p)$ models – stricktly speaking the case $(1, p)$ should correspond to (41) with a trivial potential and *non-trivial* kinetic term, but the corresponding integral do not depend on kinetic (or D-) term due to $N = 2$ bosonic-fermionic cancellation. Eqs. (45) demonstrate that actually the path integral is not still the most effective description for those models – it can be reduced to a more simple object. Such objects are directly related to integrable systems and we will pass to their description below.

4 Matrix models

To understand better the effective description of $2d$ gravity and string models let us for a moment trivialize the situation and return back from strings to particles, i.e. from surfaces to lines. A natural question is what is the analog of topological string models in the one-dimensional case and the answer should be very simple. Indeed, for the topological one-dimensional theory the only thing which can appear is something related to the points at the end of paths and their permutations, so these should be combinatorial numbers attached to the ends of Feynman diagrams.

The module space for one-dimensional theories consists of the lengths of world-lines, so inclusion of topological one-dimensional gravity should somehow take this into account. For the simplest case of the propagator one should get

$$G_{\alpha\beta} = \int_0^\infty dT \, f_{\alpha\beta}(T) \tag{46}$$

with T being the length of the world-line while α and β are indices running over the space attached to each point - end of the line, i.e. over the Hilbert space of the corresponding theory. The objects $G_{\alpha\beta}$ can be considered as building blocks for the theory.

In the case of absence of the target space the only choice for $f_{\alpha\beta}(T)$ is $\delta_{\alpha\beta}f(T)$, so instead of nontrivial propagators one gets just a number G and the "theory" reduces to a "generation function" via the one-dimensional integral

$$\int d\phi \exp\left(-\frac{\phi^2}{2G^2} + t\phi + \sum g_n\phi^n\right) \qquad (47)$$

where one should fix by hands what sort of one-dimensional "branches" - i.e. geometries is allowed. This is a typical "counting diagram" integral and it should be considered as a one-dimensional analog of generating function below.

From this point of view, two-dimensional topological gravity should naturally bring to "fat graphs" where generation function has a nice prescription to be computed via matrix models. A simple analog of (47) would look like

$$Z_N = \int DM_{N\times N} \exp\left(-TrV(M)\right) \qquad (48)$$

which was proven (in the limit $N \to \infty$) to be an effective way to compute the integral over two-dimensional metrics, including the sume over topologies. The continuum integration (1) is approximated by triangulations of world-sheet in (48).

Below, we will concentrate mostly to a slightly different version of the integral (48) which rather has an interpretation of the target space theory. The exact expression is [5]

$$Z^{(N)}[V|M] \equiv C^{(N)}[V|M]e^{TrV(M)-TrMV'(M)} \int DX \ e^{-TrV(X)+TrV'(M)X} \qquad (49)$$

where the integral is taken over $N \times N$ "Hermitean" matrices, with the normalizing factor given by Gaussian integral

$$C^{(N)}[V|M]^{-1} \equiv \int DY \ e^{-TrU_2[M,Y]},$$

$$U_2 \equiv \lim_{\epsilon\to 0} \frac{1}{\epsilon^2} Tr[V(M + \epsilon Y) - V(M) - \epsilon YV'(M)] \qquad (50)$$

Including an external matrix field, which can be considered as a source ($\equiv$ coupling constants) it allows us do assign more or less concrete potential to a theory.

The formula (49) has in fact a lot of similiarities with the Landau-Ginzburg model we discussed in the previous section. Both theories are determined by a potential and as we will see below there exists a simple relation between the potential in (49) and the superpotential of the Landau-Ginzburg model $W(X)$, namely:

$$W(X) = V'(X) \qquad (51)$$

4.1 From matrix models to integrable systems

Now we are going to demonstrate that matrix models being an adequate formulation for certain very simple string theories naturally lead to appearance of the *classical* integrable systems describing the exact solutions for such strings. Namely, we will show that introduced in the previous section model (49) is a particular solution to the KP (Toda lattice) hierarchy. That is:

(*A*) The partition function $Z_N^V[M]$ (49), if considered as a function of time-variables

$$T_k = \frac{1}{k} Tr\, M^{-k}, \, k \geq 1 \tag{52}$$

is a KP τ-function for *any* value of N and *any* potential $V[X]$.

(*B*) As soon as $V[X]$ is homogeneous polynomial of degree $p+1$, $Z_N^{\{V\}}[M] = Z_N^{\{p\}}[M]$ is in fact a τ-function of p-reduced KP hierarchy. [8]

In order to prove these statements, first, we rewrite (49) in terms of determinant formula

$$Z_N^{\{V\}}[M] = \frac{\det_{(ij)} \phi_i(\mu_j)}{\Delta(\mu)} \qquad i,j = 1, ..., N. \tag{53}$$

Then, we show that *any* KP τ-function in the Miwa parameterization does have the same determinant form.

The main thing which distinguishes matrix models from the point of view of solutions to the KP-hierarchy is that the set of functions $\{\phi_i(\mu)\}$ in (53) is not arbitrary. This is the origin of $\mathcal{L}_{-1}$ and other $\mathcal{W}$- constraints (which in the context of KP-hierarchy may be considered as implications of $\mathcal{L}_{-1}$).

The fact that the classical integrable system appear in string theory, of course has more deep reason that this simple illustration for low-dimensional models.

4.2 Integrability from the determinant formula

We begin with an evaluation of the integral [5]:

$$\mathcal{F}_N^{\{V\}}[\Lambda] \equiv \int DX\, e^{-Tr[V(X)-Tr\Lambda X]} \tag{54}$$

The integral over the "angle" $U(N)$-matrices can be easily taken with the help of [40] and if eigenvalues of X and Λ are denoted by $\{x_i\}$ and $\{\lambda_i\}$ respectively, the result is

$$\frac{1}{\Delta(\Lambda)} \left[\prod_{i=1}^{N} \int dx_i e^{-V(x_i)+\lambda_i x_i} \right] \Delta(X) \tag{55}$$

---

[8]Moreover, actually, $\dfrac{\partial Z^{\{p\}}}{\partial T_{np}} = 0$.

$\Delta(X)$ and $\Delta(\Lambda)$ are Van-der-Monde determinants, $e.g.$ $\Delta(X) = \prod_{i>j}(x_i - x_j)$. The $r.h.s.$ of (55) can be rewritten as

$$\Delta^{-1}(\Lambda)\Delta(\frac{\partial}{\partial \Lambda}) \prod_i \int dx_i e^{-V(x_i)+\lambda_i x_i} =$$

$$= \Delta^{-1}(\Lambda)\det_{(ij)} F_i(\lambda_j) \tag{56}$$

with

$$F_{i+1}(\lambda) \equiv \int dx\, x^i e^{-V(x)+\lambda x} = (\frac{\partial}{\partial \lambda})^i F_1(\lambda). \tag{57}$$

Note that

$$F_1(\lambda) = \mathcal{F}_{N=1}^{\{V\}}[\lambda] . \tag{58}$$

If we recall that

$$\Lambda = V'(M) = W(M) \tag{59}$$

and denote the eigenvalues of M through $\{\mu_i\}$, then:

$$\mathcal{F}_N^{\{V\}}[W(M)] = \frac{\det \tilde{\Phi}_i(\mu_j)}{\prod_{i>j}(W(\mu_i) - W(\mu_j))} , \tag{60}$$

with

$$\tilde{\Phi}_i(\mu) = F_i(W(\mu)). \tag{61}$$

Proceed now to the normalization (50). Indeed, it is given by the Gaussian integral:

$$C^{(N)}[V|M]^{-1} \equiv \int DX\, e^{-U_2(M,X)}. \tag{62}$$

Then for evaluation of (62) it remains to use the obvious rule of Gaussian integration,

$$\int DX\, e^{-\sum_{i,j}^N U_{ij} X_{ij} X_{ji}} \sim \prod_{i,j}^N U_{ij}^{-1/2} \tag{63}$$

and substitute the explicit expression for $U_{ij}(M)$. If potential is represented as a formal series,

$$V(X) = \sum \frac{v_n}{n} X^n$$

$$W(X) = \sum v_n X^n \tag{64}$$

we have

$$U_2(M,X) = \sum_{n=0}^{\infty} v_{n+1} \left\{ \sum_{a+b=n-1} Tr M^a X M^b X \right\},$$

and

$$U_{ij} = \sum_{n=0}^{\infty} v_{n+1} \left\{ \sum_{a+b=n-1} \mu_i^a \mu_j^b \right\} = \sum_{n=0}^{\infty} v_{n+1} \frac{\mu_i^n - \mu_j^n}{\mu_i - \mu_j} = \frac{W(\mu_i) - W(\mu_j)}{\mu_i - \mu_j}.$$

Coming back to (49), we conclude that

$$Z_N^{\{V\}}[M] = e^{Tr[V(M)-MW(M)]} C^{(N)}[V|M] \mathcal{F}_N[W(M)] \sim$$

$$\sim [\det \tilde{\Phi}_i(\mu_j)] \prod_{i>j}^{N} \frac{U_{ij}}{(W(\mu_i) - W(\mu_j))} \prod_{i=1}^{N} s(\mu_i) = \frac{[\det \tilde{\Phi}_i(\mu_j)]}{\Delta(M)} \prod_{i=1}^{N} s(\mu_i) . \quad (65)$$

$$s(\mu) = [W'(\mu)]^{1/2} e^{V(\mu)-\mu W(\mu)} \qquad (66)$$

The product of s-factors at the *r.h.s.* of (65) can be absorbed into $\tilde{\Phi}$-functions:

$$Z_N^{\{V\}}[M] = \frac{\det \Phi_i(\mu_j)}{\Delta(M)}, \qquad (67)$$

where

$$\Phi_i(\mu) = s(\mu) \tilde{\Phi}_i(\mu) \underset{\mu \to \infty}{\longrightarrow} \mu^{i-1}(1 + \mathcal{O}(\frac{1}{\mu})). \qquad (68)$$

where the asymptotic is crucial for the determinant (67) to be a solution to the KP hierarchy in the sense of [42].

The Kac-Schwarz operator [35, 36]. From eqs.(61),(66) and (68) one can deduce that $\Phi_i(\mu)$ can be derived from the basic function $\Phi_1(\mu)$ by the relation

$$\Phi_i(\mu) = [W'(\mu)]^{1/2} \int x^{i-1} e^{-V(x)+xV'(\mu)} dx = A_{\{V\}}^{i-1}(\mu) \Phi_1(\mu) , \qquad (69)$$

where $A_{\{V\}}(\mu)$ is the first-order differential operator

$$A_{\{V\}}(\mu) = s \frac{\partial}{\partial \lambda} s^{-1} = \frac{e^{V(\mu)-\mu W(m)}}{[W'(\mu)]^{1/2}} \frac{\partial}{\partial \mu} \frac{e^{-V(\mu)+\mu W(\mu)}}{[W'(\mu)]^{1/2}} =$$

$$= \frac{1}{W'(\mu)} \frac{\partial}{\partial \mu} + \mu - \frac{W''(\mu)}{2[W'(\mu)]^2} . \qquad (70)$$

In the particular case of $V(x) = \frac{x^{p+1}}{p+1}$

$$A_{\{p\}}(\mu) = \frac{1}{p\mu^{p-1}} \frac{\partial}{\partial \mu} + \mu - \frac{p-1}{2p\mu^p} \qquad (71)$$

coincides (up to the scale transformation of μ and $A_{\{p\}}(\mu)$) with the operator which determines the finite dimensional subspace of the Grassmannian in ref.[35] We emphasize that the property

$$\Phi_{i+1}(\mu) = A_{\{V\}}(\mu)\Phi_i(\mu) \quad \left(F_{i+1}(\lambda) = \frac{\partial}{\partial \lambda} F_i(\lambda)\right) \tag{72}$$

is exactly the thing which distinguishes partition functions of GKM from the expression for generic τ-function in Miwa's coordinates,

$$\tau_N^{\{\phi_i\}}[M] = \frac{[\det \phi_i(\mu_j)]}{\Delta(M)}, \tag{73}$$

with arbitrary sets of functions $\phi_i(\mu)$. In the next section we demonstrate that the quantity (73) is exactly a KP τ-function in Miwa coordinates, and we return to the Kac-Schwarz operator in sect.5.

4.3 KP τ-function in Miwa parameterization

A generic KP τ-function is a correlator of a special form [41]:

$$\tau^G\{T_n\} = \langle 0| : e^{\sum T_n J_n} : G|0\rangle \tag{74}$$

with

$$J(z) = \tilde{\psi}(z)\psi(z); \ G = : \exp \mathcal{G}_{mn}\tilde{\psi}_m\psi_n : \tag{75}$$

in the theory of free 2-dimensional fermionic fields $\psi(z)$, $\tilde{\psi}(z)$ with the action $\int \tilde{\psi}\bar{\partial}\psi$. The vacuum states are defined by conditions

$$\psi_n|0\rangle = 0 \ n < 0 \,, \ \tilde{\psi}_n|0\rangle = 0 \ n \geq 0 \tag{76}$$

where $\psi(z) = \sum_{\mathbf{Z}} \psi_n z^n \ dz^{1/2}$, $\tilde{\psi}(z) = \sum_{\mathbf{Z}} \tilde{\psi}_n z^{-n-1} \ dz^{1/2}$.

The crucial restriction on the form of the correlator, implied by (75) is that the operator $: e^{\sum T_n J_n} : G$ is *Gaussian* exponential, so that the insertion of this operator may be considered just as a modification of $\langle \tilde{\psi}\psi \rangle$ *propagator*, and the Wick theorem is applicable. Namely, the correlators

$$\langle 0| \prod_i \tilde{\psi}(\mu_i)\psi(\lambda_i)G|0\rangle \tag{77}$$

for *any* relevant G are expressed through the pair correlators of the same form:

$$(77) = \det_{(ij)} \langle 0|\tilde{\psi}(\mu_i)\psi(\lambda_j)G|0\rangle \tag{78}$$

The simplest way to understand what happens to the operator $e^{\sum T_n J_n}$ after the substitution of (52) is to use the free-*boson* representation of the current $J(z) = \partial\varphi(z)$. Then $\sum T_n J_n = \sum_i \left\{ \sum_n \frac{1}{n \cdot \mu_i^n} \varphi_n \right\} = \sum_i \varphi(\mu_i)$, and

$$: e^{\sum_i \varphi(\mu_i)} := \frac{1}{\prod_{i<j}(\mu_i - \mu_j)} \prod_i : e^{\varphi(\mu_i)} : . \tag{79}$$

In fermionic representation it is better to start from

$$T_n = \frac{1}{n} \sum_i \left(\frac{1}{\mu_i^n} - \frac{1}{\tilde{\mu}_i^n}\right) \tag{80}$$

instead of (52). Then

$$: e^{\sum T_n J_n} := \frac{\prod_{i,j}^N (\tilde{\mu}_i - \mu_j)}{\prod_{i>j}(\mu_i - \mu_j) \prod_{i>j}(\tilde{\mu}_i - \tilde{\mu}_j)} \prod_i \tilde{\psi}(\tilde{\mu}_i)\psi(\mu_i) . \tag{81}$$

In order to come back to (52) it is necessary to shift all $\tilde{\mu}_i$'s to infinity. This may be expressed by saying that the left vacuum is substituted by

$$\langle N| \sim \langle 0| \tilde{\psi}(\infty)\tilde{\psi}'(\infty)...\tilde{\psi}^{(N-1)}(\infty).$$

The τ-function now can be represented in the form:

$$\tau_N^G[M] = \langle 0| : e^{\sum T_n J_n} : G|0\rangle = \Delta(M)^{-1}\langle N| \prod_i : e^{\varphi(\mu_i)} : G|0\rangle =$$

$$= \lim_{\tilde{\mu}_j \to \infty} \frac{\prod_{i,j}(\tilde{\mu}_i - \mu_j)}{\prod_{i>j}(\mu_i - \mu_j) \prod_{i>j}(\tilde{\mu}_i - \tilde{\mu}_j)} \langle 0| \prod_i \tilde{\psi}(\tilde{\mu}_i)\psi(\mu_i)G|0\rangle \tag{82}$$

applying the Wick's theorem (77), (78) and taking the limit $\tilde{\mu}_i \to \infty$ we obtain:

$$\tau_N^G[M] = \frac{\det \phi_i(\mu_j)}{\Delta(M)} \tag{83}$$

with functions

$$\phi_i(\mu) \sim \langle 0| \tilde{\psi}^{(i-1)}(\infty)\psi(\mu)G|0\rangle \underset{\mu \to \infty}{\to} \mu^{i-1}(1 + \mathcal{O}(\frac{1}{\mu})). \tag{84}$$

Thus, we proved that KP τ-function in Miwa coordinates (52) has exactly the determinant form (53), or is a τ-function of KP hierarchy.

4.4 Universal $\mathcal{L}_{-1}$-constraint and string equation

Let us return to the question of specifying particular "stringy" solutions to the KP hierarchy which we already demostrated considering basis vectors (70). We will show that the matrix version of the Kac-Schwarz operator which is almost

$$Tr\frac{\partial}{\partial \Lambda_{tr}} = Tr\frac{1}{W'(M)}\frac{\partial}{\partial M_{tr}} \tag{85}$$

acting on τ-function gives the string equation. Therefore it is natural to examine, how this operator acts on

$$Z^{\{V\}}[M] = \frac{\det \tilde{\Phi}_i(\mu_j)}{\Delta(M)} \prod_i s(\mu_i), \tag{86}$$

$$s(\mu) = (W'(\mu))^{1/2} e^{V(\mu) - \mu W(\mu)}, \tag{87}$$

$$\tilde{\Phi}_i(\mu) = F_i(\lambda) = (\partial/\partial\lambda)^{i-1} F_1(\lambda)$$

First of all, if $Z^{\{V\}}$ is considered as a function of T-variables,

$$\frac{1}{Z^{\{V\}}} Tr \frac{\partial}{\partial \Lambda_{tr}} Z^{\{V\}} = -\sum_{n \geq 1} Tr[\frac{1}{W'(M)M^{n+1}}] \frac{\partial log Z^{\{V\}}}{\partial T_n}. \tag{88}$$

On the other hand, if we apply (85) to explicit formula (86), we obtain:

$$\frac{1}{Z^{\{V\}}} Tr \frac{\partial}{\partial \Lambda_{tr}} Z^{\{V\}}$$

$$= -Tr\ M + \frac{1}{2} \sum_{i,j} \frac{1}{W'(\mu_i)W'(\mu_j)} \frac{W'(\mu_i) - W'(\mu_j)}{\mu_i - \mu_j} + Tr \frac{\partial}{\partial \Lambda_{tr}} \log \det F_i(\lambda_j), \tag{89}$$

We can prove that

$$\frac{1}{Z^{\{V\}}} \mathcal{L}_{-1} Z^{\{V\}} = -\frac{\partial}{\partial T_1} \log\ Z^{\{V\}} + TrM - Tr \frac{\partial}{\partial \Lambda_{tr}} \log \det F_i(\lambda_j). \tag{90}$$

can be used in order to suggest the formula for the universal operator $\mathcal{L}_{-1}$.
Here

$$\mathcal{L}_{-1} = \sum_{n \geq 1} Tr[\frac{1}{W'(M)M^{n+1}}] \frac{\partial}{\partial T_n} + + \frac{1}{2} \sum_{i,j} \frac{1}{W'(\mu_i)W'(\mu_j)} \frac{W'(\mu_i) - W'(\mu_j)}{\mu_i - \mu_j} - \frac{\partial}{\partial T_1}, \tag{91}$$

So, in order to prove the $\mathcal{L}_{-1}$-constraint, one should prove that the *r.h.s.* of (90) vanishes, i.e.

$$\frac{\partial}{\partial T_1} \log\ Z_N^{\{V\}} = TrM - Tr \frac{\partial}{\partial \Lambda_{tr}} \log \det F_i(\lambda_j), \tag{92}$$

This is possible to prove only if we remember that $Z_N^{\{V\}} = \tau_N^{\{V\}}$. In this case the *l.h.s.* may be represented as residue of the ratio

$$res_\mu \frac{\tau_N^{\{V\}}(T_n + \mu^{-n}/n)}{\tau_N^{\{V\}}(T_n)} = \frac{\partial}{\partial T_1} \log\ \tau_N^{\{V\}}(T_n). \tag{93}$$

However, if expressed through Miwa coordinates, the τ-function in the numerator is given by the same formula with one *extra* parameter μ , *i.e.* is in fact equal to $\tau_{N+1}^{\{V\}}$. This idea is almost enough to deduce (92). For example, if $N = 1$

$$\tau_1^{\{V\}}(T_n) = \tau_1^{\{V\}}[\mu_1] = e^{V(\mu_1) - \mu_1 W(\mu_1)}[W'(\mu_1)]^{1/2} F(\lambda_1),$$

$$\tau_1^{\{V\}}(T_n + \mu^{-n}/n) = \tau_2^{\{V\}}[\mu_1, \mu] =$$

$$= e^{V(\mu_1) - \mu_1 W(\mu_1)} e^{V(\mu) - \mu W(\mu)} \frac{[W'(\mu_1) W'(\mu)]^{1/2}}{\mu - \mu_1} \times$$

$$\times [F(\lambda_1) \partial F(\lambda)/\partial \lambda - F(\lambda) \partial F(\lambda_1)/\partial \lambda_1] =$$

$$= \frac{e^{V(\mu) - \mu W(\mu)}[W'(\mu)]^{1/2} F(\lambda)}{\mu - \mu_1} \tau_1^{\{V\}}[\mu_1] \cdot [-\partial log F(\lambda_1)/\partial \lambda_1 + \partial log F(\lambda)/\partial \lambda].$$

$$(94)$$

The function

$$F(\lambda) = \int dx \, e^{-V(x) + \lambda x} \sim e^{V(\mu) - \mu W(\mu)}[W'(\mu)]^{-1/2}\{1 + O(\frac{W'''}{W'W'})\}. \quad (95)$$

If $W(\mu)$ grows as μ^p when $\mu \to \infty$, then $W'''/(W')^2 \sim \mu^{-p-1}$, and for our purposes it is enough to have $p > 0$, so that in the braces at the *r.h.s.* stands $\{1 + o(1/\mu)\}(\mu \cdot o(\mu) \to 0$ as $\mu \to \infty)$. Then numerator at the *r.h.s.* of (94) is $\sim 1 + o(1/\mu)$, while the second item in square brackets behaves as $\partial log F(\lambda)/\partial \lambda \sim \mu(1 + o(1/\mu))$. Combining all this, we obtain:

$$\frac{\partial}{\partial T_1} \log \tau_1^{\{V\}} = res_\mu \left\{ \frac{1 + o(1/\mu)}{\mu - \mu_1} [-\partial log F(\lambda_1)/\partial \lambda_1 + \mu(1 + o(1/\mu))] \right\} =$$

$$= \mu_1 - \partial log F(\lambda_1)/\partial \lambda_1. \quad (96)$$

i.e. (92) is proved for the particular case of $N = 1$.

In the particular case of monomial potential $V = \frac{X^{p+1}}{p+1}$ (91) turns into more common form [2, 3]:

$$\mathcal{L}_{-1}^{\{p\}} = \frac{1}{p} \sum_{n \geq 1}(n + p)T_{n+p}\frac{\partial}{\partial T_n} + + \frac{1}{2p} \sum_{\substack{a+b=p \\ a,b \geq 0}} aT_a bT_b - \frac{\partial}{\partial T_1}, \quad (97)$$

5 Canonical quantization and p-q duality

5.1 General ideology

Now, let us turn to somewhat more general question of how a generic string theory (first- quantized or second quantized) should look like. In the simplest case of topological string we can reduce ourselves to the question of basic module space. In the frames of this ideology module spaces corresponding to topological theories should be considered as a background for the first-quantized theory while the second-quantized theory should be related to the quantization of module space.

In the well-known case of pure topological gravity we should expect *nothing* since that theory does not have any target space at all. This is somehow consistent with the observation that the partition function can be made trivial just by a choice of gauge (polarization).

We are going to demonstrate that the matrix model solution can be obtained within the frames of second quantization on a kind of "module space" for these theories (see [43] for more detailed information on this point).

Finally we will make some comments on the considered problem in the framework of mirror symmetry (see for example [47]). The important remark is that mirror manifolds should be distinguished classically and this effect is very closely related to that one we have in the case of (p, q) models.

5.2   String equation and Heisenberg algebra

Now we are going directly to a problem of description of a particular representation of the Heisenberg algebra. One should start from [43] where the "phase space" for (p, q) models is considered as a certain "generalized" module space for the Riemann surfaces with punctures. In the simplest case of sphere with the only puncture one might take the phase space with a symplectic structure

$$\{W, Q\} = 1 \tag{98}$$

which is actually generated by

$$\{z, t_1\} = 1$$
$$\{\tilde{z}, \tilde{t}_1\} = 1 \tag{99}$$

(where $z^p = W(\mu)$ and $\tilde{z}^q = Q(\mu)$). For trivial $(1, p)$ topological theories $\tilde{z} = \mu$.

From this point of view what we consider is a quantization of a symplectic manifold

$$\omega = \delta W \wedge \delta Q = \delta z \wedge \delta t_1 \tag{100}$$

and we can perform it by standard methods.

The corresponding action is

$$S = \int W \, dQ + S_0$$
$$dS = \delta W \wedge \delta Q \tag{101}$$

and S_0 parameterizes an "initial point". Now, it is obvious that in the proposed quantization scheme the set of coupling constants depends on the way of quantization, so does the solutions (potentials) of the hierarchy, τ- or the BA function *etc.*

Now the quantization gives the representation of the Heisenberg operators, satisfying the string equation

$$[\hat{P}, \hat{Q}] = 1 \tag{102}$$

in the "momentum" (spectral) space

$$\hat{P} = \lambda$$
$$\hat{Q} = \frac{\partial}{\partial \lambda} + Q(\lambda) \tag{103}$$

From the point of view of the KP hierarchy, we will also add some additional requirements on the "spectral parameter" implying that

$$\lambda = W(\mu) = \mu^p \tag{104}$$

then (p, q) models correspond to the case where $Q(\lambda)$ should be a *polynomial* of μ of degree q [2], (while the corresponding wave functions should have specific asymptotics when $\mu \to \infty$).

Wave functions of this problem appear to be the Baker-Akhiezer functions of the corresponding integrable system and when acting on wave functions conditions (103) get the form of the Kac-Schwarz equations [35, 36]:

$$\lambda \varphi_i(\mu) = \sum_j W_{ij} \varphi_j(\mu)$$
$$\hat{A} \varphi_i(\mu) = \sum_j A_{ij} \varphi_j(\mu) \tag{105}$$

where

$$\lambda = W(\mu) \sim \mu^p$$
$$A^{(W,Q)} \equiv s^{(W,Q)}(\mu) \frac{1}{W'(\mu)} \frac{\partial}{\partial \mu} [s^{(W,Q)}(\mu)]^{-1} =$$
$$= \frac{1}{W'(\mu)} \frac{\partial}{\partial \mu} - \frac{1}{2} \frac{W''(\mu)}{W'(\mu)^2} + Q(\mu) \tag{106}$$

The standard way to construct wave functions of the theory is to define the Fock vacuum by

$$\hat{A} \Psi_0 = 0 \tag{107}$$

with an obvious solution

$$\Psi_0 = \sqrt{W'(\mu)}\exp\int QdW \qquad (108)$$

and the corresponding τ-function is a determinant projection of higher states

$$\Psi_n \sim W^n\Psi_0 \qquad (109)$$

to the states with a *canonical* asymptotics

$$\varphi_i(\mu) \underset{\mu\to\infty}{\to} \mu^{i-1} \qquad (110)$$

forming the conventional basis in the space of wave functions – the point of infinite-dimensional Grassmannian.

The only simple case arises when the Kac-Schwarz equations (105) have trivial solution, *i.e.* when $p = 1$. Starting from normalization $\varphi_1(\mu) = 1$ (corresponding to $\Psi_0 = \exp\int Qd\mu$), and using first of eqs.(105) one can always get $\Psi_n = \mu^n \exp\int Qd\mu \to \varphi_i(\mu) = \mu^{i-1}$ *exactly*. Then the second condition of (105) is fulfilled *automatically* for *any* $Q(\mu)$.

However, one can see that the corresponding solutions are related to topological models by a kind of Fourier transformation. Indeed, it has been observed [33, 34] that the system of equations (105) posseses a *duality* symmetry which relates (p, q) to (q, p) solution. The duality transformation for the Baker-Akhiezer functions looks like

$$\psi^{(P,Q)}(z) = [P'(z)]^{1/2}\int dQ\, e^{P(z)Q(x)}\psi^{(Q,P)}(x)[Q'(x)]^{-1/2} \qquad (111)$$

and it can be also written for the basis vectors in the Grassmannian

$$\phi_i(\mu) = [W'(\mu)]^{1/2}\exp(-S_{W,Q}|_{x=\mu})\int d\mathcal{M}_Q(x)f_i(x)\exp S_{W,Q}(x,\mu) \qquad (112)$$

with

$$d\mathcal{M}_Q(x) = dx\sqrt{Q'(x)}$$

$$S_{W,Q}(x,\mu) = -\int^x WdQ + W(\mu)Q(x) \qquad (113)$$

and for the partition functions

$$\tau^{(W,Q)}[M] =$$

$$= C[V,M]\int DX\tau^{(Q,W)}[X]\exp\left\{Tr[1/2\log Q'(X) + \int_M^X W(z)dQ(z) + \right.$$

$$\left. +W(M)Q(X)]\right\} \qquad (114)$$

(here, better to consider *normalized* partition function $\tau^{(W,Q)} \rightarrow Z^{(W,Q)} \rightarrow \Psi_{BA}^{(W,Q)}(t_k - \frac{1}{k}TrM^{-k})$. It makes possible to obtain solutions for nontrivial models – topological $(p, 1)$ models [5] and their Landau-Ginzburg deformations [32].

$$\varphi_i(\mu) = \sqrt{p\mu^{p-1}} \exp\left(-\sum t_k \mu^k\right) \int dx \; x^{i-1} \exp(-V(x) + x\mu^p) \tag{115}$$

which are *dual* to $(1, p)$ model in the above sense.

Here, we immediately run into a puzzle: how to interpret this from the point of view of quantization theory. Indeed, the duality transformation (112) is nothing but a transformation from $\hat{p}$ to $\hat{q}$ quantization procedure or from one to another representation of quantum algebra and as it is well-known the quantization should be independent of this. It means that $(p, 1)$ and $(1, p)$ or trivial theory are in fact equivalent as string theories, i.e. the nontrivial partition functions for $(p, 1)$ theories corresponding to some well-known topological theories (twisted $N = 2$ Landau-Ginzburg theories) give nothing from a physical point of view [9]. Thus, the first puzzle is that $\tau^{(1,p)} \equiv 1$ seems to contain all the "topologcal" information as a "dual" partition function does. Second, the topological numbers perhaps should not be considered as "real observables" of the theory – they rather correspond to a sort of combinatorial factors for Feynman diagramms in particle theories.

This is actually a new feature of string theory if we compare it to quantum field theory – i.e. even trivial target-space model can possess rich and nontrivial structure. The Virasoro action in these theories naturally follows from (103), (105).

Let us finally add few comments about holomorphic anomaly. The "quasi-classical" τ-function obeys a homogeneous relation

$$\sum t_j \frac{\partial}{\partial t_j} \log \tau_0 = 2 \log \tau_0 \tag{116}$$

spoilt by the contribution of the one-loop correction, having the form, for example, for the $(2, 1)$ theory

$$\sum t_j \frac{\partial}{\partial t_j} \log \tau - 2 \log \tau = -\frac{1}{24} \tag{117}$$

The similiar expressions appear when one considers the logariphm of the partition function for the higher-dimensional theories [31] and this should mean that the expression (117) should have a similiar nature.

---

[9]$(2, 1)$ model corresponds to pure topological gravity and generates intersection indices on module spaces of Riemann surfaces with punctures - it appears that the intersection indices in topological gravity are just a "physical artefact"

6   Conclusion

Now let us briefly summarize the main ideas presented above. We have tried to demonstrate that appearing in the context of matrix models effective target-space description of string theory can be a useful tool for constructing a nonperturbative string field theory. Indeed, the space of coupling constants $\{T_k\}$ may be considered for simplest string models as a space of background fields and one might hope to get a second-quantized theory by quantization of appearing there structures. It has been shown by Krichever [43] that the "small phase space" in fact can be considered as a certain module space for a spectral surface with marked points if one restricts the order of singularities in these marked points. Then it is natural to consider the quantization of (98) as a quantization of this module space. In fact we have shown above that the particular example of $(p, 1)$ models rather leads to a trivial theory – topological gravity (W-gravity) which is not too much interesting as a target space theory. However, the natural question that appears is a generalization of this approach to more interesting module spaces.

For example, there exists a quite interesting scheme of quantization of module spaces of flat connections and projective structures on Riemann surfaces with punctures [44]. This is not far from what we need in the case of string models: in fact module spaces of flat connections already appeared in the context of two-dimensional Yang-Mills theory and its relation to string theory [45, 46]. It is natural to think that the related string models should have partition (generating) functions more simple than the discussed above theories, being related thus from the point of view of integrable hierarchies with the, say, rational τ-functions. The appeerence of such τ-functions can be interpreted in the way that a restricted amount of world-sheet topologies give contribution to the partition function. In fact [48] there exists another, so-called "character" phase of GKM considered above which is closely related to the Yang-Mills theory and rational τ-functions.

7   Acknowledgements

I would like to thank V.Fock, A.Gerasimov, A.Gorsky, R.Kashaev, S.Kharchev, I.Krichever, A.Levin, A.Losev, A.Mironov, A.Morozov, A.Niemi, K.Palo, A.Rosly and V.Rubtsov for fruitful discussions.

The work was in part supported by Fundamental Research Foundation of Russia, contract No 93–02–3379 and by NFR-grant No F-GF 06821-305 of the Swedish Natural Science Research Council. I am grateful for warm hospitality to the organizers of the Third Baltic Rim student seminar and to the Institute of Theoretical Physics of Uppsala University when this paper has been completed.

References

[1] M.Douglas *Phys.Lett.*, **238B** (1990) 176

[2] M.Fukuma, H.Kawai, and R.Nakayama *Int.J.Mod.Phys.*, **A6** (1991) 1385

[3] R.Dijkgraaf, H.Verlinde, and E.Verlinde, *Nucl.Phys.*, **B348** (1991) 435.

[4] A.Gerasimov *et al. Nucl.Phys.*, **B357** (1991) 565.

[5] S.Kharchev et al. *Nucl.Phys.*, **B380** (1992) 181.

[6] E.Witten , *Chern-Simons theory as a string theory*, Preprint IASSNS-HEP-92-45.

[7] B.Zwiebach *Nucl.Phys.* **B390** (1993) 33.

[8] A.Marshakov, *On string field theory for $c \leq 1$*, preprint *FIAN/TD/08-92* (June, 1992), hepth/9208022; in *Pathways to Fundamental Theories*, World Scientific, 1993.

[9] E.Witten, Nucl.Phys. **B340** (1990) 281;
Surveys Diff.Geom. **1** (1991) 243
R.Dijkgraaf and E.Witten, *Nucl.Phys* **B342** 486

[10] J. Distler, *Nucl.Phys* **B342** (1990) 523

[11] A.Gerasimov et al *Phys.Lett.* **B242** (1990) 345.

[12] M.Kontsevich, *Comm. Math.Phys.* **147** (1992) 1.

[13] E.Witten *On the Kontsevich model and other models of 2d gravity*, Preprint IASSNS-HEP-91-24.

[14] A.Marshakov, A.Mironov, and A.Morozov, *Phys.Lett.*, **B274** (1992) 280.

[15] A.Mikhailov *Int.J.Mod.Phys.* **A9** (1994) 873

[16] R.Dijkgraaf, E. Verlinde and H. Verlinde, *Nucl.Phys* **B352** (1991) 59

[17] R.Dijkgraaf, *Intersection theory, integrable hierarchies and topological field theory*, preprint IASSNS-HEP-91/91

[18] A.Lossev, *Descendans constructed from matter fields and K.Saito higher residue pairing in Landau-Ginzburg theories coupled to topological gravity*, preprint *ITEP/TPI-MINN* (May, 1992).
A.Losev, I.Polyubin, *On connection between topological Landau-Ginzburg gravity and integrable systems*, preprint ITEP/Uppsala, (1993)

[19] K.Li, Nucl.Phys. **B354** (1991) 711

[20] J.Duistermaat and G.Heckman, *Invent.Math.* **69** (1982) 259
M.Atiyah, R.Bott, *Topology* **23**, (1984) 1
M.Blau, E.Keski-Vakkuri and A.Niemi, *Phys.Lett.* **B246** (1990) 92;
E.Keski-Vakkuri, A.Niemi, G.Semenoff and O.Tirkkonen, *Phys.Rev.* **D44** (1991) 3899;

A.Hietamaki, A.Morozov, A.Niemi, K.Palo *Phys.Lett.* **263B** (1991) 417;
A.Morozov, A.Niemi, K.Palo *Phys.Lett.* **271B** (1991) 365; *Nucl.Phys.*
B377 (1992) 295,
A.Niemi and O.Tirkkonen, *On Exact Evaluation of Path Integrals*, Preprint
UU-ITP, 3/93

[21] M.Kontsevich A_∞-*algebras in mirror symmetry*, Preprint (1994)

[22] M.Kontsevich, Yu.Manin *Gromov-Witten classes, quantum cohomology and enumerative geometry*, Preprint (1994)

[23] E.Witten *Quantum background independence in string theory*, preprint *IASSNS-HEP-93/29*

[24] S.Kharchev, A.Marshakov *Quantization of string theory for $c \leq 1$*, Preprint FIAN/TD-14/93.

[25] J.-L.Gervais *Nucl.Phys.* **B391** (1993) 287

[26] J.-L.Gervais, J.Schnittger *Nucl.Phys.* **B413** (1994) 433
Phys.Lett. **B315** (1993) 258.

[27] V.Knizhnik, A.Polyakov, A.Zamolodchikov *Mod.Phys.Lett.* **A3** (1988) 819

[28] J.Distler, H.Kawai, *Nucl.Phys.* **B312** (1989) 509

[29] F.David, *Mod.Phys.Lett.* **A3** (1988) 1651

[30] Vl.Dotsenko *Mod.Phys.Lett.* **A7** (1992) 2505.

[31] M.Bershadsky, S.Cecotti, H.Ooguri and C.Vafa *Kodaira-Spencer theory of gravity and exact results for quantum string amplitudes*, preprint HUTP-93/A025.

[32] S.Kharchev et al. *Landau-Ginzburg topological theories in the framework of GKM and equvalent hierarchies*, preprint FIAN/TD/7-92, ITEP-M-5/92 (July, 1992), hepth/9208046.

[33] S.Kharchev, A.Marshakov, *Topological versus non-topological theories and $p-q$ duality in matrix models*, preprint FIAN/TD/15-92 (September, 1992), hepth/9210072; in *String Theory, Quantum Gravity and the Unification of the Fundamental Interactions*, World Scientific, 1993.

[34] S.Kharchev, A.Marshakov, *On $p - q$ duality and explicit solutions in $c \leq 1$ 2d gravity models*, preprint NORDITA-93/20 P, FIAN/TD-04/93

[35] V.Kac, A.Schwarz, *Phys.Lett.* **B257** (1991) 329

[36] A.Schwarz, *Mod.Phys.Lett.* **A6** (1991) 611; 2713

[37] A.Gerasimov et al *Phys.Lett.* **B236** (1990) 269

[38] A.Gerasimov *et al. Int.J.Mod.Phys.*, **A5** (1990) 2495

[39] G.Moore, N.Seiberg, M.Staudacher *Nucl.Phys.* **B362** (1991) 665

[40] C.Itzyzson and J.-B.Zuber, *J.Math.Phys.*, **21** (1980) 411
M.L.Mehta, *Commun.Math.Phys.*, **79** (1981) 327

[41] E.Date, M.Jimbo, M.Kashiwara, and T.Miwa, *Transformation group for soliton equation: III, preprint RIMS-358* (1981),
E.Date, M.Jimbo, M.Kashiwara, and T.Miwa. In: *Proc.RIMS symp.Nonlinear integrable systems — classical theory and quantum theory*, page 39, Kyoto, 1983.

[42] G.Segal and G.Wilson, *Publ.I.H.E.S.*, **61** (1985) 1.

[43] I.Krichever, *Comm.Math.Phys.*, **143** (1992) 415; *The tau-function of the universal Whitham hierarchy, matrix models and topological field theories*, preprint *LPTENS-92/18.*
also talk at *IV Conference on Mathematical Physics, Rakhov, 1994* and private communication.

[44] V.Fock, A.Rosly *Poisson structure on moduli of flat connections on Riemann surfaces and r-matrix*, preprint *ITEP-72-92*;
V.Fock, A.Rosly *Flat connections and polyubles*, Theor. and Math. Phys., **95** (1993) 228.

[45] D.Gross, W.Taylor, *Twists and Wilson loops in the string theory of two dimensional QCD*, Preprint CERN-TH.6827/93, PUPT-1382, LBL-33767, UCB-PTH-93/09

[46] A.Gorsky, N.Nekrasov, *Hamiltonian systems of Calogero type and two dimensional Yang-Mills theory*, Preprint UUITP-6/93, ITEP-20/93

[47] C.Vafa, *Mirror transform and string theory*

[48] S.Kharchev et al *Generalized Kazakov-Migdal-Kontsevich model: group theory aspects*, Preprint UUITP-10/93, FIAN/TD-07/93, ITEP-M4/93.

On Background Independence in String Theory

Samson L. Shatashvili † #

School of Natural Sciences
Institute for Advanced Study
Olden Lane
Princeton, NJ 08540

Abstract. The problems with background independence are discussed in the example of open string theory. Based on the recent proposal by Witten I calculate the String Field Theory action in conformal perturbation theory to second order and demonstrate that the proper treatment of contact terms leads to non-trivial equations of motion. I conjecture the form of the field theory action to all orders.

1 Introduction

In contrast with difficult problems in realistic 4d Field Theory models, where the theory is defined and an explicit analytic solution is not yet known, String Theory isn't yet even defined. In many cases what we have is just a number of S-matrices for the processes when the background is fixed by our choice of conformal field theory in the first quantized formulation for amplitudes. Satisfactory formulation of String Theory would have been a formulation where we don't need to refer to any particular classical background and these "classical backgrounds" are given by solutions of some equations. The latter statement is very vague, because unfortunately it is not even clear (at least to the author) what should be the right terminology to address the question. It is believed, by analogy with the second quantized description of ordinary quantum field theory, that the understanding of vacuum structure of string theory as well as the non-perturbative character, can be achieved by developing the field theory language

† Research supported by NSF grant PHY92-45317.

On leave of absence from St. Petersburg Branch of Mathematical Institute (LOMI), Fontanka 27, St. Petersburg 191011, Russia.

to describe the target space theory. It might well be that the procedure that allows us to construct second quantized field theory from Feynman sum over trajectories directly applied to string theory is not the best way to approach the problem and some other new ideas should be introduced. One of the most important ingredients of any construction has to be a background independence.

In this paper I will address the question of background independent formulation of string theory in the example of open string theory recently suggested by Witten [1][1]. I can't claim that at present every point is understood for the case of open string; this paper should be considered as an attempt to single out main problems and find a correct language based on this experience. The calculations and observations presented in section 3, together with final result (see below) might serve as a proper guide. This explains the title.

I'll show that the integrals of total derivatives do not decouple inside the correlation functions that defines the String Field Theory action in the formalism of [1] due to contact term contributions. I'll explain that these contributions have universal character and can't be removed by change of renormalization scheme (this statement has the same origin as the one for gauge anomalies in field theory). This fact leads to slight modifications of the assumptions made in original paper [1] and also in [3]. It was shown in [3] that under the key assumption of decoupling of total derivatives, plus the requirement that BRST operator on the boundary is coupling constant independent, theory has a linear character. I'll demonstrate here that including the contribution of total derivatives one also has to properly define "BRST" operator on the boundary, which now necessarily should depend on couplings to satisfy the consistency condition. I'll discuss the ambiguities related to this issue and will make a particular choice. In this setup I'll calculate the field theory action in the lines of [3] using the conformal perturbation theory around some fixed point and demonstrate the existence of following relation :

$$S = -\beta^i \frac{\partial}{\partial t^i} Z(t) + Z(t), \qquad (1.1)$$

up to the second order in coupling constant. Here β is the world-sheet β-function and Z is a partition function. I think that (1.1) is true to all orders, but calculations beyond second order, as usual, are very complicated. It was discovered in [4] from very general arguments that action should have the above form with first term in (1.1) given by some vector field. Thus, this vector field is identified with β-function in our approach. This identification is consistent with the statement of [4], that zeros of vector field are the classical equations of motion (CFT on world-sheet).[2] We will give an alternative way of explaining this statement in the section 3.

---

[1] For the discussion of the same problem in the case of closed string field theory in a different formalizm see [2] and references therein.

[2] The fact that the string field theory action on the classical equations of motion is given by the world-sheet bosonic partition function was previously suspected in [5] [6]

For the reasons that we are dealing with interacting field theory, we are forced to loose the background independence during the calculation of the action in the perturbation theory, so this formalism doesn't achieve the final goal; the approach is also coordinate system dependent. In fact, the latter makes it difficult to reconstruct the final answer globally, once it is computed in the perturbation theory. But, if (1.1) is correct to all orders, formal background independence is preserved. At the same time any approach to write down the expressions for β and Z should appeal to perturbation expansion.

I'll not discuss the important issue of gauge symmetries for (1.1), but as it follows from (1.1), all symmetries of partition function are automatically the symmetries of action S. I think that the general transformation properties could be written using the results of Section 3 in the lines of [1].

The form of (1.1) is quite general one could conjecture that it should be true also for the closed string theory even the analog of [1] (the corresponding space of 2d field theories together with background independent formulation) is not yet known. The expression (1.1) is simple, but the objects that enter, β and Z, usually are impossible to calculate in closed form. As a result this formalism doesn't avoid the usual technical difficulties that are present in any other formulation of string field theory, although it is formally background independent and contains all string modes.

2 Boundary Problem and Open String Field Theory

In the beginning of this section we first will describe the construction of [1] with the emphasis of places where some assumptions are made. The idea of the construction in [1] is based on BV formalism. Let M be a supermanifold which is equipped with closed, nondegenerate odd simplectic structure ω and $U(1)$ symmetry, called ghost number U. This means that in Darboux coordinates ψ, θ on M with ψ fermionic and θ bosonic $\omega = d\psi d\theta$ and ω has ghost number 1. In analogy with ordinary (bosonic) simplectic manifolds one can define the Poisson Brackets, antibracket, with

$$\{A, B\} = \frac{\partial_r A}{\partial t^k} \omega^{kj} \frac{\partial_l B}{\partial t^j}. \tag{2.1}$$

One can show that the following two simple facts take place:

i. If V is a vector field that generates the symmetries of ω, which means that $(di_V + i_V d)\omega = 0$, then there exists a function S that

$$dS = i_V \omega. \tag{2.2}$$

ii. Vector field defined by (2.2) generates a symmetry of ω for any function S.

From the above immediately follows that the Poisson brackets of function S, $\{S, S\}$, defined by (2.2) is annihilated by d and thus the function $\{S, S\}$ is a constant

$$d\{S, S\} = 0. \tag{2.3}$$

The equation

$$\{S, S\} = 0 \tag{2.4}$$

is called the BV master equation and S is the action functional if it solves the master equation. Every solution of (2.4) is automatically gauge invariant [1],[7]. ; the variation of the action under any symplectic transformation

$$\delta t^i = \{t^i, K\}, \tag{2.5}$$

generated by hamiltonian function K (odd function), is given by $\delta S = \{S, K\}$ and for $\{S, K\} = 0$, it vanishes; trivial transformations are given with $K = \{S, \Lambda\}$.

In the quantization of gauge theories the action S is given on subspace of M with $U = 0$, S_0, and we have to find S. That is in fact what the Faddeev-Popov procedure does in the case of Gauge Theories. In the case of String Field Theory, the idea of [1] was to identify the antibracket and vector field in terms of the world-sheet theory and thus identify S as the action of the corresponding target space theory. It was claimed in [1] that this can be done in the case of Open String in a background independent way. The following identifications were proposed:

$$\omega : \omega(\delta O, \delta O) = \int_{\partial \Sigma} d\sigma_1 \int_{\partial \Sigma} d\sigma_2 < \delta O(\sigma_1) \delta O(\sigma_2) > \tag{2.6}$$

$$V : \delta_V O = \{Q, O\} \tag{2.7}$$

where $< ... >$ formally is defined through the world-sheet theory given by path integral corresponding to the 2d action

$$L_a = \int_{\Sigma} d^2 z (\frac{1}{8\pi} g^{\alpha\beta} \partial_\alpha X^i \partial_\beta X^j \eta_{ij} + \frac{1}{2\pi} b^{ij} D_i c_j) + \int_{\partial \Sigma} d\theta V(X, b, c, t) \tag{2.8}$$

Here, the first term is the closed string background and the second term describes an arbitrary boundary interaction, parametrised by coupling constants t^i (in general there are infinite number of coupling constants), with the condition that the boundary operator V has the form

$$V = b_{-1} O, \tag{2.9}$$

with O being a general operator of ghost number 1. Q is a BRST operator defined by BRST current: $Q = \int_C d\sigma J_{BRST}$ with contour C approaching the boundary $\partial \Sigma$ [3] and $b_{-1} = \int_C b(v)$, $b(v) = v^i b_{ij} \epsilon^j_k dx^k$ with v^i being the killing vector that

---

[3] We will not worry about generality and assume that Σ is just a disc.

generates the rotation of disc. This world-sheet action is also equipped with an ultraviolet cutoff a.

From the above identifications we have the definition of string field theory action:

$$dS = \frac{1}{2} \int_0^{2\pi} d\theta_1 d\theta_2 < dO(\theta_1)\{Q, O\}(\theta_2) >, \qquad (2.10)$$

where $d = dt^i d/dt^i$ and $< ... >$ again denotes un-normalized correlation function. Witten has shown that (2.6) gives a closed form and it is invariant under (2.7). We need for future use to repeat his arguments and stress the points where some assumptions are made. [4]

The fact that ω defined by (2.6) is closed follows from the identity:

$$0 = < b_{-1}(A_1(\theta_1)...A_n(\theta_n)) > \qquad (2.11)$$

Here we use the definition of b_{-1} and take two limits: first we shrink the contour C to a point and get zero; second we take the limit when the contour approaches the boundary $\partial\Sigma$ and get the right hand side in (2.11). Thus, in the notation $\delta_i O = \frac{\partial}{\partial t^i} O$ we have:

$$d\omega(\delta_i O, \delta_j O, \delta_k O) = \frac{\partial}{\partial t_k}\omega(\delta_i O, \delta_j O) -$$

$$- \text{cyclic permutations} = < (b_{-1}\delta_k O)\delta_i O \delta_j O > - \text{cyclic perm.} = 0 \qquad (2.12)$$

and the last step follows from (2.11).

BRST invariance of (2.6) is equivalent to exactness of the right hand side in (2.10). This follows from the simple observation that because the transformation law of ω is $\omega' = \omega + \epsilon(i_V d + d i_V)\omega$ and we already have shown that $d\omega = 0$, what we have to show is that $d i_V \omega = 0$. We have

$$d < dO\{Q, O\} > = < (b_{-1}dO)dO\{Q, O\} > -$$

$$- < dO\{dQ, O\} > - < dO\{Q, dO\} > \qquad (2.13)$$

If we use the identity (2.11) for the first term in (2.13) and the definition (2.9) we get:

$$< (b_{-1}dO)dO\{Q, O\} > - < dOb_{-1}dO\{Q, O\} > +$$

$$+ < (dO)^2[L_0, O] > - < (dO)^2[Q, V] > = 0. \qquad (2.14)$$

We are considering a deformation of the critical string, so we can drop all terms of the type $< \{Q, ...\} >_0$, where subscript 0 means the expectation value in the unperturbed theory of some number of operators, using the argument of the

---

[4] One should note that the action defined by (2.10) differs from Zamolodchikov's c-function [8], but like a c-function, it has to have a local minimum at points where world-sheet theory is conformal invariant. Probably (2.10) could be considered as a boundary problem version of c-function.

contour deformation in the definition for Q, and thus in the last term of (2.14) we can integrate by parts in the path integral to obtain $+ < \{Q, (dO)^2\} >$; the same is true for the last term in (2.13), which leads to $+\frac{1}{2} < \{Q, (dO)^2\} >$. The first two terms in (2.14) are equal and contribute as $2 < (b_{-1}dO)dO\{Q, O\} >$. Combining all the terms we get after some cancellations:

$$d < dO\{Q, O\} > = - < dO\{dQ, O\} > +\frac{1}{2} < (dO)^2[L_0, O] > = 0 \qquad (2.15)$$

and we see that ω, defined by (2.6) is BRST invariant only if the right hand side in the equation (2.15) is identically zero.

It was concluded in [1] that (2.6) is BRST invariant because of the following two assumptions:

$$\frac{\partial Q}{\partial t^i} = 0 \qquad (2.16)$$

$$[L_0, O] = \int \frac{\partial}{\partial \theta} O(\theta) d\theta = 0 \qquad (2.17)$$

here, in (2.17), the first identity means that L_0 is a generator of the rotation of circle, and the second identity assumes that total derivatives decouple inside the correlation functions.[5]

Comment: the second correlator in (2.15), the one with total derivative inside the integral, generically is not zero and might receive the contribution from the boundary of moduli space (position of operators on the circle are moduli). Thus, we have to treat such terms and include their contribution. Or, if we want to set up such a scheme during the evaluation of correlation functions in (2.10), when (2.16) is satisfied once inside the correlator, we have to make sure that our regularization scheme leads to decoupling of total derivatives in the second term in (2.15). The latter is a nontrivial statement and in the next section we are going to address this question in detail. The identity in (2.15) should be considered as the requirement for operator Q; so, Q, when the contour approaches the boundary $\partial\Sigma$ should depend on the couplings according to equation (2.15) and this leads to consistency condition on the construction. From the point of view of conformal perturbation theory the above requirement means that we have to use the parallel transport of Q, consistent with (2.15) when we move away from the critical point t_{CFT}. It happens that only the $PSL(2, R)$ subalgebra of Virasoro algebra is relevant, thus what one needs is to deform this subalgebra by including the contributions of boundary term. In the next section we will evaluate the right hand side in (2.10) and formulate this consistency condition in more clear terms for the case when the boundary interaction doesn't mixes ghosts and matter.

At the end of this section as an illustration I would like to discuss a known example of perturbation of a conformal field theory (closed string) by dimension

---

[5] I would like to thank K. K. Li and Erik Verlinde for discussions on importance of (2.16) in [1] (see also [3] and [9]) and E. Witten, who insisted that the whole perturbation theory should be used for proper definition of BRST commutator in (2.10).

one operator, where the decoupling of total derivatives doesn't takes place and the obstruction is a β-function [10]. Similar calculations will be performed in the next section for open strings.

Consider some CFT perturbed by a dimension one operator $V_i t^i$. [6] We will denote the correlation functions in the perturbed theory by $<< ... >>$ and those in the unperturbed theory by $< .. >$; so, the partition function for can be written as $<< 1 >>$, or $< exp(i \int V) >$. One can calculate the trace of stress-tensor in the perturbed theory in the following way. We start from the expectation value of the holomorphic part of stress tensor $<< T(z, \bar{z}) >>$ and use the operator expansion algebra

$$T(z, \bar{z}) V_i(w) = \frac{1}{(z-w)^2} + \frac{1}{z-w} \frac{\partial}{\partial w} V(\omega) + ... = \frac{\partial}{\partial w} (\frac{1}{z-w} V + ...) \quad (2.18)$$

(here we used the fact that V has dimension one in the CFT that we are perturbing; we are speaking about closed string) to deduce that expectation value of the stress tensor is given by total derivatives, integrated over the points where the operator is inserted:

$$<< T(z, \bar{z}) >> =$$
$$= \sum_i < \int d^2 w_i \frac{\partial}{\partial w_i} (\frac{1}{z-w_i} V(w_i) + ...) \sum_n \frac{1}{(n-1)!} (\int V)^{n-1} > . \quad (2.19)$$

If we claim that the contribution of the total derivative in the right hand side is zero, we will conclude that the expectation values of stress tensor is zero; but the latter is wrong – we know that the following relation holds:

$$\frac{\partial}{\partial \bar{z}} << T(z, \bar{z}) >> = \frac{\partial}{\partial z} << trT(z, \bar{z}) >> = \beta^i \frac{\partial}{\partial z} << V_i(z, \bar{z}) >> \quad (2.20)$$

Thus the obstruction for decoupling is the β-function, so we have to calculate the contribution of contact terms. We have to use the operator expansion algebra for V_i to proceed further:

$$V_i(w_1, \bar{w}_1) V_j(w_2, \bar{w}_2) = \frac{g_{ij}}{|w_1 - w_2|^4} + \frac{C_{ij}^k}{|w_1 - w_2|^2} V_k(w_2) + ... \quad (2.21)$$

Now we see that because of the poles in (2.19) and (2.21), there is nonzero boundary contribution in the integral over w_i in (2.19). For this we substitute (2.21) in (2.19). After integration over w_i in (2.19) we are left with the boundary

---

[6] I would like to thank A. Polyakov for pointing out to me the following example.

contour integral, with small contour surrounding each point w_j; denoting $w_i = w_j + \rho e^{i\theta}$ with small ρ, we get:

$$<< T(z, \bar{z}) >>= \sum_{ij} t^i t^j < \int d^2 w_j \int d\theta \rho e^{-i\theta} \frac{1}{z - w_j - \rho e^{i\theta}} (\frac{g_{ij}}{\rho^4} + $$
$$+ \frac{C_{ij}^k}{\rho^2} V_k(w_j) + ...) \sum_n \frac{1}{(n-2)!} (\int V)^{n-2} > . \tag{2.22}$$

If we expand the denominator in (2.22) in the powers of ρ and integrate over θ, we see that only second term contributes, with final answer:

$$<< T(z, \bar{z}) >>= C_{ij}^k t^i t^j \int d^2 w \frac{1}{(z-w)^2} << V_k(w) >> + ... \tag{2.23}$$

Thus, we obtain the desired formula (2.20) in the second order for β-function.

consider simplicity that enters in and replace it is which will be inserted. We

3 Modifications and Perturbative Calculation

In this section we consider the situation when ghosts and matter are decoupled. This means that the boundary interaction V is a functional purely of the matter coordinates X and the operators O are just $O = cV$. The following calculations were largely already presented in [3], but it was assumed that (2.16) and (2.17) are valid assumptions. As a result the calculations have captured only the linear part of β in (1.1) and the conclusion was that the theory has a linear equation of motion. Below I will modify the construction by using the parallel transport of Q for to satisfy the consistency condition (2.15). This procedure is generally ambiguous and I will suggest the possible criteria. I would like to note that there has to be a relation of this procedure with the issues discussed in [11].

On the world-sheet we are dealing with interacting field theory and thus only the way to perform the calculations is to use the perturbation theory. This means that in any calculation we will loose background independence (formally), but the goal is to get final expressions in the invariant terms that don't appeal to a particular background. Thus, we will consider the perturbation theory near a fixed point t_0, which corresponds to some conformal field theory with stress tensor T and corresponding BRST operator Q. When the contour C approaches the boundary of the disc the operator

$$Q = \int d\theta c(\theta) [T_m + \frac{1}{2} T_{gh}]. \tag{3.1}$$

is ambiguous, or it is better to say, it needs to be defined. We have to understand what the correct Q is and make sure that the consistency condition is satisfied.

We could write the left hand side of equation (2.10) in the following form:

$$dS = \frac{1}{2} \int_0^{2\pi} d\theta_1 d\theta_3 << c(\theta_1) dV(\theta_1) c(\theta_3) \partial_{\theta_3} c(\theta_3) V(\theta_3) >>$$

$$+ \frac{1}{2} \int_0^{2\pi} d\theta_1 d\theta_2 d\theta_3 << c(\theta_1) dV(\theta_1) c(\theta_2) c(\theta_3) [T_m(\theta_2), V(\theta_3)] >> . \tag{3.2}$$

The ghost correlation functions in (3.2) is easy to evaluate and it is given by a standard formula, because we consider the case when ghosts and matter are decoupled. For the general 3-point function we have:

$$<< c(\theta_1) c(\theta_2) c(\theta_3) >> = 2(\sin(\theta_1 - \theta_2) - \sin(\theta_1 - \theta_3) + \sin(\theta_2 - \theta_3)). \tag{3.3}$$

Let us treat (3.2) term by term. The expression for the first term is simple:

$$\frac{1}{2} \int_0^{2\pi} d\theta_1 d\theta_2 (2\cos(\theta_1 - \theta_2) - 2) << dV(\theta_1) V(\theta_2) >> =$$

$$= \int d\theta_1 d\theta_2 \cos(\theta_1 - \theta_2) << dV(\theta_1) V(\theta_2) >> \tag{3.4}$$

$$- d << \int d\theta V(\theta) >> + d << 1 >> .$$

If we use the notation $L_n = \int_0^{2\pi} d\theta e^{in\theta} T_m(\theta)$, the second term in (3.2) can be written as a combination of three expressions:

$$\frac{1}{2}[i(- << [L_{-1}, \int d\theta_2 V(\theta_2)] \int d\theta_1 e^{i\theta_1} dV(\theta_1) >> -c.c.)+$$

$$+ i(<< [L_{-1}, \int d\theta_2 e^{i\theta_2} V(\theta_2)] \int d\theta_1 dV(\theta_1) >> -c.c)- \tag{3.5}$$

$$-2 \int d\theta_1 d\theta_2 \sin(\theta_1 - \theta_2) << dV(\theta_1)[L_0, V(\theta_2)] >>].$$

One can simplify (3.5) first noting that:

$$<< \int d\theta_1 e^{i\theta_1} dV(\theta_1)[L_{-1}, \int d\theta_2 V(\theta_2)] >> =$$

$$= < \int d\theta e^{i\theta} dV(\theta)[L_{-1}, exp(iL^{int})] >, \tag{3.6}$$

$$<< \int d\theta_1 dV(\theta_1)[L_{-1}, \int d\theta_2 e^{i\theta_2} V(\theta_2)] >> =$$

$$= < d(exp(iL^{int}))[L_{-1}, \int d\theta e^{i\theta} V(\theta)] > . \tag{3.7}$$

We need two more identities:

$$< d(exp(iL^{int}))[L_{-1}, \int d\theta e^{i\theta} V(\theta)] >= d << [L_{-1}, \int d\theta e^{i\theta} V(\theta)] >> -$$

$$- << [L_{-1}, \int d\theta e^{i\theta} dV(\theta)] >> - << [dL_{-1}, \int d\theta e^{i\theta} V(\theta)] >>, \tag{3.8}$$

and finally for the second term in (3.8)

$$<< [L_{-1}, \int d\theta e^{i\theta} dV(\theta)] >>=$$

$$= - < [L_{-1}, exp(iL^{int})] \int d\theta e^{i\theta} dV(\theta) > + < [L_{-1}, exp(iL^{int})] \int d\theta e^{i\theta} dV(\theta) > . \tag{3.9}$$

Here we used the notation where L^{int} is the boundary interaction term, and $dL_{-1} = dt^i \frac{\partial}{\partial t^i} L_{-1}$, assuming that L_{-1} might depend on the couplings.

Now we see that the difference between (3.6) and (3.7), which enters in (3.5) is given by:

$$<< \int d\theta_1 e^{i\theta_1} dV(\theta_1)[L_{-1}, \int d\theta_2 V(\theta_2)] >> -$$

$$- << \int d\theta_1 dV(\theta_1)[L_{-1}, \int d\theta_2 e^{i\theta_2} V(\theta_2)] >>=$$

$$= d << [L_{-1}, \int d\theta e^{i\theta} V(\theta)] >> - << [dL_{-1}, \int d\theta e^{i\theta} V(\theta)] >> -$$

$$< [L_{-1}, exp(iL^{int}) \int d\theta e^{i\theta} dV(\theta) > . \tag{3.10}$$

The last term in (3.10) could be dropped because it is an expectation value of operator $[L_{-1}, ...]$ in a critical theory, thus we can integrate by parts in path integral and this term is identically zero. So, if we combine (3.2), (3.4) and (3.10) we get:

$$dS = d\frac{i}{2}(<< [L_{-1}, \int d\theta e^{i\theta} V(\theta)] >> -c.c.) + dZ - d << \int d\theta V(\theta) >> +$$

$$+ X, \tag{3.11}$$

where we denote by one form X the following expression:

$$X = \int d\theta_1 d\theta_2 \frac{\partial}{\partial \theta_2} sin(\theta_1 - \theta_2) << dV(\theta_1)V(\theta_2) >> -$$

$$- \int d\theta_1 d\theta_2 sin(\theta_1 - \theta_2) << dV(\theta_1)[L_0 - \frac{\partial}{\partial \theta_2}, V(\theta_2)] >> - \tag{3.12}$$

$$- \frac{i}{2}(<< [dL_{-1}, \int d\theta e^{i\theta} V(\theta)] >> -c.c.).$$

Now I would like to require that under proper definition of renormalization scheme and the generators of $SL(2, R)$ subalgebra, the object X is identically zero. One should note that the consistency condition (2.15) requires that X is just an exact form, thus there is an ambiguity in the definition of Q and equivalently the symmetries of (2.6). Moreover it is not guaranteed that in the deformed theory the stress tensor, that enters in the definition of Q, is deformed accordingly to this requirement. Below I will give arguments that it is indeed the case and that the vanishing of X is a natural choice. They couldn't be considered as a rigorous proof to all orders in t; they are just arguments and most likely they can lead to such a proof. Before I turn to this very important question let us evaluate the first term in (3.12), to be sure that the total derivative doesn't make it zero. We have in the lowest order in t:

$$\int d\theta_1 d\theta_2 \frac{\partial}{\partial \theta_2} sin(\theta_1 - \theta_2) << dV(\theta_1)V(\theta_2) >>=$$

$$= 2dt^i t^j C_{ij}^k \int d\theta_1 \frac{sin a}{sin \frac{1}{2}a^{\Delta_i + \Delta_j - \Delta_k}} << V_k(\theta_1) >> + ... = \qquad (3.13)$$

$$= 4dt^i t^j C_{ij}^k(a) << \int d\theta V_k(\theta) >> + ...,$$

with $c_{ij}^k(a) = c_{ij}^k(a/2)^{1+\Delta_k-\Delta_i-\Delta_j}$. Here we had used the operator expansion algebra for V_i,

$$V_i(\theta_1)V_j(\theta_2) = \frac{c_{ij}^k}{|sin\frac{1}{2}(\theta_1 - \theta_2)|^{\Delta_i + \Delta_j - \Delta_k}} V_k(\theta_1) + ..., \qquad (3.14)$$

with Δ_i being the dimension of V_i in the CFT corresponding to the coupling constant t_0, and integrated over θ_2 from $\theta_1 + a$ to $\theta_1 + 2\pi - a$. The term in c with

$$\Delta_i + \Delta_j - \Delta_k = 1, \qquad (3.15)$$

the "resonance term", is convergent in the limit when we remove the cutoff, and others diverge. These divergent terms can be removed by redefinition of couplings or the same, by a subtraction procedure (see below), while the constant term is universal and can't be removed. Also, there should be a higher order correction in couplings in (3.13).

We see that, like in the example for closed string at the end of the previous section, total derivative in X doesn't decouple and is proportional to β-function coefficient C_{ij}^k; thus (2.17) fails, so does (2.16).

Until now we had avoided the question about the transformation properties of boundary operator. To proceed further it is necessary to know the action of L_{-1}, L_0 and L_1 on V's. What we need to cancel the contribution of (3.13) in X is:

$$\frac{\partial}{\partial t^i} \delta_\epsilon V_j = 4C_{ij}^k V_k \partial \epsilon + ... \qquad (3.16)$$

with $\epsilon = e^{-i\theta}, 1, e^{i\theta}$.

This leads to a suggestion (that has to be verified) that the proper deformation of the action of $SL(2, R)$ algebra is given by:

$$\delta_\epsilon V_i = \epsilon \partial V_i + \gamma_i^j(t) \partial \epsilon V_j, \tag{3.17}$$

and γ is the matrix of anomalous dimensions, which for operators V are simply given by [8]

$$\gamma_i^j(t) = \delta_i^j - \frac{\partial \beta^j}{\partial t^i} = \delta_i^j \Delta_i + 4 c_{ik}^j t^k + ..., \tag{3.18}$$

and β^j is the β-function for operator V_j. This deformation is a very natural one.[7] In fact what we need is (3.17) in the first nontrivial order in couplings to compare with (3.13), because the latter we are able to calculate only up to this order. Here are the arguments in support of (3.17): consider the auotomorphizms of disc:

$$z' = \alpha \frac{z - z_0}{1 - z_0 z}, \tag{3.19}$$

or infinitesimally

$$z' = \alpha(z - z_0 + z_0 z^2) \tag{3.20}$$

with $|z_0| < 1$ and $|\alpha| = 1$. The requirement, that corresponding $PSL(2, R)$ subalgebra of Virasoro algebra is not broken in perturbed theory, fixes the transformation law (3.17). In the case of closed string, the transformation properties under constant shift is standard, and under dilatation is controlled by the Callan-Symanzik equation, that leads to change of classical dimension to anomalous dimension [8]; so these two elements are universal, and the third one is fixed by the requirement, mentioned above. The open string version is given by (3.17) and leads to (3.16). We see from this expression that they are compatible in the lowest order in couplings and guarantee that X vanishes. What is needed to complete the proof is that one has to show the compatibility of (3.17), in particular regularization scheme (note that higher order terms in β-function are scheme dependent), with the vanishing of X in (3.12) and (3.13) to all orders..

Thus, the first term in equation (3.11) defines the string field theory action

$$S = \frac{i}{2}(<< [L_{-1}, \int d\theta e^{i\theta} V(\theta)] >> -c.c.)- << \int d\theta V(\theta) >> + << 1 >>; \tag{3.21}$$

and we know the transformation of boundary operator with respect to $SL(2, R)$ from $X = 0$ or (3.17). We had derived this expression in the second order for

---

[7] There are similarities between (3.17) and the boundary problem analog of anomaly equation (2.20), recently derived by Zamolodchikov [12], see also [13]. I would like to thank A. Zamolodchikov for sharing his insight on the problem of boundary deformations with me.

conformal perturbation theory. At present it is difficult to make any rigorous statement to all orders, but I think that (3.21) should be correct up to all orders. My believe is based on important check which is provided by the expression (3.9); if one calculates the left and right hand sides of this identity using (3.17), or (3.16), he will find out that these transformation laws are consistent with (3.9). For direct proof we need to calculate X in all orders and there are two problems involved: first, it is difficult to perform the calculation beyond second order; second, we have to calculate the parallel transport of Virasoro generators and derive (3.17) in the perturbation theory (or its modification in higher orders) and make sure that X vanishes to all orders. In fact, there are ambiguities related to both calculations, caused by renormalization scheme dependence beyond second order.

Our final expression (3.21) can be written in the form announced in the introduction:

$$S = -\beta^i \frac{\partial}{\partial t^i} Z(t) + Z(t). \tag{3.22}$$

If we remember that structure constants in (3.13) were cutoff dependent and this dependence we removed by a subtraction, we might have kept it up to (3.22). The reason is that as it follows form Poincare-Dulac Theorem[8] about vector fields, every vector field can be linearized by appropriate redefinition of coordinates up to the resonant terms, and the resonant condition is related to the zero modes of linear part $\alpha_1, ..., \alpha_n$. The N-th order term can not be removed by this redefinition if and only if there exists the integer relation of the form:

$$\alpha_s = \sum_1^N m_i \alpha_i \tag{3.23}$$

with $(m_1, ...)$ integers and $m_k \geq 0$, $\sum m_k \geq 2$ and (3.23) is called the resonance relation. The linear term for β^i is given by $(1 - \Delta_i)t^i$, thus the resonance condition in the second order is the one we had written before (3.15): $1 - \Delta_k = 1 - \Delta_i + 1 - \Delta_j$. They correspond to finite terms in (3.13) and can't be removed by coordinate transformation. This in fact proves that the non-vanishing of total derivative term in S is universal.

The expression (3.22) shows that for exactly marginal deformations of base point (which is a particular CFT) action S is the same as partition function and this confirms the statement of [4]. Obviously, assuming that $\beta = 0$ ($\Delta_i = 1$ and $c_{ij}^k = 0$ in the perturbation theory) from the begining and going through our calculations again we get $dS = 0$. Because any attempt to calculate string field theory action in the present approach should use the perturbation theory we can't make any statement about global properties of action. Also, it is difficult to check above statement about equation of motion directly from final expression (3.22) without going to world-sheet and using the identities described in this Section; it would be very interesting to find such procedure. Last important

---

[8] The relevance of Poincare-Dulac theorem to β-function related issues was stressed many times by Zamolodchikov, see [14].

comment related to these questions: we couldn't compare two actions (note that dependance on t's enter through β) if they are calculated in perturbation theory arround different points in space of t's. For this we will need the natural paralel transport in the space of theories that we unfortunately don't have; we only know the deformed relations for $SL(2, R)$ generators in perturbation theory. Thus, the result is not truly background independent even it looks so formally.

Acknowledgments: I would like to thank A. Zamolodchikov, E. Verlinde and E. Witten for very useful discussions.

References

[1] E. Witten, Phys.Rev., D46 (1992) 5446, hep-th - 9208027.

[2] A. Sen, B. Zwiebach, Quantum Background Independence of Closed String Theory, Preprint MIT-CTP-2244, TIFR-TH-93-37, hep-th/9311009.

[3] S. Shatashvili, Phys. Lett. B311 (1993) 83, hep-th - 9303143.

[4] E. Witten, Phys. Rev., D47 (1993) 3405, hep-th - 9210065.

[5] E. Fradkin and A. Tseytlin, Phys. Lett. B 163 (1985) 123.

[6] A. Abouelsaood, C. Callan, C. Nappi and S. Yost, Nucl. Phys. B 280 (1987) 559.

[7] A. Sen and B. Zwiebach, A note on gauge transformations in Batalin-Vilkovisky theory, Preprint MIT-CTP2240, TIFR-TH-93-38, hep-th/93099027.

[8] A. B. Zamolodchikov, Yad. Fiz. 46 (1987) 1819, Sov. J. Nucl. Phys. 46 (1987) 1091.

[9] K. K. Li and E. Witten, Phys. Rev. D, 48 (1993) 853, hep-th - 9303067.

[10] A. Polyakov, Unpublished, A. M. Polyakov, Gauge Fields and Strings, Harwood, 1991.

[11] K. Ranganathan, H. Sonoda, B. Zwiebach, Connection on the state space over Conformal Field Theories, Preprint MIT-CTP-2206, hep-th - 9304053.

[12] A. Zamolodchikov, Unpublished

[13] S. Ghoshal and A. Zamolodchikov, Boundary S-matrix and boundary state in two-dimensional integrable quantum field theory, Rutgers Preprint RU-93-20, hep-th - 9306002.

[14] A. Zamolodchikov, Adv. St. in Pure Math., 19 (1989) 641.

Lectures on Mirror Symmetry

S. Hosono[1], A. Klemm[2] and S. Theisen[3]

[1] Department of Mathemathics, Toyama University
Toyama 930, Japan

[2] Department of Mathematics, Harvard University
Cambridge, MA 02138, U.S.A.

[3] Sektion Physik der Universität München
Theresienstraße 37, D - 80333 München, FRG

Abstract. We give an introduction to mirror symmetry of strings on Calabi-Yau manifolds with an emphasis on its applications e.g. for the computation of Yukawa couplings. We introduce all necessary concepts and tools such as the basics of toric geometry, resolution of singularities, construction of mirror pairs, Picard-Fuchs equations, etc. and illustrate all of this on a non-trivial example.

Contents:

1 Introduction

Almost ten years after the revival of string theory we are still a far cry away from being able to convince the critics of the viability of string theory as a unified description of elementary particles and their interactions, *including* gravity.

A lot of work has been devoted to the construction of consistent classical string vacua and to the study of some of their features, such as their massless spectrum (i.e. gauge group, number of massless generations, their representations with respect to the gauge group, their interactions, etc.). This line of research is still continuing but a complete classification seems to be out of reach.

One particular class of string vacua which has received much attention are compactifications on so-called Calabi-Yau (CY) manifolds. These notes will deal with some aspects of strings on CY manifolds. Our goal is to enable the reader to compute the phenomenologically relevant Yukawa couplings between charged matter fields, which in a realistic model will have to be identified with quarks, leptons and Higgs fields. Even though to date there does not exist a CY model with all the qualitative features of the standard model, it is still useful to develop techniques to do explicit computations in the general case. This is where mirror symmetry enters the stage in that it allows for the (explicit) computation of a class of Yukawa couplings which are otherwise very hard (if not impossible) to obtain.

As it is the purpose of these notes to give a pedagogical introduction to mirror symmetry and its applications, we like to review in the introduction some general concepts of string theory in view of mirror symmetry. Throughout the text we have to assume some familiarity with strings theory and conformal field theory [1].

Some basic properties of closed string theory are best discussed in the geometrical approach, i.e. by looking at the classical σ-model action. It is defined by a map Φ from a compact Riemann surface Σ_g of genus g (the world-sheet with metric $h_{\alpha\beta}$) to the target space X (the space-time) $\Phi : \Sigma_g \to X$ and an action $S(\Phi, G, B)$, which may be viewed as the action of a two dimensional field theory. The latter depends on the dynamical field Φ, whereas the metric G of X and an antisymmetric tensorfield B on X are treated as background fields. As the simplest example one may take a bosonic action, which reads

$$S = \frac{1}{2\pi\alpha'} \int_{\Sigma_g} d^2\sigma \sqrt{h} \Big(h^{\alpha\beta} G_{ij}(\phi) \partial_\alpha \phi^i \partial_\beta \phi^j + \epsilon^{\alpha\beta} B_{ij}(\phi) \partial_\alpha \phi^i \partial_\beta \phi^j + \ldots \Big) \quad (1.1)$$

where ϕ^i $(i = 1, \ldots, \dim(X))$ and σ^α $(\alpha = 1, 2)$ are local coordinates on Σ_g and X respectively. The dots indicate further terms, describing the coupling to other background fields such as the dilaton and gauge fields. The first quantized string theory can then be perturbatively defined in terms of a path integral as[1]

$$S(X) = \sum_g \int_{\mathcal{M}_g} \int D\Phi e^{iS(\Phi, G, B, \ldots)} \quad (1.2)$$

---

[1] We always assume that we quantize in the critical dimension. Integration over the world-sheet metric can then be converted to an integration over the moduli space $\mathcal{M}_g$ with suitable measure which, for general correlation functions, depends on the number of operator insertions [2]. By $S(X)$ we mean the generating function of all correlators of the string theory on X.

For a particular background to provide a classical string vacuum, the sigma model based on it has to be conformally invariant [3]. This means that the energy-momentum tensor, including corrections from σ-model loops, must be traceless, or, equivalently, that the β-functions must vanish. Vanishing of the dilaton β-function demands that we quantize in the critical dimension, whereas the β-functions associated to the metric and the anti-symmetric tensor impose dynamical equations for the background, in particular that is has (to lowest order in α') to be Ricci flat, i.e. that the metric satisfy the vacuum Einstein equations[2].

Since we are dealing with strings, it is not the classical geometry (or even topology) of X which is relevant. In fact, path integrals such as (1.2) are related to the loop space of X. Much of the attraction of string theory relies on the hope that the modification of the concept of classical geometry to string geometry at very small scales will lead to interesting effects and eventually to an understanding of physics in this range. At scales large compared to the scale of the loops (which is related to α') a description in terms of point particles should be valid and one should recover classical geometry. The limit in which the classical description is valid is referred to as the large radius limit.

One particular property of strings as compared to ordinary point particles is that there might be more than one manifold X which leads to identical theories; i.e. $\mathcal{S}(X_1) = \mathcal{S}(X_2)$[3]. The case where two manifolds have just different *geometry* is usually referred to as duality symmetry [5]. Mirror symmetry [6] [7] [8] [9]. relates, in the generic case, identical string theories on *topologically* different manifolds. These symmetries are characteristic features of string geometry. For the case of mirror symmetry, which is the central topic of these lectures, this will become evident as we go along.

So far the analysis of (parts of) $\mathcal{S}(X)$ can be explicitly performed only for very simple target spaces X, such as the torus and orbifolds. Much of our understanding about the relation between classical and string geometry is derived from these examples. As a simple example we want to discuss compactification of the bosonic string on a two dimensional torus [10]; for review, see also [11]. This also allows us to introduce some concepts which will appear in greater generality later on.

The torus $T^2 = \mathbb{R}^2/\Gamma$ is defined by a two-dimensional lattice Γ which is generated by two basis vectors e_1 and e_2. The metric, defined by $G_{ij} = e_i \cdot e_j$, has three independent components and the antisymmetric tensor $B_{ij} = be_{ij}$ has one component. For any values of the altogether four real components does

---

[2] In fact, this point is subtle, as for X a Calabi-Yau manifold (cf. section two), higher (≥ 4) σ-model loop effects modify the equations of motion for the background. It can however be shown [4] that σ-models for Calabi-Yau compactifications are conformally invariant and that by means of a (non-local) redefinition of the metric one can always obtain $R_{ij} = 0$ as the equations of motion for the background.

[3] This resembles the situation of quantized point particles on so called isospectral manifolds. However in string theory the invariance is more fundamental, as no experiment can be performed to distinguish between X_1 and X_2.

one get a consistent string compactification. We thus have four real moduli for strings compactified on T^2. We can combine them into two complex moduli as follows: $\sigma = \frac{|e_1|}{|e_2|}e^{i\phi}$ and $\tau = 2(b + iA)$ where ϕ is the angle between the two basis vectors which we can, without loss of generality, choose to be $0 \le \phi \le \pi$ and $A = \sqrt{|\det G|} > 0$ is the area of the unit cell of Γ. σ parameterizes different complex structures on the torus and is usually called the Teichmüller parameter. The imaginary part of τ parameterizes the Kähler structure of the torus. We have used the antisymmetric tensor field to complexify the Kähler modulus. The role the two moduli play is easily recognized if one considers $ds^2 = G_{ij}dx^i dx^j \equiv \tau_2 dz d\bar{z}$ where the relation between the real coordinates x_i and the complex coordinates $z, \bar{z}$ only involves σ but not τ.

If one now considers the spectrum of the theory, one finds various symmetries. They restrict the moduli space of the compactification which is naively just two copies of the upper half complex plane. With the definition of the left and right momenta

$$p_L^2 = \frac{1}{\sigma_2\tau_2}|(m_1 - \sigma m_2) - \tau(n_2 + \sigma n_1)|^2, \quad p_R^2 = \frac{1}{\sigma_2\tau_2}|(m_1 - \sigma m_2) - \bar{\tau}(n_2 + \sigma n_1)|^2$$

the spectrum of the energy and conformal spin eigenvalues can be written as

$$m^2 = p_L^2 + p_R^2 + N_L + N_R - 2, \quad s = p_L^2 - p_R^2 + N_L - N_R \qquad (1.3)$$

where n_i and m_i are winding and momentum quantum numbers, respectively, $N_{L,R}$ are integer oscillator contributions and the last term in m^2 is from the zero point energy. The symmetries of the theory are due to invariance of (1.3) under the transformation $(\sigma, \tau) \mapsto (\tau, \sigma)$, $(\sigma, \tau) \mapsto (-\bar{\sigma}, -\bar{\tau})$, $(\sigma, \tau) \mapsto (\sigma + 1, \tau)$ and $(\sigma, \tau) \mapsto (-\frac{1}{\sigma}, \tau)$ accompanied by a relabeling and/or interchange of the winding and momentum quantum numbers. The transformation which interchanges the two types of moduli generates in fact what we call mirror symmetry. The torus example is however too simple to exhibit a change of topology as it is its own mirror. The transformations which reflect the string property are those which require an interchange of momentum and winding modes. The last three of the generators given above are not of this type and they are also present for the point particle moving on a torus (then $n_1 = n_2 = 0$). It is the addition of the mirror symmetry generator which introduces the stringy behavior. Interchange of the two moduli must be accompanied by $n_2 \leftrightarrow m_2$. Composing the mirror transformation with some of the other generators given above, always involves interchanges of winding and momentum quantum numbers. E.g. for the transformation $\tau \to -\frac{1}{\tau}$ we have to redefine $m_1 \to n_2, m_2 \to -n_1, n_2 \to -m_1, n_1 \to m_2$ and if we set $b = 0$ then this transformation identifies compactification on a torus of size A to compactification on a torus of size $1/A$. Integer shifts of τ are discrete Peccei-Quinn symmetries. One can show that the interactions (correlation functions) are also invariant under these symmetries.

These lectures deal with mirror symmetry of strings compactified on Calabi-Yau spaces. In section two we will review some of the main features of Calabi-Yau compactifications, in particular the correspondence of complex structure and Kähler moduli with elements of the cohomology groups $H^{2,1}$ and $H^{1,1}$, respectively. For X and X^* to be a mirror pair of Calabi-Yau manifolds (we will use this notation throughout) one needs that $h^{p,q}(X) = h^{3-p,q}(X^*)$ (for three dimensional Calabi-Yau manifolds this is only non-trivial for $p, q = 1, 2$). The *mirror hypothesis* is however much more powerful since it states that the string theory on X and X^* are identical, i.e. $\mathcal{S}(X) = \mathcal{S}(X^*)$. In particular it implies that one type of couplings on X can be interpreted as another type of couplings on X^* after exchanging the role of the complex structure and the Kähler moduli.

In section three we give a description of Calabi-Yau compactification in terms of symmetric $(2,2)$ superconformal field theory. The moduli of the Calabi-Yau space correspond to exactly marginal deformations of the conformal field theory. They come in two classes. Mirror symmetry appears as a trivial statement, namely as the change of relative sign, which is pure convention, of two $U(1)$ charges [6][7]. By this change the two classes of marginal perturbations get interchanged. This does not change the conformal field theory and thus leads to the same string vacuum. In the geometrical interpretation this is however non-trivial, as it entails the mirror hypothesis which implies the existence of pairs of topologically different manifolds with identical string propagation.

We will apply mirror symmetry to the computation of Yukawas couplings of charged matter fields. They come in two types, one easy to compute and the other hard to compute. On the mirror manifold, these two couplings change role. What one then does is to compute the easy ones on either manifold and then map them to one and the same manifold via the so-called mirror map. In this way one obtains both types of couplings. This will be explained in detail in section six. Before getting there we will show how to construct mirror pairs and how to compute the easy Yukawa couplings. This will be done in sections four and five. In section seven we present an example in detail, where the concepts introduced before will be applied. In the final section we draw some conclusions.

Before continuing to section two, let us give a brief guide to the literature. The first application of mirror symmetry was given in the paper by Candelas, de la Ossa, Green and Parkes [12] where the simplest Calabi-Yau manifold was treated, the quintic in $\mathbb{P}^4$ which has only one Kähler modulus. Other one-moduli examples were covered in [13] (for hypersurfaces) and in [14] (for complete intersections). Models with several moduli were examined in refs. [15] [16] (two and three moduli models) and [17]. Other references, especially to the mathematical literature, will be given as we go along. A collection of papers devoted to various aspects of mirror symmetry is [9]. Some of the topics and results to be discussed here are also contained in [15][18] [19] [20]. These notes draw however most heavily from our own papers on the subject.

2   Strings on Calabi-Yau Manifolds

One of the basic facts of string theory is the existence of a critical dimension, which for the heterotic string, is ten. To reconciliate this with the observed four-dimensionality of space-time, one makes the compactification Ansatz that the ten-dimensional space-time through which the string moves has the direct product form $X_{10} = X_4 \times X_6$ where X_6 is a six-dimensional compact internal manifold, which is supposed to be small, and X_4 is four-dimensional Minkowski space. If one imposes the 'phenomenologically' motivated condition that the theory has $N = 1$ supersymmetry in the four uncompactified dimensions, it was shown in [21] that X_6 has to be a so-called Calabi-Yau manifold [22] [23].

Def.[4]: *A Calabi-Yau manifold is a compact Kähler manifold with trivial first Chern class.*

The condition of trivial first Chern class on a compact Kähler manifold is, by Yau's theorem, equivalent to the statement that they admit a Ricci flat Kähler metric. The necessity is easy to see, since the first Chern class $c_1(X)$ is represented by the 2-form $\frac{1}{2\pi}\rho$ where ρ is the Ricci two form, which is the 2-form associated to the Ricci tensor of the Kähler metric: $\rho = R_{i\bar{\jmath}}dz^i \wedge d\bar{z}^{\bar{\jmath}}$. Locally, it is given by $\rho = -i\partial\bar{\partial}\log\det((g_{i\bar{\jmath}}))$. One of the basic properties of Chern classes is their independence of the choice of Kähler metric; i.e. $\rho(g') = \rho(g) + d\alpha$. If now $\rho(g) = 0$, $c_1(X)$ has to be trivial. That this is also sufficient was conjectured by Calabi and proved by Yau [24].

Ricci flatness also implies that the holonomy group is contained in $SU(3)$ (rather than $U(3)$; the $U(1)$ part is generated by the Ricci tensor $R_{i\bar{\jmath}} = R_{i\bar{\jmath}k}{}^k$). If the holonomy is $SU(3)$ one has precisely $N = 1$ space-time supersymmetry. This is what we will assume in the following. (This condition e.g. excludes the six-dimensional torus T^6, or $K_3 \times T^2$, which would lead to extended space-time supersymmetries.)

Another consequence of the CY conditions is the existence of a unique nowhere vanishing covariantly constant holomorphic three form, which we will denote by $\Omega = \Omega_{ijk}dz^i \wedge dz^j \wedge dz^k$ $(i, j, k = 1, 2, 3)$, where z^i are local complex coordinates of the CY space. Since Ω is a section of the canonical bundle[5], vanishing of the first Chern class is equivalent to the triviality of the canonical bundle.

A choice of complex coordinates defines a complex structure. The transition functions on overlaps of coordinate patches are holomorphic functions. There are in general families of possible complex structures on a given CY manifold. They are parameterized by the so-called complex structure moduli. Using Kodaira-Spencer defomation theory [25], it was shown in [26] that for Calabi-Yau manifolds this parameter space is locally isomorphic to an open set in $H^1(X, T_X)$. For algebraic varieties the deformation along elements of $H^1(X, T_X)$ can often be described by deformations of the defining polynomials (cf. section four).

---

[4] Here and below we restrict ourselves to the three complex-dimensional case.

[5] The canonical bundle is the highest (degree $\dim_{\mathbb{C}}(X)$) exterior power of the holomorphic cotangent bundle T_X^*.

In addition to the complex structure moduli there are also the Kähler moduli. They parameterize the possible Kähler forms. A Kähler form is a real closed $(1,1)$ form $J = \omega_{i\bar{j}}dz^i \wedge d\bar{z}^{\bar{j}}$ (with the associated Kähler metric $g_{i\bar{j}} = i\omega_{i\bar{j}}$) which satisfies the positivity conditions

$$\int_C J > 0, \quad \int_S J^2 > 0, \quad \int_X J^3 > 0 \qquad (2.1)$$

for all curves C and surfaces S on the CY manifold X. Since $\frac{1}{3!}J^3$ is the volume form on X, one concludes that the Kähler form cannot be exact and consequently one has $\dim(H^{1,1}(X)) \equiv h^{1,1} \geq 1$ for X Kähler. If there are more than one harmonic $(1,1)$ forms on X, i.e. if $h^{1,1} > 1$, then $\sum_{a=1}^{h^{1,1}} t'_a h_a$, $t' \in \mathbb{R}$, with $h_a \in H_{\bar{\partial}}^{1,1}(X)$ will define a Kähler class, provided the Kähler moduli lie within the so-called Kähler cone, i.e. (2.1) is satisfied.

From the local expression of the Ricci form it follows that it depends on the complex structure and on the volume form of the Kähler metric. The question now arises whether by changing the Kähler form and the complex structure Ricci flatness is preserved. This means that the moduli of the CY manifold must be associated with deformations of the *Ricci flat* Kähler metric: $\delta g_{i\bar{j}}$ with Kähler deformations and δg_{ij} and $\delta g_{\bar{i}\bar{j}}$ with deformations of the complex structure. If one examines the condition $\rho(g + \delta g) = 0$ one finds that $\delta g_{i\bar{j}}dz^i \wedge d\bar{z}^{\bar{j}}$ is harmonic, i.e. we can expand it as $\delta g_{i\bar{j}}dz^i \wedge d\bar{z}^{\bar{j}} = \sum_{a=1}^{h^{1,1}} \delta t'_a h_a$. Likewise, $\Omega_{ij}{}^{\bar{l}}\delta g_{\bar{l}\bar{k}}dz^i \wedge dz^j \wedge d\bar{z}^{\bar{k}} = \sum_{\alpha=1}^{h^{2,1}} \delta\lambda_\alpha b_\alpha$, with $b_a \in H^{2,1}(X)$. Here we have employed the unique Ω.

One can show that the only independent non trivial Hodge numbers of CY manifolds are $h^{0,0} = h^{3,0} = 1$ and $h^{1,1}$ and $h^{2,1}$ depending on the particular manifold. In addition we have $h^{p,q} = h^{q,p}$ (complex conjugation), $h^{p,q} = h^{3-p,3-q}$ (Poincaré duality) and $h^{0,p} = h^{0,3-p}$ (isomorphism via Ω). For Kähler manifolds the Betti numbers are $b_r = \sum_{p+q=r} h^{p,q}$ and the Euler number is thus $\chi(X) = \sum_{r=0}^{\dim(X)} b_r$.

If we were geometers we would only be interested in the deformations of the *metric* and the number of (real) moduli would be $h^{1,1} + 2h^{2,1}$. However, in string theory compactified on CY manifolds we have additional massless scalar degrees of freedom from the non-gauge sector, namely those coming from the (internal components of the) antisymmetric tensor field $B_{i\bar{j}}$. As a result of the equations of motion it is a harmonic $(1,1)$ form and its changes can thus be parameterized as $\delta B_{i\bar{j}}dz^i \wedge d\bar{z}^{\bar{j}} = \sum_{a=1}^{h^{1,1}} \delta t''_a h_a$ where t''_a are real parameters. One combines $(i\delta g_{i\bar{j}} + \delta B_{i\bar{j}})dz^i \wedge dz^{\bar{j}} = \sum_{a=1}^{h^{1,1}} \delta t_a h_a$ where now the $t_a = t''_a + it'_a$ are complex parameters. This is referred to as the complexification of the Kähler cone.

Recall (and see below) that for strings on CY manifolds, the massless sector of the theory consists of a universal sector, containing the graviton, an antisymmetric tensor field (by duality related to the axion) and a dilaton, and a matter sector with $h^{1,1}$ 27-plets and $h^{2,1}$ $\overline{27}$-plets of E_6 and a certain number of E_6 singlets. E_6 invariance restricts the possible Yukawa couplings to the following four types: $\langle 27^3 \rangle$, $\langle \overline{27}^3 \rangle$, $\langle 27 \cdot \overline{27} \cdot 1 \rangle$ and $\langle 1^3 \rangle$. In the following we will only treat the former two couplings [27]. Not much is known about the remaining two[6].

One considers the coupling of two fermionic and one bosonic field (Yukawa coupling) in the ten-dimensional field theory. All these fields are in the fundamental (248) representation of E_8, the gauge group of the uncompactified heterotic string[7]. (The bosonic field is the E_8 gauge field.) One then expands the fields in harmonics on the internal CY manifold and arrives at couplings which factorize into two terms: one is a cubic coupling of three fields on the four-dimensional Minkowski space and the other an overlap integral over the internal manifold of three zero-modes (we are interested in massless fields) of the Dirac and the wave operators, respectively. The second factor is the effective Yukawa coupling of the four-dimensional field theory. Under $E_8 \supset E_6 \times SU(3)$ we have the decomposition $248 = (27,3) \oplus (\overline{27}, \overline{3}) \oplus (1,8) \oplus (78,1)$. The $(78,1)$ gives the E_6 gauge fields and the $(1,8)$ is the spin connection which has been identified with the $SU(3)$ part of the E_8 gauge connection [21].

The four-dimensional matter fields transform as 27 and $\overline{27}$ of E_6 and the zero modes in the internal CY manifold carry a $SU(3)$ index in the 3 and $\overline{3}$ representations, respectively. Group theory then tells us that there are two different kinds of Yukawa couplings among the charged matter fields: $\langle 27^3 \rangle$ and $\langle \overline{27}^3 \rangle$. The zero modes on the internal manifold can be related to cohomology elements of the CY space and one finds [27] that the two types of Yukawa couplings are of the form:

$$\kappa_{abc}^{0(27)}(X) \equiv \kappa_{abc}^0(X) = \int_X h_a \wedge h_b \wedge h_c, \qquad a,b,c = 1,\ldots,h^{1,1}$$

$$\kappa_{\alpha\beta\gamma}^{(\overline{27})}(X) \equiv \bar{\kappa}_{\alpha\beta\gamma}(X) = \int_X \Omega \wedge b_\alpha^i \wedge b_\beta^j \wedge b_\gamma^k \, \Omega_{ijk}, \qquad \alpha,\beta,\gamma = 1,\ldots,h^{2,1}$$

$$(2.2)$$

where h_a are the harmonic $(1,1)$ forms and $b_\alpha^i = (b_\alpha)_{\bar{j}}^i d\bar{z}^{\bar{j}}$ are elements of $H^1(X,T_X)$ which are related to the harmonic $(2,1)$ forms via the unique element of $H^3(X)$: $(b_\alpha)_{\bar{j}}^i = \frac{1}{2\|\Omega\|^2} \Omega^{ikl}(b_\alpha)_{kl\bar{j}}$, $\|\Omega\|^2 = \frac{1}{3!} \bar{\Omega}^{ijk}\Omega_{ijk}$. Note that while the former couplings are purely topological the latter do depend on the complex structure (through Ω). Both types of cubic couplings are totally symmetric. Note also that by the discussion above there is a one-to-one correspondence between charged matter fields and moduli: $27 \leftrightarrow (1,1)$ moduli and $\overline{27} \leftrightarrow (2,1)$ moduli[8].

---

[6] except for cases in which the corresponding conformal field theory can be treated exactly, e.g. for Calabi-Yau spaces with an toroidal orbifold limit they can be calculated for the untwisted sector at the orbifold point [28] and for Calabi-Yau spaces with Gepner [29] model interpretation at one point (the Gepner point) in moduli space.

[7] We do not consider the second E_8 factor here. It belongs to the so-called hidden sector.

[8] This identification is a matter of convention. Here we have identified the $\overline{3}$ of $SU(3)$ with a holomorphic tangent vector index. The 3 of $SU(3)$ is a holomorphic cotangent vector index and one uses (Dolbeault theorem) $H^1(X,T_X^*) \simeq H^{1,1}(X)$.

These results for the couplings have been derived in the (classical) field theory limit and do not yet incorporate the extended nature of strings. This issue will be taken up next.

In general, to compute the string Yukawa couplings, one has to take into account sigma model and string perturbative and non-perturbative effects. One can show that both types of Yukawa couplings do not receive corrections from sigma model loops and string loops. The couplings $\bar{\kappa}$ are, in fact, also unmodified by non-perturbative effects on the world-sheet, which, due to absence of string loop corrections, is just the sphere [30]. The couplings κ^0 do however receive corrections from world-sheet instantons [31]. These are non-trivial holomorphic embeddings of the world-sheet $\Sigma_0 \simeq \mathbb{P}^1$ in the CY manifold. In algebraic geometry they are known as rational curves C on X. Then there are still non-pertubative string-effects, i.e. possible contributions from infinite genus world-sheets. We do not know anything about them and will ignore them here. The possibility to incorporate them in the low-energy effective action has been discussed in [32].

The couplings in eq.(2.2) are computable in classical algebraic geometry, and, were they the whole truth to the Yukawa couplings of strings on CY manifolds, they would blatantly contradict the mirror hypothesis: κ^0 is independent of moduli whereas $\bar{\kappa}$ depends on the complex structure moduli. In fact, the mirror hypothesis states that the full $\langle 27^3 \rangle$ couplings on the manifold X depend on the Kähler moduli in such a way that they are related to the $\langle \overline{27}^3 \rangle$ couplings on the mirror manifold X^* via the mirror map. The main topic of these lectures is to explain what this means and to provide the tools to carry it through. The dependence on the Kähler moduli is the manifestation of string geometry and is solely due to the extended nature of strings.

When computing the Yukawa-couplings in conformal field theory as correlation functions of the appropriate vertex operators, inclusion of the nonperturbative σ-model effects means that in the path-integral one has to sum over all holomorphic embeddings of the sphere in X. This is in general not feasible since it requires complete knowledge of all possible instantons and their moduli spaces. In fact, this is where mirror symmetry comes to help.

We have thus seen that, modulo the remark on non-perturbative string effects, the Yukawa couplings are $\bar{\kappa}(X)$ and for the instanton corrected couplings $\kappa(X)$ one expects an expansion of the form $(\imath : C \hookrightarrow X)$

$$\kappa_{abc} = \kappa^0_{abc} + \sum_C \int_C \imath^*(h_a) \int_C \imath^*(h_b) \int_C \imath^*(h_c) \frac{e^{2\pi i \int_C \imath^*(J(X))}}{1 - e^{2\pi i \int_C \imath^*(J(X))}}, \qquad (2.3)$$

which generalizes the Ansatz made in [12], which led to a successfull prediction for the numbers of instantons on the quintic hypersurface in $\mathbb{P}^4$, to the multimoduli case. This ansatz was justified in ref. [33] in the framework of topological sigma models [34]. The sum is over all instantons C of the sigma model based on X and the denominator takes care of their multiple covers. J is the Kähler form on X.

One sees from (2.3) that as we go 'far out' in the Kähler cone to the 'large radius limit', the instanton corrections get exponentially supressed and one recovers the classical result.

3 Superconformal Field Theory and CY Compactification

Let us now turn to an alternative view of string compactification. We recall that the existence of a critical dimension is due to the requirement that the total central charge (matter plus ghost) of the Virasoro algebra vanishes. The critical dimensions for the bosonic and fermionic strings then follow from the central charges of the Virasoro algebras generated by the reparametrization and local $n = 1$ world-sheet supersymmetry ghost systems, which are $c = -26$ and $\hat{c} = \frac{2}{3}c = -10$, respectively. If we want a four-dimensional Minkowski space-time, we need four left-moving free bosonic fields and four right-moving free chiral superfields, contributing $(\bar{c}, c) = (4, 6)$ to the central charge. (Barred quantities refer to the left-moving sector.) Compactification might then be considered as an internal conformal field theory with central charge $(\bar{c}_{int}, c_{int}) = (22, 9)$. There are however internal conformal field theories which satisfy all consistency requirements (e.g. modular invariance, absence of dilaton tadpoles, etc.) which do not allow for a geometric interpretation as compactification. In the case of CY compactification, the internal conformal field theory is of a special type. The left moving central charge splits into a sum of two contributions, $\bar{c}_{int} = 22 = 13 + 9$, where the first part is due to a $E_8 \times SO(10)$ gauge sector (at level one; $E_8 \times SO(10)$ is simply laced of rank 13). The remaining contribution combines with the right-moving part to a symmetric (i.e. the same for both left and right movers) $(\bar{n}, n) = (2, 2)$ superconformal field theory with central charges $(9, 9)$. A right moving (global) $n = 2$ extended algebra is necessary for space-time supersymmetry [35], whereas the symmetry between left and right movers are additional inputs which allow for the geometrical interpretation as CY compactification with the spin connection embedded in the gauge connection [21].

The fact that we have a left as well as a right moving extended superconformal symmetry will be crucial for mirror symmetry. Before explaining this, let us briefly mention the relevant features of the $n = 2$ superconformal algebra [36]. It has four generators, the bosonic spin two energy momentum tensor T, two fermionic spin $3/2$ super-currents $T_F^\pm$ and a bosonic spin one current J which generates a $U(1)$ Kac-Moody algebra. If we expand the fields in modes as $T(z) = \sum L_n z^{-n-2}$, $T_F^\pm(z) = \sum G_r^\pm z^{-r-3/2}$ and $J(z) = \sum J_n z^{-n-1}$ the algebra takes the form $(G_r^+ = (G_{-r}^-)^\dagger)$

$$[L_n, L_m] = (n - m)L_{n+m} + \frac{c}{12}(n^3 - n)\delta_{n+m,0}$$

$$\{G_r^\pm, G_s^\mp\} = 2L_{r+s} \pm (r - s)J_{r+s} + \frac{c}{3}(r^2 - \frac{1}{4})\delta_{r+s,0}, \quad \{G_r^\pm, G_s^\pm\} = 0$$

$$[L_n, G_r^\pm] = (\frac{n}{2} - r)G_{n+r}^\pm, \quad [L_n, J_m] = -mJ_{n+m} \tag{3.1}$$

$$[J_n, J_m] = \frac{c}{3}n\delta_{m+n,0}, \quad [J_n, G_r^\pm] = \pm G_{n+r}^\pm$$

The moding of the fermionic generators is $r \in \mathbf{Z}$ in the Ramond (R) and $r \in \mathbf{Z}+\frac{1}{2}$ in the Neveu-Schwarz (NS) sector. The finite dimensional subalgebra in the NS

sector, generated by $L_{0,\pm1}$, J_0 and $G_{\pm1/2}^{\pm}$ is $OSp(2|2)$. In a unitary theory we need

$$\langle\phi|\{G_r^{\pm}, G_{-r}^{\mp}\}|\phi\rangle = 2h \pm 2rq + \frac{c}{3}(r^2 - \frac{1}{4}) \geq 0 \qquad (3.2)$$

for any state with $U(1)$ charge q (i.e. $J_0|\phi\rangle = q|\phi\rangle$) and conformal weight h (i.e. $L_0|\phi\rangle = h|\phi\rangle$). Setting $r = 0$ we thus find that in the R sector $h \geq \frac{c}{24}$ and in the NS sector (setting $r = \frac{1}{2}$) $h \geq \frac{|q|}{2}$.

There is a one-parameter isomorphism of the algebra, generated by $\mathcal{U}_\theta$, called spectral flow [37]

$$\mathcal{U}_\theta L_n \mathcal{U}_\theta^{-1} = L_n + \theta J_n + \frac{c}{6}\theta^2 \delta_{n,0} \quad \rightarrow \Delta h = \frac{3}{2c}\Delta(q^2)$$

$$\mathcal{U}_\theta J_n \mathcal{U}_\theta^{-1} = J_n + \frac{c}{3}\theta\delta_{n,0} \quad \rightarrow \Delta q = +\frac{c}{3}\theta \qquad (3.3)$$

$$\mathcal{U}_\theta G_r^{\pm} \mathcal{U}_\theta^{-1} = G_{r\pm\theta}^{\pm}$$

States transform as $|\phi\rangle \rightarrow \mathcal{U}_\theta|\phi\rangle$. For $\theta \in \mathbf{Z} + \frac{1}{2}$ the spectral flow interpolates between the R and the NS sectors and for $\theta \in \mathbf{Z}$ it acts diagonally on the two sectors.

We have two commuting copies of the $n = 2$ algebra. The left moving $U(1)$ current combines with the $SO(10)$ Kac-Moody algebra to form an E_6 algebra at level one. Hence the gauge group E_6 for CY compactifications. (The E_8 factor, which is also present, will play no role here.)

Let us first consider the R sector[9]. Ramond ground states $|i\rangle_R$ satisfy $G_0^{\pm}|i\rangle_R = 0$, i.e. $\{G_0^+, G_0^-\}|i\rangle_R = 0$. From (3.2) it follows that R ground states have conformal weight $h = \frac{c}{24}$. Under spectral flow by $\theta = \mp\frac{1}{2}$, the R ground states flow into *chiral/anti-chiral* primary states of the NS sector. They are primary states that satisfy the additional constraint $G_{-1/2}^{\pm}|i\rangle_{NS} = 0$. (Recall that primary states are annihilated by all positive modes of all generators of the algebra.) It follows from (3.3) that chiral/anti-chiral primary states satisfy $h = \pm\frac{1}{2}q$. The $OSp(2|2)$ invariant NS vacuum $|0\rangle$ is obviously chiral and anti-chiral primary. Under spectral flow by $\theta = \pm1$ it flows to a unique chiral (anti-chiral) primary field $|\rho\rangle$ ($|\bar\rho\rangle$) with $h = \frac{c}{6}$ and $q = \pm\frac{c}{3}$. It follows from (3.2) that for chiral primary fields $h \leq \frac{c}{6}$.

We now look at the operator product of two chiral primary fields $\phi_i(z)\phi_j(w) \sim \sum_k (z-w)^{h_k - h_i - h_j} \psi_k(w)$ where the ψ_k are necessarily chiral but not necessarily primary. $U(1)$ charge conservation requires $q_k = q_i + q_j$ and due to the inequality $h \geq \frac{|q|}{2}$ with equality for primary fields we conclude that in the limit $z \rightarrow w$ only the chiral primaries survive. They can thus be multiplied pointwise and therefore form a ring $\mathcal{R}$ under multiplication: $\phi_i\phi_j = \sum_k c_{ij}{}^k \phi_k$ where the structure constants are functions of the moduli (cf. below). This ring is called chiral primary ring. The same obviously holds for anti-chiral fields forming the anti-chiral ring.

---

[9] The following paragraphs draw heavily from ref.[7].

Note that spectral flow by θ is merely a shift of the $U(1)$ charge by $\frac{c}{3}\theta$ and the accompanying change in the conformal weight. Indeed, in terms of a canonically normalized boson $(\phi(z)\phi(w) = -\ln(z-w) + \dots)$ we can express the $U(1)$ current as $J(z) = \sqrt{\frac{c}{3}}\partial\phi(z)$. Any field[10] with $U(1)$ charge q can be written as $\phi_q = e^{i\sqrt{\frac{3}{c}}q\phi}\mathcal{O}$ where $\mathcal{O}$ is neutral under $U(1)$. The conformal weight of ϕ_q is $\frac{3}{2c}q^2 + h_{\mathcal{O}}$ and of $\phi_{q+\frac{c}{3}\theta}$ it is $\frac{3}{2c}(q+\frac{c}{3}\theta)^2 + h_{\mathcal{O}}$, in agreement with (3.3). We thus find that the spectral flow operator $\mathcal{U}_\theta$ can be written as $\mathcal{U}_\theta = e^{i\theta\sqrt{\frac{c}{3}}\phi}$ and also $\rho(z) = \mathcal{U}_1(z) = e^{i\sqrt{\frac{c}{3}}\phi(z)}$.

The foregoing discussion of course applies separately to the left and right moving sectors and we in fact have four rings: (c,c) and (a,c) and their conjugates (a,a) and (c,a). Here c stands for chiral and a for anti-chiral.

We will now make the connection to our discussion of CY compactification and set $(\bar{c}, c) = (9, 9)$. We have already mentioned that $N = 1$ space-time supersymmetry for the heterotic string requires $n = 2$ superconformal symmetry for the right movers. This is however not sufficient. The additional requirement is that for states in the right-moving NS sector $q_R \in \mathbb{Z}$. The reason for this is the following [35]. The operator $\mathcal{U}_{\frac{1}{2}}(z)$ takes states in the right-moving NS sector to states in the right-moving R sector i.e. it transforms space-time bosons into space-time fermions (and vice versa). In fact, $\mathcal{U}_{\frac{1}{2}}(z) = e^{i\frac{\sqrt{3}}{2}\phi(z)}$ is the internal part of the gravitino vertex operator, which, when completed with its space-time and super-conformal ghost parts, must be local with respect to all the fields in the theory. When considering space-time bosons this leads to the requirement of integer $U(1)$ charges[11] (which have to be in the range $-3, \dots, +3$)[12]. Since we are dealing with *symmetric* $(2,2)$ superconformal theories both q_L and q_R must be integer for states in the (NS, NS) sector.

We now turn to the discussion of the moduli of the Calabi-Yau compactification. In an effective low energy field theory they are neutral (under the gauge group $E_6 \times E_8$) massless scalar fields with vanishing potential (perturbative and non-perturbative) whose vacuum expectation values determine the 'size' (Kähler moduli) and 'shape' (complex structure moduli) of the internal manifold. In the conformal field theory context they parameterize the perturbations of a given conformal field theory by exactly marginal operators. Exactly marginal operators can be added to the action without destroying $(2,2)$ superconformal invariance of the theory. Their multi-point correlation functions all

---

[10] Recall the correspondence between states and fields: $|\phi\rangle = \lim_{z \to 0} \phi(z)|0\rangle$.

[11] To see this, take the operator product of the gravitino vertex operator (e.g. in the 1/2 ghost picture) $\psi_\alpha(z) = e^{-\varphi/2}S^\alpha e^{i\frac{\sqrt{3}}{2}\phi}(z)$ (S^α is a $SO(4)$ spin field) with a space-time boson (in the zero ghost picture) with vertex operator $V = (\text{s.t.})e^{i\frac{q}{\sqrt{3}}\phi}$. The operator products of the spin field with the space-time parts (s.t) are either local, in which case we need $q = 2\mathbb{Z} + 1$, or have square root singularities, and we need $q = 2\mathbb{Z}$.

[12] Space-time supersymmetry and the existence of the unique state $|\rho\rangle$ with $q = \frac{c}{3}$ thus requires that c be an integer multiple of 3.

vanish. In fact, one can show [31][6][38] that there is a one-to-one correspondence between moduli of the (2,2) superconformal field theory and chiral primary fields with conformal weight $(\bar{h}, h) = (\frac{1}{2}, \frac{1}{2})$. The chiral primary fields are (left and right) (anti)chiral superfields whose upper components (they survive the integral over chiral superspace) have conformal weight (1,1) and are thus marginal. The lower components with weights $(\frac{1}{2}, \frac{1}{2})$ provide the internal part of the charged matter fields (27 and $\overline{27}$ of E_6). The gauge part on the left moving side and the space-time and superconformal ghost parts on the right-moving side account for the remaining half units of conformal weight for the massless matter fields. We have thus a one-to-one correspondence between charged matter fields and moduli: extended world-sheet supersymmetry relates the 27's of E_6 to the marginal operators in the (c, c) ring and the $\overline{27}$'s of E_6 to the marginal operators in the (a, c) ring[13].

We have encountered the four chiral rings. The fields in the four rings can all be obtained from the (R, R) ground states via spectral flow. The additive structure of the rings is therefore isomorphic, not however their multiplicative structures. They are in general very different. The (c, c) ring contains fields with $U(1)$ charges $(q_L, q_R) = (+1, +1)$ whereas the (a, c) contains fields with $(q_L, q_R) = (-1, +1)$, both with conformal weights $(h_L, h_R) = (\frac{1}{2}, \frac{1}{2})$. The latter are related, via spectral flow, to states in the (c, c) ring with charges $(2, 1)$.

We now turn to a comparison between the chiral primary states and the cohomology of the CY manifold. We expect a close relationship since the (2,2) super-conformal field theories we are considering correspond to conformally invariant sigma models with CY target space. Let us first look at the (c, c) ring. In the conformal field theory there is a unique chiral primary state with $(q_L, q_R) = (3, 0)$ whereas there is the unique holomorphic three form Ω in the cohomology of the CY space. By conjugation we have the state with charge $(0, 3)$ and $\bar{\Omega} \in H^{0,3}(X)$. Also $(q_L, q_R) = (3, 3) \leftrightarrow \Omega \wedge \bar{\Omega}$. The fields with $(q_L, q_R) = (1, 1)$ are marginal and correspond to the complex structure moduli, whereas the fields with $(q_L, q_R) = (2, 1)$ are related (via spectral flow by $(-1, 0)$) to marginal states in the (a, c) ring with charges $(-1, 1)$. They correspond to the complex structure deformations. In general, if we identify the left and right $U(1)$ charges with the holomorphic and anti-holomorphic form degrees, respectively, one is tempted to establish a one-to-one correspondence between elements of the (c, c) ring with charges (q_L, q_R) and elements of $H^{q_L, q_R}(X)$. This becomes even more suggestive if we formally identify the zero modes of the supercurrents with the holomorphic exterior differential and co-differential as $G_0^+ \sim \partial$, $\bar{G}_0^+ \sim \bar{\partial}$. Via the spectral flow (by $(\theta_L, \theta_R) = (-\frac{1}{2}, -\frac{1}{2})$) we can also identify $G_{-\frac{1}{2}}^+ \sim \partial$ and $\bar{G}_{-\frac{1}{2}}^+ \sim \bar{\partial}$. Furthermore, one can show [7] that each NS state has a chiral primary representative in the sense that there exists a unique decomposition $|\phi\rangle = |\phi_0\rangle + G_{-\frac{1}{2}}^+ |\phi_1\rangle + G_{\frac{1}{2}}^- |\phi_2\rangle$ with $|\phi_0\rangle$ chiral primary. For $|\phi\rangle$ itself primary,

---

[13] Note that this identification is again a matter of convention. The arbitrariness here is due to a trivial symmetry of the theory under the flip of the relative sign of the left and right $U(1)$ charges (cf. below).

$|\phi_2\rangle$ is zero. This parallels the Hodge decomposition of differential forms. In fact, the one-to-one correspondence between the cohomology of the target space of supersymmetric sigma models and the Ramond ground states has been established in [39]. Let us now compare the (c, c) ring with the cohomology ring, whose multiplicative structure is defined by taking wedge products of the harmonic forms. We know from (2.2) that in the large radius limit the Yukawa couplings are determined by the cohomology of the Calabi-Yau manifold X. However, once string effects are taken into account, the Yukawa couplings are no longer determined by the cohomology ring of X but rather by a deformed cohomology ring. This deformed cohomology ring coincides with the (c, c) ring of the corresponding super-conformal field theory. For discussions of the relation between chiral rings and cohomology rings in the context of topological σ-model we refer to Witten's contribution in [9].

On the conformal field theory level, mirror symmetry is now the following simple observation: the exchange of the relative sign of the two $U(1)$ currents is a trivial symmetry of the conformal field theory, by which the (c, c) and the (a, c) rings of chiral primary fields are exchanged. On the geometrical level this does however have highly non-trivial implications. It suggests the existence of two topologically very different geometric interpretations of a given $(2, 2)$ internal superconformal field theory. The deformed cohomology rings are isomorphic to the (c, c) and (a, c) rings, respectively. Here we associate elements in the (a, c) ring with charge (q_L, q_R) with elements of H^{3+q_L, q_R}.

If we denote the two manifolds by X and X^* then one simple relation is in terms of their Hodge numbers:

$$h^{p,q}(X) = h^{3-p,q}(X^*) \tag{3.4}$$

The two manifolds X and X^* are referred to as a mirror pair. The relation between the Hodge numbers alone is not very strong. A farther reaching consequence of the fact that string compactification on X and X^* are identical is the relation between the deformed cohomology rings. This in turn entails a relation between correlation functions, in particular between the two types of Yukawa couplings.

It is however not clear whether a given $(2, 2)$ theory always allows for two topologically distinct geometric descriptions as compactifications [14]. In fact, if one considers so-called rigid manifolds, i.e. CY manifolds with $h^{2,1} = 0$, then it is clear that the mirror manifold cannot be a CY manifold, which is Kähler, i.e. has $h^{1,1} \geq 1$. The concept of mirror symmetry for these cases has been exemplified on the $\mathbb{Z}_3 \times \mathbb{Z}_3$ orbifold in [40] and was further discussed in [41].

To close this section we want to make some general comments about the structure of the moduli space of Calabi-Yau compactifications which will be useful later on (see [11] for a review of these issues). In this context it is useful to note that instead of compactifying the heterotic string on a given Calabi-Yau manifold, one could have just as well taken the type II string. This would result

---

[14] Mirror symmetric manifolds are excluded if $\chi \neq 0$.

in $N = 2$ space-time supersymmetry. In the conformal field theory language this means that we take the same $(2,2)$ super-conformal field theory with central charge $(\bar{c}, c) = (9, 9)$ but this time without the additional gauge sector that was required for the heterotic string. This results in one gravitino on the left and on the right moving side each. The fact that the identical internal conformal field theory might also be used to get a $N = 2$ space-time supersymmetric theory leads to additional insight into the structure of the moduli space which is the same for the heterotic as for the type II string and which has to satisfy additional constraints coming from the second space-time supersymmetry. This was used in [42] to show that locally the moduli manifold has the product structure

$$\mathcal{M} = \mathcal{M}_{h^{1,1}} \times \mathcal{M}_{h^{2,1}} \tag{3.5}$$

where $\mathcal{M}_{h^{1,1}}, \mathcal{M}_{h^{2,1}}$ are two Kähler manifolds with complex dimensions $h^{1,1}$ and $h^{2,1}$, respectively. The same result was later derived in refs. [43] [44] [45]. In ref.[45] it was shown to be a consequence of the $(2,2)$ super-conformal algebra. $N = 2$ space-time supersymmetry or super-conformal Ward identities can be used to show that each factor of $\mathcal{M}$ is a so-called special Kähler manifold. Special Kähler manifolds are characterized by a prepotential F from which the Kähler potential (and thus the Kähler metric) and also the Yukawa couplings can be computed. Locally on special Kähler manifolds there exist so-called special coordinates t_i which allow for simple expressions of the Kähler potential K and the Yukawa couplings κ in terms of the prepotential as[15] [46]

$$K = -\ln\left[2(F - \bar{F}) - (t_i - \bar{t}_i)(F_i + \bar{F}_i)\right], \qquad \left(F_i = \frac{\partial F}{\partial t_i}\right)$$
$$\kappa_{ijk} = F_{ijk} \tag{3.6}$$

with one set for each factor of $\mathcal{M}$[16]. To reproduce the classical Yukawa couplings (2.2) $F^{1,1}(X)$ is simply a cubic polynomial whereas $F^{2,1}$ is a complicated function of the complex structure moduli. Instanton corrections will modify $F^{1,1}$ such as to reproduce the couplings (2.3). Mirror symmetry then relates the prepotentials on X and X^*. If λ_i are the complex structure moduli on X^* then one can find a local map (the mirror map) $\mathcal{M}_{2,1}(X^*) \to \mathcal{M}_{1,1}(X) : \lambda_i \mapsto t_i(\lambda)$ such that $F^{2,1}(\lambda)(X^*) = F^{1,1}(t)(X)$ and similarly for $F^{1,1}(X^*)$ and $F^{2,1}(X)$. We will find below that $\lambda_i(t)$ are transzendental functions containing exponentials. It is an interesting fact that $\lambda(q^i)$, with $q = e^{2\pi i t}$, is always an infinite series with integer coefficients.

---

[15] From this it follows immediately that the $\langle \overline{27}^3 \rangle$ couplings receive no instanton corrections since they would lead to a dependence on the Kähler moduli.

[16] We will denote the special coordinates on $\mathcal{M}_{h^{1,1}}$ by t and the ones on $\mathcal{M}_{h^{2,1}}$ by λ.

4   Construction of Mirror Pairs

We have already alluded to the fact that a classification of consistent string vacua is still out of reach. In fact, even a classification of three-dimensional CY manifolds, providing just a subset of string vacua, is still lacking. (One only knows how to classify the homotopy types of the manifolds by virtue of Wall's theorem; [23], p. 173.) What has been achieved so far is to give complete lists of possible CY manifolds within a given construction. But even here one does not have criteria to decide which of these manifolds are diffeomorphic.

The constructions that have been completely searched for are hypersurfaces in four-dimensional weighted projective space [47] and complete intersections of k hypersurfaces in $3 + k$ dimensional products of projective spaces [48]. In these notes we will limit ourselves to the discussion of hypersurfaces. Mirror symmetry for complete intersections has been discussed in [49] and [17].

Weighted n-dimensional complex projective space $\mathbb{P}^n[\vec{w}]$ is simply $\mathbb{C}^{n+1} \setminus \{0\}/\mathbb{C}^*$ where $\mathbb{C}^* = \mathbb{C} \setminus \{0\}$ acts as $\lambda \cdot (z_0, \ldots, z_n) = (\lambda^{w_0} z_0, \ldots, \lambda^{w_n} z_n)$. We will denote a point in $\mathbb{P}^n[\vec{w}]$ by $(z_0 : \cdots : z_n)$. The coordinates z_i are called homogeneous coordinates of $\mathbb{P}^n[\vec{w}]$ [17] and $w_i \in \mathbb{Z}_+$ their weights. For $\vec{w} = (1, \ldots, 1)$ one recovers ordinary projective space. Note however that $\mathbb{P}^n[k\vec{w}] \sim \mathbb{P}^n[\vec{w}]$. In fact, due to this and other isomorphisms (see Prop. 1.3.1 in [50]) one only needs to consider so-called well-formed weighted projective spaces. $\mathbb{P}^n[\vec{w}]$ is called well-formed if each set of n weights is co-prime. Weighted projective $\mathbb{P}^n[\vec{w}]$ can be covered with $n+1$ coordinate patches U_i with $z_i \neq 0$ in U_i. The transition functions between different patches are then easily obtained. The characteristic feature to note about the transition functions of projective space is that in overlaps $U_i \cap U_j$ they are given by Laurent monomials. For example, consider $\mathbb{P}^2$ with homogeneous coordinates $(z_0 : z_1 : z_2)$ and the three patches U_i, $i = 0, 1, 2$ with inhomogeneous coordinates $\varphi_0(z_0 : z_1 : z_2) = (u_0, v_0) = (\frac{z_1}{z_0}, \frac{z_2}{z_0})$, $\varphi_1(z_0 : z_1 : z_2) = (u_1, v_1) = (\frac{z_0}{z_1}, \frac{z_2}{z_1})$ and $\varphi_2(z_0 : z_1 : z_2) = (u_2, v_2) = (\frac{z_0}{z_2}, \frac{z_1}{z_2})$. The transition functions on overlaps are then Laurent monomials; e.g. on $U_0 \cap U_1$ we have $(u_1, v_1) = (\frac{1}{u_0}, \frac{v_0}{u_0})$, and we have $\varphi_i(U_i \cap U_j) \simeq \mathbb{C} \times \mathbb{C}^*$, $\varphi_i(U_1 \cap U_2 \cap U_3) = (\mathbb{C}^*)^2$. The reason for dwelling on these well-known facts is that below we will treat this example also in the language of toric geometry and that it is the fact that transition functions between patches are Laurent monomials, that characterizes more general toric varieties.

Weighted projective spaces are generally singular. As an example, consider $\mathbb{P}^2[1, 1, 2]$, i.e. (z_0, z_1, z_2) and $(\lambda z_0, \lambda z_1, \lambda^2 z_2)$ denote the same point and for $\lambda = -1$ the point $(0 : 0 : z_2) \equiv (0 : 0 : 1)$ is fixed but λ acts non-trivially on its neighborhood: we thus have a $\mathbb{Z}_2$ orbifold singularity at this point.

A hypersurface X in (weighted) projective space is defined as the vanishing locus of a (quasi)homogeneous polynomial, i.e. of a polynomial in the homogeneous coordinates that satisfies $p(\lambda^{w_0} z_0, \ldots, \lambda^{w_n} z_n) = \lambda^d p(z_0, \ldots, z_n)$ where $d \in \mathbb{Z}_+$ is called the degree of $p(z)$; i.e. we have

$$X_{\vec{w}} = \left\{ (z_0 : \ldots : z_n) \in \mathbb{P}^n[\vec{w}] \,\middle|\, p(z) = 0 \right\} \tag{4.1}$$

---

[17] We will sometimes denote them by $z_0, \ldots, z_n$ and othertimes by $z_1, \ldots, z_{n+1}$. We are confident that this will cause no confusion.

In order for the hypersurface to be a CY manifold one has to require that its first Chern class vanishes. This can be expressed in terms of the weights and the degree of the defining polynomial as[23]

$$c_1(X) = \left(\sum_{i=0}^{n} w_i - d\right) J \tag{4.2}$$

where J is the Kähler form of the projective space the manifold is embedded in. A necessary condition for a hypersurface in $\mathbb{P}^4[\vec{w}]$ to be a three-dimensional CY manifold is then that the degree of the defining polynomial equals the sum of the weights of the projective space. However one still has to demand that the embedding $X \hookrightarrow \mathbb{P}^n[\vec{w}]$ be smooth. This means that one has to require the transversality condition: $p(z) = 0$ and $dp(z) = 0$ have no simultaneous solutions other than $z_0 = \ldots = z_n = 0$ (which is not a point of $\mathbb{P}^n[\vec{w}]$).

There exists an easy to apply criterium [51] which allows one to decide whether a given polynomial satisfies the transversality condition. This criterium follows from Bertini's theorem (c.f. e.g. Griffiths and Harris in [22]) and goes as follows. For every index set $J = \{j_1, \ldots, j_{|J|}\} \subset \{0, \ldots, n\} = N$ denote by $z_J^{m^{(k)}}$ monomials $z_{j_1}^{m_{i_1}^{(k)}} \cdots z_{i_{|J|}}^{m_{i_{|J|}}^{(k)}}$ of degree d. Transversality is then equivalent to the condition that for every index set J there exists either (a) a monomial z_J^m, or (b) $|J|$-monomials $z_J^{m^{(k)}} z_k$ with $|J|$ distinct $k \in N \setminus J$.

Analysis of this condition shows that there are 7555 projective spaces $\mathbb{P}^4[\vec{w}]$ which admit transverse hypersurfaces. They were classified in [47]. In case that $q_i = d/w_i \in \mathbb{Z}$, $\forall i = 0, \ldots, k$ one gets a transverse polynomial of Fermat type: $p(z) = \sum_{i=0}^{4} z_i^{q_i}$.

If the hypersurfaces X meet some of the singularities of the weighted projective space, they are themselves singular. Let us first see what kind of singular sets one can get. If the weights w_i for $i \in I$ have a common factor N_I, the singular locus S_I of the CY space is the intersection of the hyperplane $H_I = \{z \in \mathbb{P}^4[\vec{w}] | z_i = 0 \text{ for } i \notin I\}$ with $X_{\vec{w}}$. We will see that the singular locus consists either of points or of curves.

As $\mathbb{P}^4[\vec{w}]$ is wellformed we have $|I| \leq 3$. Consider $|I| = 3$ and apply the transversality criterum to $J = I$. Obviously, only transversality condition (a) can hold, which implies that p will not vanish identically on H_I, hence $\dim(S_I) = 1$. It is important for the following to consider the $\mathbb{C}^*$-action on the normal bundle[18] to this curve. We write the $c_1(X) = 0$ condition as $\sum_{i \in I} w_i' + \sum_{j \notin I}(w_j/N_I) = (d/N_I)$, with $w_i' \in \mathbb{Z}_>$ and $(w_j/N_I) \notin \mathbb{Z}$. Because of (a) one has $d = \sum_{i \in I} m_i w_i = N_I \sum_{i \in I} m_i w_i'$, for $m_i \in \mathbb{Z}_{\geq}$, from which we can conclude $\sum_{j \notin I}(w_i/N_I) \in \mathbb{Z}$. Locally we can then choose (z_{k_1}, z_{k_2}) with $k_i \notin I$ as the coordinates normal to the curve. The $\mathbb{C}^*$-action which fixes S_I will therefore be generated by $(z_{k_1}, z_{k_2}) \mapsto (\lambda z_{k_1}, \lambda^{-1} z_{k_2})$, where we define $\lambda = e^{2\pi i/N_I}$. That is, locally the singularity in the normal bundle is of type $\mathbb{C}^2/\mathbb{Z}_{N_I}$.

---

[18] The normal bundle on C is the quotient bundle $N_C = Tx|_C/T_C$.

Finally for $|I| \leq 2$ clearly $\dim(S_I) \leq 1$. From the analysis of the divisibility condition imposed by transversality and $c_1(X) = 0$ we can summarize that the $\mathbb{C}^*$ action in the normal bundle of singular curves or the neighborhood of singular points is in local coordinates always of the form $\mathbb{C}^2/\mathbb{Z}_{N_I}$ and $\mathbb{C}^3/\mathbb{Z}_{N_I}$ with

$$\lambda \cdot (z_1, z_2) = (\lambda z_1, \lambda^{-1} z_2)$$
$$\lambda \cdot (z_1, z_2, z_3) = (\lambda z_1, \lambda^a z_2, \lambda^b z_3) \quad \text{with} \quad 1 + a + b = N_I, \, a, b \in \mathbb{Z} \tag{4.3}$$

Note that for the case of fixed curves invariant monomials are $z_1^{N_I}, z_2^{N_I}$ and $z_1 z_2$, i.e. we can describe the singularity as $\{(u, v, w) \in \mathbb{C}^3 | uv = w^{N_I}\}$. This type of singularity is called a rational double point of type A_{N_I-1}. The relation with the Lie-algebras from the A series and the discussion of the resolution of the singularities within toric geometry will be given below.

Let us mention that the types of singularities encountered here are the same as the ones in abelian toroidal orbifolds, discussed in [52] [53]. The condition on the exponents of the $\mathbb{C}^*$-action there is related to the fact that one considers only subgroups of $SU(3)$ as orbifold groups. This was identified as a necessary condition to project out of the spectrum three gravitinos in order to obtain exactly $N = 1$ spacetime supersymmetry [54]. As we will briefly explain below, it also ensures $c_1(\hat{X}) = 0$, i.e. triviality of the canonical bundle of the resolved manifold $\hat{X}$. Desingularisations with this property are referred to as minimal desingularizations [55]. In ref.[47] the Hodge numbers of the minimal desingularizations of all 7555 transverse hypersurfaces were evaluated.

Among the singular points we have to distinguish between isolated points and exceptional points, the latter being singular points on singular curves or the points of intersection of singular curves. The order N_I of the isotropy group I of exceptional points exceeds that of the curve. In order to get a smooth CY manifold $\hat{X}$, these singularities have to be resolved by removing the singular sets and replacing them by smooth two complex dimensional manifolds which are then called exceptional divisors. Each exceptional divisor D provides, by Poincaré duality, a harmonic $(1,1)$ form h_D: $\int_D \alpha = \int_{\hat{X}} \alpha \wedge h_D$ for every closed $(2,2)$ form α and $h^{1,1}(\hat{X}) = \#$ exceptional divisors $+1$ where the last contribution counts the restriction of the Kähler form from to embedding space to $\hat{X}$ (to which one can also associate a divisor).

There are only few hypersurfaces, namely the quintic (i.e. $d = 5$) in $\mathbb{P}^4$, the sextic in $\mathbb{P}^4[1, 1, 1, 1, 2]$, the octic in $\mathbb{P}^4[1, 1, 1, 1, 4]$ and finally the dectic in $\mathbb{P}^4[1, 1, 1, 2, 5]$, which do not require resolution of singularities and the Kähler form they inherit from the embedding space is in fact the only one, i.e. for these CY spaces $h^{1,1} = 1$. One easily sees that these Fermat hypersurfaces do not meet the singular points of their respective embedding spaces. They were analyzed in view of mirror symmetry in [13].

If one considers the lists of models of ref.[47] one finds already on the level of Hodge numbers that there is no complete mirror symmetry within this construction. Also, if one includes abelian orbifolds of the hypersurfaces [56] the situation does not improve.

One can however get a mirror symmetric set of CY manifolds if one generalizes the construction to hypersurfaces in so-called toric varieties. Since they are not (yet) familiar to most physicists but relevant for describing the resolution of the above encountered singularities and for the discussion of mirror symmetry, we will give a brief description of toric varieties. For details and proofs we have to refer to the literature [57]. Toric methods were first used in the construction of Calabi-Yau manifolds in [55] and [52]. They have entered the discussion of mirror symmetry through the work of V. Batyrev [58] [59] [60] [49][61].

Toric varieties are defined in terms of a lattice $N \simeq \mathbb{Z}^n$ and a fan Σ. Before explaining what a fan is, we first have to define a (*strongly convex rational polyhedral*) cone (or simply cone) σ in the real vector space $N_{\mathbb{R}} \equiv N \otimes_{\mathbb{Z}} \mathbb{R}$:

$$\sigma = \left\{ \sum_{i=1}^{s} a_i n_i; a_i \geq 0 \right\} \tag{4.4}$$

where n_i is a finite set of *lattice* vectors (hence rational), the generators of the cone. We often write simply $\langle n_1, \ldots, n_s \rangle$. Strong convexity means that $\sigma \cap (-\sigma) = \{0\}$, i.e. the cone does not contain lines through the origin. A cone σ is called *simplicial* if it is generated by linearly independent (over $\mathbb{R}$) lattice vectors. A cone generated by part of a *basis* of the lattice N is called a basic cone. If we normalize the unit cell of the lattice to have volume one, then a simplicial cone $\langle n_1, \ldots n_m \rangle$ is basic if $\det(n_1, \ldots, n_m) = 1$ (here n_i are the generators of minimal length). If a cone fails to be basic, not all lattice points within the cone can be reached as linear combinations of the generators with positive integer coefficients.

To every cone we can define the dual cone $\sigma^{\vee}$ as

$$\sigma^{\vee} = \left\{ x \in M_{\mathbb{R}}; \langle x, y \rangle \geq 0 \text{ for all } y \in \sigma \right\} \tag{4.5}$$

where M is the lattice dual to N and $\langle , \rangle : M \times N \to \mathbb{Z}$, which extends to the $\mathbb{R}$ bilinear pairing $M_{\mathbb{R}} \times N_{\mathbb{R}} \to \mathbb{R}$. $\sigma^{\vee}$ is rational with respect to M but strongly convex only if $\dim(\sigma) = \dim(M_{\mathbb{R}})$. (For instance the cones dual to one-dimensional cones in $\mathbb{R}^2$, are half-planes.). Also $(\sigma^{\vee})^{\vee} = \sigma$. Given a cone we define

$$S_{\sigma} = \sigma^{\vee} \cap M \tag{4.6}$$

which is finitely generated by say $p \geq \dim M \equiv d_M$ lattice vectors n_i. In general (namely for $p \geq d_M$) there will be non-trivial linear relations between the generators of S_{σ} which can be written in the form

$$\sum \mu_i n_i = \sum \nu_i n_i \tag{4.7}$$

with μ_i and ν_i non-negative integers. A cone then defines an affine toric variety as

$$U_{\sigma} = \{(Z_1, \ldots, Z_p) \in \mathbb{C}^p | Z^{\mu} = Z^{\nu}\} \tag{4.8}$$

where we have used the short hand notation $Z^\mu = Z_1^{\mu_1} \cdots Z_p^{\mu_d M}$. In the mathematics literature one often finds the notation $U_\sigma = \mathrm{Spec}(\mathbb{C}[S_\sigma]) = \mathrm{Spec}(\mathbb{C}[Z_1, \ldots, Z_p]/I)$ where the ideal I is generated by all the relations $Z^\mu = Z^\nu$ between the gerators of S_σ.

To illustrate the construction, let us look at a simple example. Consider the cone σ generated by $(N+1)e_1 - Ne_2$ and e_2. Then S_σ is easily shown to be generated by $n_1 = e_1^*$, $n_2 = Ne_1^* + (N+1)e_2^*$ and $n_3 = e_1^* + e_2^*$ which satisfy $(N+1)n_3 = n_1 + n_2$. This leads to

$$U_\sigma = \{(Z_1, Z_2, Z_3) \in \mathbb{C}^3 | Z_1 Z_2 = Z_3^{N+1}\} \tag{4.9}$$

This is just the A_N rational double point discussed above.

A face τ of a cone σ is what one expects; it can be defined as $\sigma \cap u^\perp = \{v \in \sigma : \langle u, v \rangle = 0\}$ for some $u \in \sigma^\vee$. The constructive power of toric geometry relies in the possibility of gluing cones to fans.

A fan Σ is a family of cones σ satisfying:
(i) any face of a cone in Σ is itself a cone in Σ;
(ii) the intersection of any two cones in Σ is a face of each of them.

After having associated an affine toric variety U_σ with a cone σ, we can now construct a toric variety $\mathbb{P}_\Sigma$ associated to a fan Σ by glueing together the U_σ, $\sigma \in \Sigma$:

$$\mathbb{P}_\Sigma = \bigcup_{\sigma \in \Sigma} U_\sigma \tag{4.10}$$

The U_σ are open subsets of $\mathbb{P}_\Sigma$. The glueing works because $U_{\sigma \cap \sigma'}$ is an open subset of both U_σ and U'_σ i.e. $U_{\sigma \cap \sigma'} = U_\sigma \cap U_{\sigma'}$. On the other hand $U_{\sigma^\vee \cap (\sigma')^\vee} \neq U_{\sigma^\vee} \cap U_{(\sigma')^\vee}$, hence gluing is natural for the cones σ. A last result we want to quote before demonstrating the above with an example is that a toric variety $\mathbb{P}_\Sigma$ is compact iff the union of all its cones is the whole space $\mathbb{R}^n$. Such a fan is called complete.

To see what is going on we now give a simple but representative example. Consider the fan Σ whose dimension one cones τ_i are generated by the vectors $n_1 = e_1$, $n_2 = e_2$ and $n_3 = -(e_1 + e_2)$. Σ also contains the dimension two cones $\sigma_1 : \langle n_1, n_2 \rangle$, $\sigma_2 : \langle n_2, n_3 \rangle$, $\sigma_3 : \langle n_3, n_1 \rangle$ and of course the dimension zero cone $\{0\}$. We first note that Σ satisfies the compactness criterium. The two-dimensional dual cones are $\sigma_1^\vee : \langle e_1^*, e_2^* \rangle$, $\sigma_2^\vee : \langle -e_1^*, -e_1^* + e_2^* \rangle$, $\sigma_3^\vee : \langle -e_2^*, e_1^* - e_2^* \rangle$ and $U_{\sigma_1} = \mathrm{Spec}(\mathbb{C}[X, Y])$, $U_{\sigma_2} = \mathrm{Spec}(\mathbb{C}[X^{-1}, X^{-1}Y])$, $U_{\sigma_3} = \mathrm{Spec}(\mathbb{C}[Y^{-1}, XY^{-1}])$ each isomorphic to $\mathbb{C}^2$. These glue together to form $\mathbb{P}^2$. Indeed, if we define coordinates u_i, v_i via $U_{\sigma_i} = \mathrm{Spec}(\mathbb{C}[u_i, v_i])$ we get the transition functions for $\mathbb{P}^2$ between the three patches. Note also that $U_{\tau_i} = U_{\sigma_i \cap \sigma_j} \simeq \mathbb{C} \times \mathbb{C}^*$ and $U_{\sigma_1 \cap \sigma_2 \cap \sigma_3} = U_{\{0\}} = (\mathbb{C}^*)^2$. To get e.g. the weighted projective space $\mathbb{P}^2[1, 2, 3]$ one simply has to replace the generator n_3 by $n_3 = -(2e_1 + 3e_2)$. In fact, all projective spaces are toric varieties. This will become clear below.

We now turn to the discussion of singularities of toric varieties and their resolution. We have already given the toric description of the rational double point. The reason why the corresponding cone leads to a singular variety is because it is not basic, i.e. it can not be generated by a basis of the lattice N.

This in turn results in the need for three generators for S_σ which satisfy one linear relation. The general statement is that U_σ is smooth if and only if σ is basic. Such a cone will also be called smooth. An n-dimensional toric variety X_Σ is smooth, i.e. a complex manifold, if and only if all dimension n cones in Σ are smooth. We do however have to require more than smoothness. We also want to end up with a CY manifold, i.e. a smooth manifold with $c_1 = 0$.

It can be shown that if Σ is a smooth fan (i.e. all its cones are smooth) X_Σ has trivial canonical bundle if and only if the endpoints of the minimal generators of all one-dimensional cones in Σ lie on a hyperplane (see the proof in the appendix to [52]). The intersection of the hyperplane with the fan is called the trace of the fan. An immediate consequence of this result is that a compact toric variety, i.e. corresponding to a complete fan, can never have $c_1 = 0$. One therefore has to consider hypersurfaces in compact toric varieties to obtain CY manifolds.

The singularities we are interested in are cyclic quotient singularities. A standard result in toric geometry states that X_Σ has only quotient singularities, i.e. is an orbifold, if Σ is a simplicial fan, i.e. if all cones in Σ are simplicial. Given a singular cone one resolves the singularities by subdividing the cone into a fan such that each cone in the fan is basic.

Let us demonstrate this on the rational A_N double point. The two-dimensional cone $\sigma^{(2)}$ was generated by $(N+1)e_1 - Ne_2$ and e_2 which is not a basis of the lattice $N = \mathbb{Z}^2$. We now add the one-dimensional cones $\sigma_m^{(1)}$ generated by $me_1 - (m-1)e_2$ for $m = 1, \ldots, N$. (In this notation σ_0 and σ_{N+1} are the original one-dimensional cones.) The two-dimensional cones $\sigma_m^{(2)} : \langle me_1 - (m-1)e_2, (m+1)e_1 - me_2 \rangle, m = 1, \ldots, N$ are then basic and furthermore we have a $c_1 = 0$ resolution.

The original singular manifold has thus been desingularized by gluing exceptional divisors $D_m, m = 1, \ldots, N$. To cover the nonsingular manifold one needs $N+1$ patches. One generator of $(\sigma_i^{(2)})^\vee$ and one of $(\sigma_{i+1}^{(2)})^\vee$ are antiparallel in $\Sigma^\vee$. These generators therefore correspond to inhomogenous coordinates of one of N $\mathbb{P}^1$'s. The exceptional divisors are therefore $\mathbb{P}^1$'s. By inspection of the various patches one can see that the $\mathbb{P}^1$'s intersect pairwise transversely in one point to form a chain. The self-intersection number is obtained as the degree of their normal bundles, which is -2, so that the intersection matrix is the (negative of) the Cartan matrix of A_N; for details we refer to Fulton, ref.[57]. Such a collection of $\mathbb{P}^1$'s is called a Hirzebruch-Jung sphere-tree [62]. Recalling that the rational double points appeared in the discussion of *curve* singularities of CY manifolds we have seen that the resolved singular curves are locally the product of the curve C and a Hirzebruch-Jung sphere-tree.

Data of toric varieties depend only on linear relations and we may apply bijective linear transformations to choose a convenient shape. E.g. given the canonical basis e_1, e_2 in $\mathbb{R}^2$, we may use e_i as generators of the cone and $n_1 = \frac{1}{N+1}e_1 + \frac{N}{N+1}e_2$, $n_2 = e_2$, as generators of $N = \mathbb{R}^2$ to describe the A_N singularity. This form generalizes easily to higher dimensional cyclic singularities such as $\mathbb{C}^3/\mathbb{Z}_N$, the general form of the *point* singularities (4.3). They are described by e_i, $i = 1, 2, 3$ as generators of the cone and the lattice basis

$n_1 = \frac{1}{N}(e_1 + ae_2 + be_3)$, $n_2 = e_2$, $n_3 = e_3$. The local desingularisation process consists again of adding further generators such as to obtain a smooth fan. One readily sees that as a consequence of (4.3) all endpoints of the vectors generating the nonsingular fan lie on the plane $\sum_{i=1}^{3} x_i = 1$, as it is necessary for having a trivial canonical bundle on the resolved manifold. The exceptional divisors are in 1-1 correspondence with lattice points inside the cone on this hyperplane (trace of the fan). Their location is given by

$$\mathcal{P} = \left\{ \sum_{i=1}^{3} \vec{e}_i \frac{a_i}{N} \,\middle|\, (a_1, a_2, a_3) \in \mathbb{Z}^3, \begin{pmatrix} e^{2\pi i \frac{a_1}{N}} & & \\ & e^{2\pi i \frac{a_2}{N}} & \\ & & e^{2\pi i \frac{a_3}{N}} \end{pmatrix} \in \mathbb{Z}_N, \right.$$
$$\left. \sum_{i=1}^{3} a_i = N, a_i \geq 0 \right\} \tag{4.11}$$

The generator of the isotropy group $\mathbb{Z}_N$ on the coordinates of the normal bundle of the singular point is given in (4.3) and $\vec{e}_1, \vec{e}_2, \vec{e}_3$ span an equilateral triangle from its center. The corresponding divisors can all be described by compact toric surfaces which have been classified.

The toric diagrams for the resolution of singular points are thus equilateral triangles with interior points. If we have exceptional points then there will also be points on the edges, which represent the traces of the fans of the singular curves.

Whereas in the case of curve singularities, whose resolution was unique, this is not so for point singularities. Given the trace with lattice points in its interior, there are in general several ways to triangulate it. Each way corresponds to a different resolution. They all lead to topologically different smooth manifolds with the same Hodge numbers, differing however in their intersection numbers.

To obtain a Kähler manifold one also has to ensure that one can construct a positive Kählerform (2.1). This is guaranteed if one can construct an upper convex piecewise linear function on Σ (see Fulton, ref. [57]). The statement is then (see e.g. [60]) that the cone $K(\Sigma)$ consisting of the *classes* of all upper convex piecewise linear functions (i.e. modulo globally linear functions) on Σ is isomorphic to the Kähler cone of $\mathbb{P}_\Sigma$. This construction is reviewed in [16]. Below we will consider hypersurfaces in $\mathbb{P}_\Sigma$. The positive Kählerform on the hypersurface is then the one induced from the Kähler form on $\mathbb{P}_\Sigma$.

We will now give another description of toric varieties, namely in terms of convex integral polyhedra. The relation between the two constructions will become clear once we have demonstrated how to extract the cones and the fan from a given polyhedron. The reason for introducing them is that this will allow for a simple and appealing description of the mirror operation and of the construction of mirror pairs [59].

We start with a few definitions. We consider *rational* (with respect to a lattice N) *convex polyhedra* (or simply polyhedra) $\Delta \subset N_\mathbb{R}$ containing the origin $\nu_0 = (0, 0, 0, 0)$. Δ is called *reflexive* if its dual defined by

$$\Delta^* = \{ (x_1, \ldots, x_4) \mid \sum_{i=1}^{4} x_i y_i \geq -1 \text{ for all } (y_1, \ldots, y_4) \in \Delta \} \tag{4.12}$$

is again a rational polyhedron. Note that if Δ is reflexive, then Δ^* is also reflexive since $(\Delta^*)^* = \Delta$. We associate to Δ a complete rational fan $\Sigma(\Delta)$ whose cones are the cones over the faces of Δ with apex at ν_0.

The l dimensional faces will be denoted by Θ_l. Completeness is ensured since Δ contains the origin. The toric variety $\mathbb{P}_\Delta$ is then the toric variety associated to the fan $\Sigma(\Delta^*)$, i.e. $\mathbb{P}_\Delta \equiv \mathbb{P}_{\Sigma(\Delta^*)}$.

Denote by ν_i, $i = 0, \ldots, s$ the integral points in Δ and consider the affine space $\mathbb{C}^{s+1}$ with coordinates a_i. We will consider the zero locus Z_f of the Laurent polynomial

$$f_\Delta(a, X) = a_0 - \sum_i a_i X^{\nu_i}, \quad f_\Delta(a, X) \in \mathbb{C}[X_1^{\pm 1}, \ldots, X_4^{\pm 1}] \qquad (4.13)$$

in the algebraic torus $(\mathbb{C}^*)^4 \subset \mathbb{P}_\Delta$, and its closure $\bar{Z}_f$ in $\mathbb{P}_\Delta$. Here we have used the convention $X^\mu \equiv X_1^{\mu_1} \ldots X_4^{\mu_4}$. Note that by rescaling the four coordinates X_i and adjusting an overall normalization we can set five of the parameters a_i to one.

$f \equiv f_\Delta$ and Z_f are called Δ-regular if for all $l = 1, \ldots, 4$ the f_{Θ_l} and $X_i \frac{\partial}{\partial X_i} f_{\Theta_l}$, $\forall i = 1, \ldots, n$ do not vanish simultaneously in $(\mathbb{C}^*)^4$. This is equivalent to the transversality condition for the quasi-homogeneous polynomials p. Varying the parameters a_i under the condition of Δ-regularity, we get a family of toric varieties.

In analogy with the situation of hypersurfaces in $\mathbb{P}^4[\vec{w}]$, the more general ambient spaces $\mathbb{P}_\Delta$ and so $\bar{Z}_f$ are in general singular. Δ-regularity ensures that the only singularities of $\bar{Z}_f$ are the ones inherited from the ambient space. $\bar{Z}_f$ can be resolved to a CY manifold $\hat{Z}_f$ iff $\mathbb{P}_\Delta$ has only so-called Gorenstein singularities, which is the case iff Δ is reflexive [59].

The families of the CY manifolds $\hat{Z}_f$ will be denoted by $\mathcal{F}(\Delta)$. The above definitions proceed in an exactly symmetric way for the dual polyhedron Δ^* with its integral points ν_i^* $(i = 0, \cdots, s^*)$, leading to families of CY manifolds $\mathcal{F}(\Delta^*)$.

In ref. [59] Batyrev observed that a pair of reflexive polyhedra (Δ, Δ^*) naturally provides a pair of mirror CY families $(\mathcal{F}(\Delta), \mathcal{F}(\Delta^*))$ as the following identities for the Hodge numbers hold

$$
\begin{aligned}
h^{1,1}(\hat{Z}_{f,\Delta^*}) &= h^{2,1}(\hat{Z}_{f,\Delta}) \\
&= l(\Delta) - 5 - \sum_{\text{codim}\Theta=1} l'(\Theta) + \sum_{\text{codim}\Theta=2} l'(\Theta)l'(\Theta^*) \\
h^{1,1}(\hat{Z}_{f,\Delta}) &= h^{2,1}(\hat{Z}_{f,\Delta^*}) \\
&= l(\Delta^*) - 5 - \sum_{\text{codim}\Theta^*=1} l'(\Theta^*) + \sum_{\text{codim}\Theta^*=2} l'(\Theta^*)l'(\Theta).
\end{aligned}
\qquad (4.14)
$$

Here $l(\Theta)$ and $l'(\Theta)$ are the number of integral points on a face Θ of Δ and in its interior, respectively (and similarly for Θ^* and Δ^*). An l-dimensional face Θ can be represented by specifying its vertices $v_{i_1}, \cdots, v_{i_k}$. Then the dual face defined by $\Theta^* = \{x \in \Delta^* \mid (x, v_{i_1}) = \cdots = (x, v_{i_k}) = -1\}$ is a $(n - l - 1)$-dimensional face of Δ^*. By construction $(\Theta^*)^* = \Theta$, and we thus have a natural pairing between l-dimensional faces of Δ and $(n - l - 1)$-dimensional faces of Δ^*. The last sum in each of the two equations in (2.4) is over pairs of dual faces. Their contribution cannot be associated with a monomial in the Laurent polynomial[19]. We will denote by $\tilde{h}^{2,1}$ and $\tilde{h}^{1,1}$ the expressions (4.14) without the last terms.

A sufficient criterion for the possibility to associate to a CY hypersurface in $\mathbb{P}^4[\vec{w}]$ a reflexive polyhedron is that $\mathbb{P}^4[\vec{w}]$ is Gorenstein, which is the case if $\mathrm{lcm}[w_1, \ldots, w_5]$ divides the degree d [63]. In this case we can define a simplicial, reflexive polyhedron $\Delta(\vec{w})$ in terms of the weights, s.t. $\mathbb{P}_{\Delta^*(\vec{w})} \simeq \mathbb{P}^4[\vec{w}]$. The associated n-dimensional integral convex dual polyhedron is the convex hull of the integral vectors μ of the exponents of all quasi-homogeneous monomials z^μ of degree d, shifted by $(-1, \ldots, -1)$:

$$\Delta^*(\vec{w}) := \{(x_1, \ldots, x_5) \in \mathbb{R}^5 \mid \sum_{i=1}^{5} w_i x_i = 0, x_i \geq -1\}. \tag{4.15}$$

Note that this implies that the origin is the only point in the interior of Δ.

If the quasihomogeneous polynomial p is Fermat, i.e. if it consists of monomials z_i^{d/w_i} ($i = 1, \cdots, 5$), $\mathbb{P}^4[\vec{w}]$ is clearly Gorenstein, and (Δ, Δ^*) are thus simplicial. If furthermore at least one weight is one (say $w_5 = 1$) we may choose $e_1 = (1, 0, 0, 0, -w_1)$, $e_2 = (0, 1, 0, 0, -w_2)$, $e_3 = (0, 0, 1, 0, -w_3)$ and $e_4 = (0, 0, 0, 1, -w_4)$ as generators for Λ, the lattice induced from the $\mathbb{Z}^5$ cubic lattice on the hyperplane $H = \{(x_1, \ldots, x_5) \in \mathbb{R}^5 \mid \sum_{i=1}^{5} w_i x_i = 0\}$. For this type of models we then always obtain as vertices of $\Delta^*(\vec{w})$ (with respect to the basis $e_1, \ldots, e_4$)

$$\nu_1^* = (d/w_1 - 1, -1, -1, -1), \quad \nu_2^* = (-1, d/w_2 - 1, -1, -1),$$
$$\nu_3^* = (-1, -1, d/w_3 - 1, -1), \quad \nu_4^* = (-1, -1, -1, d/w_4 - 1), \tag{4.16}$$
$$\nu_5^* = (-1, -1, -1, -1)$$

and for the vertices of the dual simplex $\Delta(w)$ one finds

$$\nu_1 = (1, 0, 0, 0), \quad \nu_2 = (0, 1, 0, 0), \quad \nu_3 = (0, 0, 1, 0), \quad \nu_4 = (0, 0, 0, 1)$$
$$\nu_5 = (-w_1, -w_2, -w_3, -w_4) \tag{4.17}$$

It should be clear from our description of toric geometry that the lattice points in the interior of faces of Δ of dimensions $4 > d > 0$ correspond to exceptional divisors resulting from the resolution of the Gorenstein singularities of $\mathbb{P}_{\Delta^*}$. This in turn means that the corresponding Laurent monomials in f_Δ

---

[19] In the language of Landau-Ginzburg theories, if appropriate, they correspond to contributions from twisted sectors.

correspond to exceptional divisors. For those CY hypersurfaces that can be written as a quasi-homogeneous polynomial constraint in $\mathbb{P}^4[\vec{w}]$, we can then give a correspondence between the monomials, which correspond (via Kodaira-Spencer deformation theory) to the complex structure moduli, and the divisors, which correspond to the Kähler moduli. The authors of ref. [64] call this the monomial-divisor mirror map. Not all deformations of the complex structure can be represented by monomial deformations, which are also referred to as algebraic deformations. We will however restrict our analysis to those. We will now describe it for Fermat hypersurfaces of degree d, following [59].

The toric variety $\mathbb{P}_{\Delta(\vec{w})}$ can be identified with

$$
\begin{aligned}
\mathbb{P}_{\Delta(\vec{w})} &\equiv \mathbf{H}_d(\vec{w}) \\
&= \{[U_0, U_1, U_2, U_3, U_4, U_5] \in \mathbb{P}^5 | \prod_{i=1}^{5} U_i^{w_i} = U_0^d\},
\end{aligned}
\tag{4.18}
$$

where the variables X_i in eq.(4.13) are related to the U_i by

$$
[1, X_1, X_2, X_3, X_4, \frac{1}{\prod_{i=1}^{4} X_i^{w_i}}] = [1, \frac{U_1}{U_0}, \frac{U_2}{U_0}, \frac{U_3}{U_0}, \frac{U_4}{U_0}, \frac{U_5}{U_0}].
\tag{4.19}
$$

Let us consider the mapping $\phi : \mathbb{P}^4[\vec{w}] \to \mathbf{H}_d(\vec{w})$ given by[20]

$$
[z_1, z_2, z_3, z_4, z_5] \mapsto [z_1 z_2 z_3 z_4 z_5, z_1^{d/w_1}, z_2^{d/w_2}, z_3^{d/w_3}, z_4^{d/w_4}, z_5^{d/w_5}].
\tag{4.20}
$$

Integral points in $\Delta(\vec{w})$ are mapped to monomials of the homogeneous coordinates of $\mathbb{P}^4[\vec{w}]$ by

$$
\mu = (\mu_1, \mu_2, \mu_3, \mu_4) \mapsto \phi^*(X^\mu U_0) = \frac{\prod_{i=1}^{4} z_i^{\mu_i d/w_i}}{\left(\prod_{i=1}^{5} z_i\right)^{\sum_{i=1}^{4} \mu_i - 1}}
\tag{4.21}
$$

Note that the Laurent polynomial f_Δ and the quasi-homogeneous polynomial $p = \phi^*(U_0 f_\Delta)$ between which the monomial-divisor map acts correspond to a mirror pair. The point at the origin of Δ is always mapped to the symmetric perturbation $z_1 \cdots z_5$ which is always present. It represents the restriction of the Kähler form of the embedding space $\mathbf{P}_{\Delta}$. to the hypersurface.

The situation for CY hypersurfaces in non-Gorenstein $\mathbb{P}^4[\vec{w}]$'s was discussed in [16]. Here the corresponding polyhedron is still reflexive but no longer simplicial and the associated toric variety $\mathbb{P}_\Delta$ is Gorenstein.

We have already mentioned that Fermat hypersurfaces in weighted projective space do not intersect with the singular points of the embedding space. This is no longer generally true for non-Fermat hypersurfaces. What does however

---

[20] In toric geometry this mapping replaces the orbifold construction for the mirror manifolds described in [8].

hold is that hypersurfaces $Z_{f,\Delta}$ in the corresponding (Gorenstein) $\mathbb{P}_\Delta$ do not intersect singular points in $\mathbb{P}_\Delta$. The singular points of the embedding space correspond to the lattice points in the interior of faces of $\mathbb{P}_\Delta$ of codimension one. The Laurent monomials for these points thus do not correspond to complex structure deformations and we will in the following always restrict the sum in (4.13) to those lattice points of Δ which do not lie in the interior of faces of codimension one.

The final point to mention in this section is the computation of topological triple intersection numbers. They represent the classical part (field theory limit, large radius limit) of the $\langle 27^3 \rangle$ Yukawa couplings.

Given three divisors the topological triple intersection number can be computed in terms of an integral of the (by Poincaré duality) associated harmonic forms: $D_i \cdot D_j \cdot D_k = \int_{\hat{X}} h_{D_i} \wedge h_{D_j} \wedge h_{D_k}$. In toric geometry their evaluation reduces to combinatorics. The results are scattered through the (mathematics) literature [57][65] and have been collected in [16]. We will not repeat them here.

5 Periods, Picard-Fuchs Equations and Yukawa Couplings

Now that we have learned how to construct CY manifolds and even mirror pairs we can move on towards applying mirror symmetry. This, as was mentioned before, requires the knowledge of the $\langle \overline{27}^3 \rangle$ Yukawa couplings on the mirror manifold X^* in order to compute the $\langle 27^3 \rangle$ couplings on X. But in addition we need to know (at least locally) how to map the complex structure moduli space of X^* to the Kähler structure moduli space of X. This is the so-called mirror map. Our discussion will be restricted to the neighbourhood of the large complex structure limit of X^* and the large radius limit of X which will be mapped to each other by the mirror map.

For the purposes of getting the Yukawa couplings and the mirror map the Picard-Fuchs equations play a crucial role. So let us turn to them. It is quite easy to explain what they are but harder to find them explicitly. We will start with the easy part.

We know from the discussion in section two that the dimension of the third cohomology group is $\dim(H^3) = b_3 = 2(h^{2,1} + 1)$. Furthermore we know that the unique holomorphic three form Ω depends only on the complex structure. If we take derivatives with respect to the complex structure moduli, we will get elements in $H^{3,0} \oplus H^{2,1} \oplus H^{1,2} \oplus H^{0,3}$. Since b_3 is finite, there must be linear relations between derivatives of Ω of the form $\mathcal{L}\Omega = d\eta$ where $\mathcal{L}$ is a differential operator with moduli dependent coefficients. If we integrate this equation over an element of the third homology group H_3, i.e. over a closed three cycle, we will get a differential equation $\mathcal{L}\Pi_i = 0$ satisfied by the periods of Ω. They are defined as $\Pi_i(a) = \int_{\Gamma_i} \Omega(a)$. $\Gamma_i \in H_3(X, \mathbb{Z})$ and we have made the dependence on the complex structure moduli explicit. In general we will get a set of coupled linear partial differential equations for the periods of Ω. These equations are called Picard-Fuchs equations. In case we have only one complex structure modulus (i.e. $b_3 = 4$) one gets just one ordinary (in fact, hypergeometric) differential equation of order four. For the general case, i.e. if $b_3 \geq 1$, we will describe below how to set up a complete system of Picard-Fuchs equations.

For a detailed discussion of ordinary differential equations we recommend the book by Ince [66]. For the general case we have profited from the book by Yoshida [67]. There exists a vast mathematical literature on PF systems and the theory of complex moduli spaces; main results are collected in [68]. Two results which are of relevance for our discussion are that the global monodromy is completely reducible and that the PF equations have only regular singularities. The first result enables us to consider only a subset of the periods by treating only the moduli corresponding to $\tilde{h}^{2,1}$ out of $h^{2,1}$ moduli. The second result means that the PF equations are Fuchsian and we can use the theory developped for them.

A systematic, even though generally very tedious procedure to get the Picard-Fuchs equations for hypersurfaces in $\mathbb{P}[\vec{w}]$, is the reduction method due to Dwork, Katz and Griffiths. As shown in ref. [69] the periods $\Pi_i(a)$ of the holomorphic three form $\Omega(a)$ can be written as[21]

$$\Pi_i(a) = \int_{\Gamma_i} \Omega(a) = \int_\gamma \int_{\Gamma_i} \frac{\omega}{p(a)} , \quad i = 1, \ldots, 2(h^{2,1} + 1) . \tag{5.1}$$

Here

$$\omega = \sum_{i=1}^{5} (-1)^i w_i z_i dz_1 \wedge \ldots \wedge \widehat{dz_i} \wedge \ldots \wedge dz_5 ; \tag{5.2}$$

Γ_i is an element of $H_3(X, \mathbb{Z})$ and γ a small curve around the hypersurface $p = 0$ in the 4-dimensional embedding space. a_i are the complex structure moduli, i.e. the coefficients of the perturbations of the quasi-homogeneous polynomial p. The fact that $\Omega(a)$ as defined above is well behaved is demonstrated in [70].

The observation that $\frac{\partial}{\partial z_i} \left(\frac{f(z)}{p^r} \right) \omega$ is exact if $f(z)$ is homogeneous with degree such that the whole expression has degree zero, leads to the partial integration rule, valid under the integral ($\partial_i = \frac{\partial}{\partial z_i}$):

$$\frac{f \partial_i p}{p^r} = \frac{1}{r-1} \frac{\partial_i f}{p^{r-1}} \tag{5.3}$$

In practice one chooses a basis $\{\varphi_k(z)\}$ for the elements of the local ring $\mathcal{R} = \mathbb{C}[z_1, \ldots, z_{n+1}]/(\partial_i p)$. From the Poincaré polynomial associated to p [7] one sees that there are $(1, \tilde{h}^{2,1}, \tilde{h}^{2,1}, 1)$ basis elements with degrees $(0, d, 2d, 3d)$ respectively. The elements of degree d are the perturbing monomials. One then takes derivatives of the expressions $\pi_k = \int \frac{\varphi_k(z)}{p^{n+1}}$ ($n = \deg(\varphi_k)/d$) w.r.t. the moduli. If one produces an expression such that the numerator in the integrand is not one of the basis elements, one relates it, using the equations $\partial_i p = \ldots$, to

---

[21] Again, here and below we only treat the case of hypersurfaces in a single projective space. Complete intersections in products of projective spaces are covered in [17].

the basis and uses (5.3). This leads to the so called Gauss-Manin system of first order differential equations for the π_k which can be rewritten as a system of partial differential equations for the period. These are the Picard-Fuchs equations. In fact, the Picard-Fuchs equations just reflect the structure of the local ring and expresses the relations between its elements (modulo the ideal). It thus depends on the details of the ring how many equations and of which order comprise a complete system of Picard-Fuchs equations. To see this, let us consider a model with $\tilde{h}^{2,1} = 2$, i.e. we have two monomials at degree $d : \varphi_1$ and φ_2. Since the dimension of the ring at degree $2d$ is two, there must be one relation (modulo the ideal) between the three combinations φ_1^2, $\varphi_1\varphi_2$, φ_2^2. Multiplying this relation by φ_1 or φ_2 leads to two independent relations at degree three. Since the dimension of the ring at degree three is one, there must be one further relation at degree three. The system of Picard-Fuchs equations for models with $\tilde{h}^{2,1} = 2$ thus consists of one second and one third order equation [22]. For models with $\tilde{h}^{2,1} > 2$, a general statement is no longer possible. For instance, for $\tilde{h}^{2,1} = 3$, the simplest case is a system of three equations of second order. The three relations at degree $2d$ then generate all relations at degree $3d$. There are however cases where this is not the case and one has to add extra relations at degree three, leading to third order equations. Examples of this type are presented in [16].

It is clear from the discussion that the Picard-Fuchs equations we have obtained only contain those complex structure moduli for which there exists a monomial perturbation. Also, the method outlined above applies only to manifolds in projective spaces and not to manifolds embedded in more general toric varieties.

We will now describe an alternative and often more efficient way to obtain the Picard-Fuchs equations using the toric data of the hypersurfaces. The general method has been outlined, in the context of mirror symmetry, in [58] and is based on the generalized hypergeometric system of Gelf'and, Kapranov and Zelevinsky (GKZ) [71]. We will not decribe it in its generality here, since for our purposes the following simplified treatment is sufficient.

The way we will proceed is to compute one of the periods, the so-called fundamental period [19][58] directly and then set up a system of partial differential equations satisfied by this period. This system is the GKZ hypergeometric system. It is not quite yet the PF system since its solution space is larger than that of the PF system, which are the periods of the holomorphic three form, of which there are $2(\tilde{h}^{2,1} + 1)$. (As mentioned before, we are only able to treat a subset of the $2(h^{2,1} + 1)$ periods.) However the monodromy acts reducibly on the (larger) space of solutions and the periods are a subset on which it acts irreducibly. It is often (sometimes?) easy, starting with the GKZ system, to find a reduced system of differential equations which is then the PF system. This will be explained in the example that we will treat below.

---

[22] In these considerations we use the fact that the homogeneous subspace of $\mathcal{R}$ of degree nd is generated by the elements of degree d and furthermore that taking a derivative w.r.t. to the modulus parameter a_i produces one power of the corresponding monomial φ_i in the numerator of the period integral.

Before showing how to compute the fundamental period and how to extract the GKZ generalized hypergeometric system, we first have to discuss the correct choice of coordinates on moduli space. We have already mentioned that we will only discuss the mirror map in the neighbourhood of the large complex structure limit. The large complex structure limit corresponds to the point in moduli space where the periods have maximal unipotent monodromy [72]. This in particular means that the characteristic exponents of the PF equations are maximally degenerate. What this means is the following: if we make a power series ansatz for the solution of the PF equations one gets recursion relations for the coefficients. However, the condition that the lowest powers vanish, gives polynomial equations (of the order of the differential equations) for the (characteristic) exponents of the lowest order terms of the power series [23]. At the point of large complex structure the common zeros of these polynomials are all equal, and in fact, by a suitable moduli dependent rescaling of the period (this constitutes a choice of gauge, c.f. below) they can be chosen to be zero. According to the general theory of Frobenius we then get, in a neighbourhood of large complex structure, one power series solutions and the other solutions contain logarithms. We will have more to say about these solutions later. The problem now consists of finding the correct variables in which we can write a power series expansion with these properties.

To find these variables it is necessary to introduce the so-called lattice of relations. Among the $5 + \tilde{h}^{2,1}$ integer points $\nu_0, \ldots, \nu_{\tilde{h}^{2,1}+5}$ (ν_0 is the origin) in Δ which do not lie in the interior of faces of codimension one, i.e. those points to which we associate Laurent monomials in (4.13), there are relations of the form $\sum_i l_i v_i = 0$, $l_i \in \mathbb{Z}$. The vectors l generate the $\tilde{h}^{2,1}$ dimensional lattice of relations. In this lattice one defines a cone, the so-called Mori cone, whose minimal set of generators we denote by l^α, $\alpha = 1, \ldots, \tilde{h}^{2,1}$. (They are also a basis of the lattice of relations.) This cone is in fact the same as the one mentioned in our brief discussion about the Kähler condition of the resolution of singularities. We then define the extended vectors $(\bar{l}^\alpha) \equiv (-\sum_i l_i^\alpha, \{l_i^\alpha\}) \equiv (l_0^\alpha, \{l_i^\alpha\})$. In terms of the parameters appearing in the Laurent polynomial (4.13) the large complex structure limit is defined to be the point $u_1 = \ldots = u_{\tilde{h}^{2,1}} = 0$ in complex structure moduli space with $u_\alpha \equiv a^{\bar{l}^\alpha} \equiv \prod a_i^{\bar{l}_i^\alpha}$.

A systematic method to find the generators of Mori's cone has been reviewed in [16] where the construction of the upper convex piecewise linear functions was explained. An equivalent way[24] is as follows. Consider a particular 'triangulation' (i.e. decomposition into four-simplices with apex at the origin) of Δ with lattice points $\nu_0, \ldots, \nu_{\tilde{h}^{2,1}+5}$. Each four-simplex is then specified by four vertices (in addition to the origin ν_0). We then take any pair of four simplices which have a common three simplex and look for integer relations $\sum n_i \nu_i = 0$, $n_i \in \mathbb{Z}$ of the five non-trivial vertices such that the coefficients of the two vertices which

---

[23] In the case of ordinary differential equations this polynomial equation is called indicial equations; see e.g. [66].

[24] which we have learned from V. Batyrev; see also [73].

are not common to both four-simplices are positiv. This provides a set of relations. There will be $\tilde{h}^{2,1}$ independent relations in terms of which the others can be expanded with non-negative integer coefficients. They constitute the basis of Mori's cone and define the coordinates in the neighbourhood of large complex structure.

Let us now turn to the computation of the fundamental period. In the language of toric geometry the period integrals are written as [25]

$$\Pi_i(a) = \int_{\Gamma_i} \frac{a_0}{f(a,X)} \frac{dX_1}{X_1} \wedge \cdots \wedge \frac{dX_4}{X_4} \tag{5.4}$$

with $\Gamma_i \in H_4((\mathbb{C}^*)^4 \setminus Z_f)$ and $f(a,X)$ the Laurent polynomial (4.13). We get the fundamental period if we choose the cycle $\Gamma = \left\{ (X_1,\dots,X_4) \in \mathbb{C}^4 \big| |X_i| = 1 \right\}$, expand the integrand in a power series in $1/a_0$ and evaluate the integral using the residue formula. Straightforward computation gives

$$w_0(a) = \sum_{\substack{\mu_i \geq 0 \\ \sum \mu_i \nu_i = 0}} \frac{(\sum \mu_i)!}{\prod \mu_i!} \frac{\prod_i a_i^{\mu_i}}{a_0^{\sum \mu_i}} \tag{5.5}$$

i.e. the sum is over a subset of the lattice of relations. Since the generators of Mori's cone are a basis of the lattice of relations we can in fact express the relations to be summed over in the form $\sum n_\alpha l^\alpha$ and sum over the n_α. If we then introduce the variables u_α, the fundamental period becomes[26]

$$w_0(u) = \sum_{\{n_\alpha\}} \frac{(-\sum_\alpha l_0^\alpha n_\alpha)!}{\prod_{i>0} (\sum_\alpha l_i^\alpha n_\alpha)!} \prod_\alpha u_\alpha^{n_\alpha} \equiv \sum_n c(n) u^n \tag{5.6}$$

where the sum is over those n_α which leave the arguments of the factorials nonnegative. It is now straightforward to set up a system of $\tilde{h}^{2,1}$ partial differential equations which are satisfied by $w_0(u)$. Indeed, the coefficients $c(n)$ satisfy recursion relations of the form $p_\beta(n_\alpha, n_\beta+1) c(n_\alpha, n_\beta+1) - q_\beta(n_\alpha, n_\beta) c(n_\alpha, n_\beta) = 0$. Here p_α and q_α are polynomials of their respective arguments. The recursion relations translate to linear differential operators

$$\mathcal{L}_\beta = p_\beta(\Theta_\alpha, \Theta_\beta) - u_\beta q_\beta(\Theta_\alpha, \Theta_\beta) \tag{5.7}$$

where we have introduced the logarithmic derivatives $\Theta_\alpha = u_\alpha \frac{d}{du_\alpha}$. The order of the operator $\mathcal{L}_\beta$ equals the sum of the positive (or negative) components of $\bar{l}^\beta$. The system of linear differential equations $\mathcal{L}_\beta w(u) = 0$ is the generalized

[25] This in fact differs from the period defined before by a factor of a_0, which we have included for later convenience. This corresponds to a redefinition of $\Omega \rightarrow a_0 \Omega$.

[26] Here and below we will use the following notation: for a multi-index $n = (n_1,\dots,n_N)$ we define $|n| = n_1 + \dots + n_N$, $u^n = u_1^{n_1} \cdots u_N^{n_N}$ and also $n! = \prod n_i!$.

hypergeometric system of GKZ. Since it is related to the polyhedron Δ, it is also called the Δ-hypergeometric system.

The hypergeometric systems that one gets this way are not generic, but rather (semi-non)resonant in the language of [71] in which case the monodromy acts no longer irreducibly. It is in general not straightforward to extract the PF system from the GKZ system, but in simple cases the operators $\mathcal{L}_\alpha$ factorize $\mathcal{L}_\alpha = \ell_\alpha \mathcal{D}_\alpha$ and the $\mathcal{D}_\alpha$ form the complete PF system. The general situation, which might require an extension of the Δ-hypergeometric system, has been discussed in [16] and [17].

In any case, when the dust settles, one has a system of PF operators of the form

$$\mathcal{D}_\alpha = p_\alpha(\Theta) + \sum_\beta f_{\alpha\beta}(u) q_{\alpha\beta}(\Theta) \tag{5.8}$$

where $p_\alpha, q_{\alpha\beta}$ and $f_{\alpha\beta}$ are polynomials with $f_{\alpha\beta}(0) = 0$ and p_α is homogeneous and p_α and $q_{\alpha\beta}$ are of the same degree. The homogeneity of $p_\alpha(\Theta)$ follows from the characterization of the large complex structure by the requirement that the characteristic exponents of the PF differential equations should be maximally degenerate and the gauge choice which gives a power series solution that starts with a constant.

Note that the terms in (5.8) of top degree in the Θ correspond to relations between monomials of the same degree modulo the ideal, whereas the lower order terms correspond to terms in the ideal. This comment applies if we work with quasi-homogeneous polynomials.

A necessary condition for a set of period equations to be complete is that there be $2(\tilde{h}^{2,1}+1)$ degenerate characteristic exponents at $z = 0$. The polynomial ring

$$\mathcal{R} = \mathbb{C}[\Theta_1, \cdots, \Theta_{\tilde{h}^{2,1}}]/\{p_a(\Theta)\} \tag{5.9}$$

then has $(1, \tilde{h}^{2,1}, \tilde{h}^{2,1}, 1)$ elements at degrees (in Θ) $(0, 1, 2, 3)$. This in particular means that the symbols of $\mathcal{D}_\alpha$ generate the ideal of symbols [27]. This observation is important for the determination of the singular locus of the PF equations. But let us first explain what is being said [67].

The PF operators $\mathcal{D}_\alpha$ define a (left) ideal I in the ring of differential operators[28], i.e. $\mathcal{D} \in I \Leftrightarrow \mathcal{D}w = 0$. The symbol of a partial linear differential operator $\mathcal{D}$ in k variables of order m, i.e. $\mathcal{D} = \sum_{|p| \le m} a_p(u) \left(\frac{d}{du}\right)^p$, $(|p| = p_1 + \cdots + p_k)$ is defined as $\sigma(\mathcal{D}) = \sum_{|p|=m} a_p(u) \xi_1^{p_1} \cdots \xi_k^{p_k}$. $\xi_1, \ldots, \xi_k$ is a coordinate system on the fiber of the cotangent bundle T^*U at $z = 0$. The ideal of symbols is then $\sigma(I) = \{\sigma(\mathcal{D})|\mathcal{D} \in I\}$. The singular locus is $S(I) = \pi(\mathrm{Ch}(I) - U \times \{0\})$ where the characteristic variety $\mathrm{Ch}(I)$ is the subvariety in T^*U given by the ideal of symbols (i.e. given by $\sigma(\mathcal{D}) = 0, \mathcal{D} \in I$) and π the projection along the fiber, i.e. setting $\xi = 0$. The singular locus is also the discriminant of the CY hypersurface,

---

[27] This is a feature of PF systems and not generally true for generalized hypergeometric systems. A counterexample can be found in ref. [67].

[28] Here we work in a neighbourhood U of the origin of a coordinate system on $\mathbb{C}^{\tilde{h}^{2,1}}$.

i.e. the locus in moduli space where the manifold fails to be transverse [29]. We will demonstrate the method introduced here in the example in section seven.

Let us now turn to the discussion of the remaining solutions of the PF equations in the neighbourhood of the point $u = 0$. Due to the fact that the characteristic exponents are all degenerate, the fundamental period is the only power series solution. The other $(2\tilde{h}^{2,1} + 1)$ solutions contain logarithms of u. We will now show that there are $\tilde{h}^{2,1}$ solutions with terms linear in logarithms, the same number of solutions with parts quadratic and one solution with a part cubic in logarithms. This corresponds to the grading of the ring $\mathcal{R}$ (eq.(5.9)).

Extending the definition of $x! = \Gamma(x + 1)$ to $x \in \mathbb{R}$, and that of the coefficients $c(n + \rho)$ in (5.6) for arbitrary values of $\tilde{h}^{2,1}$ parameters ρ_α, we define the power series

$$w_0(u, \rho) = \sum c(n + \rho)u^{n+\rho}. \tag{5.10}$$

Clearly, setting $\rho = 0$ gives the fundamental period. By the method of Frobenius the logarithmic solutions are obtained by taking linear combinations of derivatives $D_\rho = \sum \frac{1}{(2\pi i)^n} \frac{b_n}{n!} \partial_\rho^n$ of $w_0(u, \rho)$, evaluated at $\rho = 0$. As $[\mathcal{D}_\alpha, \partial_{\rho_\alpha}] = 0$ it is then sufficient to check

$$D_\rho (\mathcal{D}_\alpha w_0(u, \rho))|_{\rho=0} = 0, \ \forall \alpha \tag{5.11}$$

to establish $D_\rho w_0(u, \rho)|_{\rho=0}$ as a solution. By consideration of the explicit form of the series (5.10) one can show that the conditions for vanishing of the constant terms in (5.11)

$$D_\rho(p_\alpha(\rho)c(0, \rho)u^\rho)|_{\rho=0} = 0, \ \forall \alpha, \tag{5.12}$$

are in fact also sufficient. A moment's thought shows that the following construction of the operators D_ρ is valid. We consider the ideal $\mathcal{I}$ in the polynomial ring $\mathbb{C}[\Theta]$, generated by the $p_\alpha(\Theta)$. We endow $\mathbb{C}[\Theta]$ with a natural vector space structure with the normalized monomials as orthonormal basis. We can then define $\mathcal{I}^\perp$ which consists of the (homogeneous) polynomials orthogonal to the elements in $\mathcal{I}$. If we denote a homogeneous element in $\mathcal{I}^\perp$ by $\sum b_n \theta^n$ then the D_ρ are simply

$$D_\rho^{(|n|)} = \sum \frac{1}{(2\pi i)^{|n|}} \frac{b_n}{n!} \partial_\rho^n. \tag{5.13}$$

Here the sum is over the n_i such that $\sum n_i = |n|$. The corresponding solutions contain up to $|n|$ powers of logarithms. Note that since the PF equations are always at least of order two, the solutions linear in logarithms are $w_i(u) = D_i^{(1)} w_0(u, \rho)\big|_{\rho=0} = \frac{1}{2\pi i} \frac{\partial}{\partial \rho_i} w_0(u, \rho)\big|_{\rho=0}$ $(i = 1, \ldots, \tilde{h}^{2,1})$. In section six we will show that as a result of mirror symmetry, the operators $D^{(3)}$ and $D^{(2)}$ are of the form $D^{(3)} = -\frac{1}{3!} \frac{1}{(2\pi i)^3} \sum \kappa_{ijk}^0 \partial_{\rho_i} \partial_{\rho_j} \partial_{\rho_k}$ and $D_i^{(2)} = \frac{1}{2} \frac{1}{(2\pi i)^2} \sum \kappa_{ijk}^0 \partial_{\rho_j} \partial_{\rho_k}$ with κ^0 the topological triple couplings in a basis to be introduced there.

---

[29] An alternative way to determine the discriminant of hypersurfaces in $\mathbb{P}_\Delta$ was given in [49].

To summarize the discussion above we collect all the solutions to the PF equations into the period vector:

$$\Pi(z) = \begin{pmatrix} w_0(u) \\ D_i^{(1)} w_0(u,\rho)|_{\rho=0} \\ D_i^{(2)} w_0(u,\rho)|_{\rho=0} \\ D^{(3)} w_0(u,\rho)|_{\rho=0} \end{pmatrix}. \tag{5.14}$$

The final point we want to discuss in this section is how to compute the $\langle \overline{27}^3 \rangle$ Yukawa couplings as functions of the complex structure moduli. In the next section we will relate them, via the mirror map, to the $\langle 27^3 \rangle$ couplings on the mirror manifold.

We have already given an expression for the $\langle \overline{27}^3 \rangle$ couplings in eq.(2.2). There is in fact a more convenient (for what is going to come) way to write the same expression in terms of bilinears of the (derivatives of the) periods of Ω [74] [75]. For this purpose we introduce an integral basis of $H^3(X, \mathbb{Z})$ with generators α_i and β^j ($i,j = 0,\ldots,h^{2,1}$) which are dual to a canonical homology basis (A^i, B_j) of $H_3(X, \mathbb{Z})$ with intersection numbers $A^i \cdot A^j = B_i \cdot B_j = 0$, $A^i \cdot B_j = \delta^i_j$. Then

$$\int_{A^j} \alpha_i = \int_X \alpha_i \wedge \beta^j = -\int_{B_i} \beta^j = \delta^j_i \tag{5.15}$$

with all other pairings vanishing. This basis is unique up to $Sp(2(h^{2,1}+1), \mathbb{Z})$ transformations.

A complex structure on X is now fixed by choosing a particular 3-form as the holomorphic (3,0) form Ω. It may be expanded in the above basis of $H^3(X, \mathbb{Z})$ as $\Omega = z^i \alpha_i - \mathcal{F}_i \beta^i$ where $z^i = \int_{A^i} \Omega$, $\mathcal{F}_i = \int_{B_i} \Omega$ are periods of Ω. As shown in [76] and [77] the z^i are local complex projective coordinates for the complex structure moduli space, i.e. we have $\mathcal{F}_i = \mathcal{F}_i(z)$. The coordinates z^i are called special projective coordinates. They are related to the special coordinates of eq.(3.6), say in a patch where $z^0 \neq 0$, i.e. $\lambda_i = \frac{z^i}{z^0}$. Under a change of complex structure Ω, which was pure (3,0) to start with, becomes a mixture of (3,0) and (2,1), i.e. $\frac{\partial}{\partial z^i}\Omega \in H^{(3,0)} \oplus H^{(2,1)}$. In fact [26] $\frac{\partial \Omega}{\partial z^i} = k_i \Omega + b_i$ where $b_i \in H^{(2,1)}$ is related to elements in $H^1(M, T_X)$ via Ω and k_i is a function of the moduli but independent of the coordinates of X. One immediate consequence is that $\int \Omega \wedge \frac{\partial \Omega}{\partial z^i} = 0$. Inserting the expression for Ω in this equation, one finds $\mathcal{F}_i = \frac{1}{2}\frac{\partial}{\partial z^i}(z^j \mathcal{F}_j)$, or $\mathcal{F}_i = \frac{\partial \mathcal{F}}{\partial z^i}$ with $\mathcal{F} = \frac{1}{2}z^i \mathcal{F}_i(z)$, $\mathcal{F}(\mu z) = \mu^2 \mathcal{F}(z)$. From $\frac{\partial^2}{\partial z^i \partial z^k}\Omega \in H^{(3,0)} \oplus H^{(2,1)} \oplus H^{(1,2)}$ it immediately follows that also $\int \Omega \wedge \frac{\partial^2}{\partial z^i \partial z^j}\Omega = 0$. In fact, this is already a consequence of the homogeneity of $\mathcal{F}$. Finally, $\frac{\partial^3}{\partial z^i \partial z^j \partial z^k}\Omega \in H^{(3,0)} \oplus H^{(2,1)} \oplus H^{(1,2)} \oplus H^{(0,3)}$ and one easily finds $\int \Omega \wedge \frac{\partial^3}{\partial z^i \partial z^j \partial z^k}\Omega = \frac{\partial^3}{\partial z^i \partial z^j \partial z^k}\mathcal{F} = (z^0)^2 \frac{\partial^3}{\partial \lambda_i \partial \lambda_j \partial \lambda_k}F$ where $\mathcal{F} = (z^0)^2 F$ (cf. (3.6)); here $i,j,k = 1,\ldots,h^{2,1}$. If one computes the (0,3) part of $\partial^3 \Omega$ explicitly [75], one recovers indeed the couplings $\bar{\kappa}_{ijk}$ in (2.2). From the discussion above it also follows that under a change of coordinates $\lambda_i \to \tilde{\lambda}_i(t)$ the Yukawa couplings transform homogeneously and thus $\bar{\kappa}_{ijk} = \int \Omega \wedge \partial_i \partial_j \partial_k \Omega$ holds in any coordinate system, whereas it can be written as the third derivative of the prepotential only in special coordinates. If we redefine $\Omega \to \frac{1}{z_0}\Omega$, the periods are $(1, \lambda_i, \frac{\partial}{\partial \lambda_i}F, 2F - \lambda_i \frac{\partial}{\partial \lambda_i}F)$. We also note that $K = -\ln \int \Omega \wedge \bar{\Omega}$, which is easily shown to be in agreement with (3.6), up to a Kähler transformation.

We are now ready to link the Yukawa couplings to the PF equations. In inhomogeneous coordinates λ_i the Yukawa couplings are

$$\bar{\kappa}_{ijk} = \int \Omega \wedge \frac{\partial^3}{\partial \lambda_i \partial \lambda_j \partial \lambda_k}\Omega = \sum_{l=0}^{\bar{h}^{2,1}}(z^l \partial_i \partial_j \partial_k \mathcal{F}_l - \mathcal{F}_l \partial_i \partial_j \partial_k z^l) \qquad (5.16)$$

where z^l and $\mathcal{F}_l$ are periods of Ω (in a symplectic basis, i.e. for the particular choice of cycles A^l, B_l as specified above).

We now define

$$W^{(k_1, \cdots, k_d)} = \sum_l [z^l \partial_{\lambda_1}^{k_1} \cdots \partial_{\lambda_d}^{k_d} \mathcal{F}_l - \mathcal{F}_l \partial_{\lambda_1}^{k_1} \cdots \partial_{\lambda_d}^{k_d} z^l]$$
$$:= \sum_l (z^l \partial^k \mathcal{F}_l - \mathcal{F}_l \partial^k z^l) \ . \qquad (5.17)$$

In this notation, $W^{(\mathbf{k})}$ with $\sum k_i = 3$ describes the various types of Yukawa couplings and $W^{(\mathbf{k})} \equiv 0$ for $\sum k_i = 0, 1, 2$.

If we now write the Picard-Fuchs differential operators in the form

$$\mathcal{D}_\alpha = \sum_{\mathbf{k}} f_\alpha^{(\mathbf{k})} \partial^{\mathbf{k}} \ , \qquad (5.18)$$

then we immediately obtain the relation

$$\sum_{\mathbf{k}} f_\alpha^{(\mathbf{k})} W^{(\mathbf{k})} = 0 \ . \qquad (5.19)$$

Further relations are obtained from operators $\partial_{\lambda_\beta} \mathcal{D}_\alpha$. If the system of PF differential equations is complete, it is sufficient for deriving linear relations among the Yukawa couplings and their derivatives, which can be integrated to give the Yukawa couplings up to an overall normalization. In the derivation, we need to use the following relations which are easily derived

$$W^{(4,0,0,0)} = 2\partial_{\lambda_1} W^{(3,0,0,0)}$$

$$W^{(3,1,0,0)} = \frac{3}{2}\partial_{\lambda_1} W^{(2,1,0,0)} + \frac{1}{2}\partial_{\lambda_2} W^{(3,0,0,0)}$$

$$W^{(2,2,0,0)} = \partial_{\lambda_1} W^{(1,2,0,0)} + \partial_{\lambda_2} W^{(2,1,0,0)}$$

$$W^{(2,1,1,0)} = \partial_{\lambda_1} W^{(1,1,1,0)} + \frac{1}{2}\partial_{\lambda_2} W^{(2,0,1,0)} + \frac{1}{2}\partial_{\lambda_3} W^{(2,1,0,0)}$$

$$W^{(1,1,1,1)} = \frac{1}{2}(\partial_{\lambda_1} W^{(0,1,1,1)} + \partial_{\lambda_2} W^{(1,0,1,1)} + \partial_{\lambda_3} W^{(1,1,0,1)} + \partial_{\lambda_4} W^{(1,1,1,0)}) \ .$$

$$(5.20)$$

By symmetry the above relations exhaust all possibilities.

We have now described all the calculation that have to be done on the mirror manifold $\hat{X}^*$, namely the computation of the couplings $\bar{\kappa}_{ijk}(\hat{X}^*)$[30]. In the next section we show how to go back to the manifold $\hat{X}$ on which we want to compute the $\langle 27^3 \rangle$ Yukawa couplings.

6 Mirror Map and Applications of Mirror Symmetry

The question now arises whether mirror symmetry is merely a hitherto unknown mathematical curiosity or whether it can also be used as a practical tool in string theory. The demonstration that this is indeed the case will be attempted in what follows. We will show how mirror symmetry allows for the computation of the otherwise difficult (if not impossible) to get Kähler moduli dependence of the $\langle 27^3 \rangle$ Yukawa couplings. The methods developped can, in principle, be applied to any CY hypersurface and incorporates the dependence of those Kähler moduli which correspond to toric divisors. Given a general model it is however in practice technically rather cumbersome to actually carry the program through. However, for models with few moduli it can and has been done successfully. A simple but non-trivial example will be given in the following section. But before turning to it, we still need a few ingredients.

In the previous section we have shown how to compute, via the PF equations, the $\langle \overline{27}^3 \rangle$ couplings on a Calabi-Yau space given as a hypersurface of a toric variety. If we now also want to compute the $\langle 27^3 \rangle$ couplings, we proceed as follows. We go to the mirror manifold $\hat{X}^*$ and compute the $\langle \overline{27}^3 \rangle$ couplings there and then use mirror symmetry to go back to $\hat{X}$. What this requires is to find the map from the complex structure moduli space with coordinates u_α, $(\alpha = 1, \ldots, \tilde{h}^{2,1}(\hat{X}^*))$ to the Kähler structure moduli space on $\hat{X}$ with coordinates t_i, $(i = 1, \ldots, \tilde{h}^{1,1}(\hat{X}) = \tilde{h}^{2,1}(\hat{X}^*))$. The mirror hypothesis then states that the two Yukawa couplings transform into each other under this transformation. What one has to take into account is the transformation properties of the Yukawas under coordinate transformations: for $t_i \to \tilde{t}_i(t)$ they transform as $\kappa_{ijk}(t) = \frac{\partial \tilde{t}_l}{\partial t_i} \frac{\partial \tilde{t}_m}{\partial t_j} \frac{\partial \tilde{t}_n}{\partial t_k} \tilde{\kappa}_{lmn}(\tilde{t}(t))$. Another point to consider is the normalization of the periods and consequently also of the Yukawas which are quadratic in the periods. In fact, a change of Kähler gauge[31] $\Omega \to f(u)\Omega$ results in an change of the Yukawa couplings $\bar{\kappa}_{ijk} \to f^2(u)\bar{\kappa}_{ijk}$. The gauge we choose is such that the fundamental period is one, i.e. $f(u) = 1/w_0(u)$. The $\tilde{h}^{2,1}$ solutions linear in logarithms are then

$$t_i(z) = \frac{w_i(u)}{w_0(u)}. \tag{6.1}$$

---

[30] In fact, if we are only interested in the couplings $\kappa_{ijk}(\hat{X})$ we do not need the $\bar{\kappa}_{ijk}(\hat{X}^*)$ explicitly, as will become clear below.

[31] This corresponds to a Kähler transformation of the Kähler potential $K = -\log\left(\int \Omega \wedge \bar{\Omega}\right)$ of moduli space.

They serve as the coordinates on the Kähler moduli space on $\hat{X}$ in the neighbourhood of infinite radius which is obtained for $\text{Im}(t_i) \to \infty$ (recall that $\text{Im}(t_i)$ are the real moduli). Equations (6.1) define the mirror map. This coordinate choice can be identified with the special coordinates of special geometry[32] [77][78]. As discussed in [79], in these coordinates the Picard-Fuchs differential equations can be written in the form

$$\sum_{i=1}^{k} \partial_j \partial_p (K_r^{-1})^{l_i} \partial_i \partial_r \, \Pi(t) = 0, \tag{6.2}$$

where $K_{ijk} = \partial_i \partial_j \partial_k F$ is derived from the prepotential $\mathcal{F} = w_0^2 F$ ($\partial_i = \frac{\partial}{\partial t_i}$). This system of fourth order equations can be rewritten as a system of linear differential equations, the Gauss-Manin system in special coordinates. The solutions of (6.2) are easily written down in terms of F: $\Pi(t) = (1, t_i, \partial_i F, 2F - t_i \partial_i F)$. Note that these are the periods in the canonical basis discussed in section five, after going to inhomogeneous coordinates. The mirror conjecture now states that $F(t)$ can also be identified with the prepotential for the Kähler structure moduli of the manifold X.

The Yukawa couplings are then

$$\kappa_{ijk}(t) = \partial_{t_i} \partial_{t_j} \partial_{t_k} F(t) = \frac{1}{w_0(u(t))^2} \frac{\partial u_\alpha(t)}{\partial t_i} \frac{\partial u_\beta(t)}{\partial t_j} \frac{\partial u_\gamma(t)}{\partial t_k} \bar{\kappa}_{\alpha\beta\gamma}(u(t)) \tag{6.3}$$

Here $\bar{\kappa}_{\alpha\beta\gamma}$ are the Yukawa couplings on the mirror manifold which we showed how to compute in the previous section. In order to espress the couplings κ_{ijk} in terms of the Kähler moduli t_i, we have to invert the expressions $t_i(u) = \frac{w_i(u)}{w_0(u)} = \frac{1}{2\pi i} \log(u_i) + O(u)$ which leads to expressions of the form $u_i = q_i(1 + O(q))$. Here we have defined $q_j = e^{2\pi i t_j}$. This then provides the Yukawa couplings as a power series in the variables q_j.

We now want to compare this with the general form for these Yukawa couplings given in eq.(2.3). For this we introduce the (multi) degree of the curve C, which is defined as $n_i = \int_C h_i \in \mathbb{Z}$ for $h_i \in H^{(1,1)}(X, \mathbb{Z})$. For the integral over the Kähler form we then get $\int_C J = \sum t_i n_i$. In terms of the degrees and the variables q_i the Yukawa coupling is

$$\kappa_{ijk} = \kappa_{ijk}^0 + \sum_{\{n_i\}} \frac{N(\{n_i\}) n_i n_j n_k \prod_l q_l^{n_l}}{1 - \prod_l q_l^{n_l}} \tag{6.4}$$

where $N(\{n_i\})$ are integers. In the simplest case of isolated non-singular rational curves C, they give the number of curves at degree $\{n_i\}$. More generally they have to be interpreted as Euler numbers of a suitably compactified moduli space of holomorphic maps of degree $\{n_i\}$ from $\mathbb{P}^1$ (the genus zero world-sheet) to the CY manifold.

---

[32] These are the so-called flat coordinates of the associated topological field theory.

Before turning to the example in the next section, we want to say a few words about the prepotential. If we introduce homogeneous coordinates on the Kähler structure moduli space of $\hat{X}$ via $t_i = z^i/z^0$, then the most general ansatz for the prepotential $\mathcal{F}(z) = (z^0)^2 F(t)$ which respects homogeneity, is

$$
\begin{aligned}
\mathcal{F} &= \frac{1}{6}\kappa^0_{ijk}\frac{z^i z^j z^k}{z^0} + \frac{1}{2}a_{ij}z^i z^j + b_i z^i z^0 + \frac{1}{2}c(z^0)^2 + \mathcal{F}_{\text{inst.}} \\
&= (z^0)^2\left(\frac{1}{6}\kappa^0_{ijk}t_i t_j t_k + \frac{1}{2}a_{ij}t_i t_j + b_i t_i + \frac{1}{2}c + F_{\text{inst.}}(t)\right)
\end{aligned}
\tag{6.5}
$$

We have split the prepotential into the classical intersection part and the instanton part ($F_{\text{inst.}}$); κ^0_{ijk} are the classical intersection numbers. The constants a_{ij}, b_i and c do not enter the Yukawa couplings $\kappa_{ijk}(t) = \partial_i\partial_j\partial_k F(t)$. Their real parts are also irrelevant for the Kähler potential. There is a continuous Peccei-Quinn symmetry $t_i \to t_i + \alpha_i$, α_i real [33], which is broken by instanton corrections to discrete shifts [80]. Requiring this symmetry in the absence of instanton corrections gives $\text{Im}(a_{ij}) = \text{Im}(b_i) = 0$.

From the function $\mathcal{F}$, viewed as the pre-potential for the Kähler structure moduli space on X, we construct the vector $(z^0, z^i, (\partial\mathcal{F}/\partial z^i), (\partial\mathcal{F}/\partial z^0)) \equiv z^0\Pi(t)$ with

$$
\Pi(t) = \begin{pmatrix} 1 \\ t_i \\ \frac{1}{2}\kappa^0_{ijk}t_j t_k + a_{ij}t_j + b_i + \partial_i(F_{\text{inst.}}) \\ -\frac{1}{6}\kappa^0_{ijk}t_i t_j t_k + b_i t_i + c + O(e^{2\pi it}) \end{pmatrix}
\tag{6.6}
$$

The mirror conjecture now says that this is the same as the period vector (5.14). Comparing the last components of these two vectors, using that $\log z_j = 2\pi it_j + O(t^2)$ we verify that, up to an overall normalization, the coefficients of the operator $D^{(3)}_\rho$ are indeed the topological couplings. We also conclude that the fully instanton corrected couplings κ_{ijk} are given by the concise expressions

$$
\kappa_{ijk}(t) = \partial_{t_i}\partial_{t_j}\frac{D^{(2)}_k w_0(u(t),\rho)|_{\rho=0}}{w_0(u(t))}
\tag{6.7}
$$

Let us finally turn to the constants a_{ij}, b_i and c in the prepotential. From what has been said we see that they can be expressed in terms as the expansion coefficient $c(0,\rho)$ in eq.(5.6): $c = D^{(3)}c(0,\rho)|_{\rho=0}$, $b_i = D^{(2)}_i c(0,\rho)|_{\rho=0}$ and $a_{ij} = 0$. One finds

$$
\frac{\partial}{\partial\rho_\beta}c(0) = -(l^\beta_0 + \sum_{i>0}l^\beta_i)\Gamma'(1) \equiv 0
$$

$$
\frac{\partial}{\partial\rho_\beta}\frac{\partial}{\partial\rho_\gamma}c(0) = \frac{\pi^2}{6}\left(l^\beta_0 l^\gamma_0 - \sum_{i>0}l^\beta_i l^\gamma_i\right)
\tag{6.8}
$$

$$
\frac{\partial}{\partial\rho_\alpha}\frac{\partial}{\partial\rho_\beta}\frac{\partial}{\partial\rho_\gamma}c(0) = 2\left(l^\alpha_0 l^\beta_0 l^\gamma_0 + \sum_{i>0}l^\alpha_i l^\beta_i l^\gamma_i\right)\zeta(3)
$$

---

[33] Under constant shifts of $\text{Re}(t_i)$ the sigma-model action (1.1) changes according to $\Delta S \sim \int_\Sigma d^2z\, b_{i\bar{j}}(\phi)(\partial\phi^i\bar{\partial}\phi^{\bar{j}} - \bar{\partial}\phi^i\partial\phi^{\bar{j}}) = \int_{\Phi(\Sigma)}b_{i\bar{j}}d\phi^i \wedge d\phi^{\bar{j}} = \int_{\Phi(\Sigma)}b$. For $\Phi(\Sigma)$ topologically trivial, $b = da$ and $\Delta S = 0$.

An interesting observation is now that for all hypersurfaces that we have treated explicitly, the following relations hold:

$$\chi(\hat{X}) = \int_{\hat{X}} c_3 = \frac{1}{3}\kappa^0_{\alpha\beta\gamma}(l_0^\alpha l_0^\beta l_0^\gamma + \sum_{i>0} l_i^\alpha l_i^\beta l_i^\gamma)$$

$$\int_{\hat{X}} c_2 \wedge h_\alpha = \frac{1}{2}\kappa^0_{\alpha\beta\gamma}(l_0^\beta l_0^\gamma - \sum_{i>0} l_i^\beta l_i^\gamma)$$

(6.9)

For the case of singular hypersurfaces we have no proof of these relations. For non-singular complete intersections in products of projective spaces equivalent formulas are derived and proven in [16]. Above results lead to the following expressions for the constants b_i and c:

$$b_i = \frac{1}{24}\int_{\hat{X}} c_2 \wedge h_i \qquad c = \frac{1}{(2\pi i)^3}\chi(\hat{X})\xi(3)$$

(6.10)

We thus find that c, being imaginary, is the only relevant contribution of a, b, c to the Kähler potential. In fact the value of c reproduce the expected contribution from the σ model loop calculation. Moreover (6.10) reproduces the values for b and c of all examples where the prepotential is derived by specifying the integral basis[12][13]. We therefore conjecture that the prepotential (6.5) describes in general the Yukawa couplings *and* the Kähler metric for $\mathcal{M}_{h^{2,1}}(\hat{X}^*)$ and therefore by mirror hypothesis also on $\mathcal{M}_{h^{1,1}}(\hat{X})$ in the region of convergence of the large complex (Kähler) structure expansion.

We now have to comment on the topological couplings $\kappa^0_{\alpha\beta\gamma}$. The indices refer to the coordinates t_α, which, via the mirror map (6.1), are related to the u_α. In the large radius limit we have $t_\alpha = \frac{1}{2\pi i}\log(u_\alpha)$. The coordinates u_α are monomials in the perturbation parameters in the Laurent polynomial f_Δ. These parameters are in one-to-one correspondence with (exceptional) divisors on $\hat{X}$. To the coordinates u_α we thus have to associate linear combinations of divisors, or, equivalently, of harmonic $(1,1)$ forms. For Fermat hypersurfaces this is done in the following way. If $h_J, h_{D_1}, \ldots, h_{D_n}$, $(n = \tilde{h}^{2,1}(\hat{X}^*) - 1 = \tilde{h}^{1,1}(\hat{X}) - 1)$ are the harmonic $(1,1)$ forms corresponding to the basis $a_0, \ldots, a_n$, the forms corresponding to the basis $u_1, \ldots, u_{\tilde{h}^{2,1}}$ are,

$$h_J = h_1 \quad h_{D_i} = \sum_\alpha l^\alpha_{i+5} h_\alpha$$

(6.11)

7 A Two Moduli Example: Hypersurface in $\mathbb{P}^4[2,2,2,1,1]$

After all the rather formal discussions, we now want to present an example for which we choose the degree eight Fermat CY hypersurface in $\mathbb{P}^4[2,2,2,1,1]$ defined by

$$p_0(z) = z_1^4 + z_2^4 + z_3^4 + z_4^8 + z_5^8 = 0$$

(7.1)

We will end up with the various Yukawa couplings between the to 27-plets of E_6, in particular their dependence on the Kähler moduli. This will also provide the number of rational curves at all degrees (instanton numbers).

Being a Fermat model, both Δ and Δ^* are simplicial and their corners can be easily written down using (4.17). Constructing the remaining lattice points in Δ and Δ^* and applying formulas (4.14) gives $h^{1,1}(\hat{X})=2$ and and $h^{2,1}(\hat{X}) = 86$ and the reversed numbers for the mirror partner. Note however that $\tilde{h}^{2,1}(\hat{X}) = 83$. This corresponds to the fact that we can only incorporate 83 complex structure deformations as monomial perturbations of the quasi-homogeneous polynomial p_0. In fact, one readily writes down the most general quasi-homogeneous polynomial of degree eight. It has 105 monomials. Using the freedom of homogeneous coordinate redefinitions, which provide 22 parameters, we are left with 83 possible monomial deformations which we might choose to be the degree eight elements of the ring $\mathcal{R} = \frac{\mathbb{C}[z_1,...,z_5]}{\{\partial_i p_0\}}$. An easy way to see this is that if we make infinitesimal homogeneous coordinate transformations p_0 changes (to first order) by terms in the ideal.

We also understand the number of Kähler moduli. The model has a singular $\mathbb{Z}_2$ curve C and the exceptional divisor is $C \times \mathbb{P}^1$. We thus have $h^{1,1} = 2$, one of the forms coming from the embedding space and the second from the exceptional divisor. Since we have a singular curve we expect a lattice point on a face of Δ of dimension one (i.e. on an edge). This point is $\nu_6 = (-1,-1,-1,0) = \frac{1}{2}(\nu_5 + \nu_4)$. Via the monomial-divisor map it corresponds to the perturbation $z_4^4 z_5^4$. The Laurent polynomial for the mirror is thus

$$f_\Delta(X) = a_0 - \left(a_1 X_1 + a_2 X_2 + a_3 X_3 + a_4 X_4 + \frac{a_5}{X_1^2 X_2^2 X_3^2 X_4} + \frac{a_6}{X_1 X_2 X_3} \right) \quad (7.2)$$

where we may use the freedom to rescale the variables X_i and the polynomial to set $a_1 = \ldots = a_5 = 1$. Equivalently we can write the homogeneous polynomial $p(z) = p_0(z) - a_0 z_1 \cdots z_5 - a_6 z_4^4 z_5^4$.

There are two independent relations between the lattice points which can be represented by the following two generators of the lattice of relations: $l^{(1)} = (-4,1,1,1,0,0,1)$ and $l^{(2)} = (0,0,0,0,1,1,-2)$. They do in fact generate the Mori cone and thus define the coordinates in the neighbourhood of the large complex structure point: $u_1 \equiv u = \frac{a_1 a_2 a_3 a_6}{a_0^4}$ and $u_2 \equiv v = \frac{a_4 a_5}{a_6^2}$. The fundamental period is

$$w_0(u,v) = \sum c(n,m) u^n v^m \quad (7.3)$$

with

$$c(n,m) = \frac{(4n)!}{(n!)^3 (m!)^2 (n-2m)!}$$

The Picard-Fuchs equations are then found to be $\mathcal{D}_{1,2} \omega(u,v) = 0$ with

$$\begin{aligned}
\mathcal{D}_1 &= \Theta_u^2 (\Theta_u - 2\Theta_v) - 4u(4\Theta_u + 3)(4\Theta_u + 2)(4\Theta_u + 1) \\
\mathcal{D}_2 &= \Theta_v^2 - v(2\Theta_v - \Theta_u + 1)(2\Theta_v - \Theta_u)
\end{aligned} \quad (7.4)$$

where $\mathcal{L}_1 = \Theta_u \mathcal{D}_1$ and $\mathcal{L}_2 = \mathcal{D}_2$.

From the Picard-Fuchs equations we can read off the symbols $\sigma(\mathcal{D}_1)$ and $\sigma(\mathcal{D}_2)$ which generate the ideal of symbols:

$$\sigma(\mathcal{D}_1) = u^2\xi_u^2\left(u\xi_u(1 - 4^4u^2) - 2v\xi_v\right)$$
$$\sigma(\mathcal{D}_2) = v^2\xi_v^2 - v(2v\xi_v - u\xi_u)^2 \qquad (7.5)$$

To get the discriminant we have to look for simultaneous solutions of $\sigma(\mathcal{D}_1) = \sigma(\mathcal{D}_2) = 0$ other than $\xi_u = \xi_v = 0$, which leads to the characteristic variety $\mathrm{Ch}(I)$. Setting $\xi_u = \xi_v = 0$ then gives the discriminant. It is straightforward to verify that $\mathrm{dis}(\hat{X}^*) = \Delta_1\Delta_2\Delta_3\Delta_4$ with

$$\Delta_1 = (1 - 512u + 65536u^2 - 262144u^2v), \Delta_2 = (1 - 4v), \Delta_3 = u, \Delta_4 = v \quad (7.6)$$

being its irreducible components.

The PF equations also determine the Yukawa couplings (up to an overall multiplicative constant). From the PF equations we derive three third order equations with operators $\Theta_u\mathcal{D}_1$, $\Theta_v\mathcal{D}_1$ and $\mathcal{D}_2$ which provide three linear relations between the four different Yukawa couplings. We can thus express all of them in terms of one for which we derive two linear first oder differential equations which can be integrated. The final result is:

$$K_{uuu} = c\frac{1}{\Delta_1\Delta_3^3}, \quad K_{uuv} = c\frac{1 - 256u}{2\Delta_1\Delta_3^2\Delta_4}$$
$$K_{uvv} = c\frac{512u - 1}{\Delta_1\Delta_2\Delta_3\Delta_4}, \quad K_{vvv} = c\frac{1 - 256u + 4v - 3072uv}{2\Delta_1(\Delta_2\Delta_4)^2} \qquad (7.7)$$

where c is an integration constant which will be fixed below. These are the $\langle\overline{27}^3\rangle$ couplings on the manifold $\hat{X}^*$. We now perform the mirror map to compute the $\langle 27^3\rangle$ couplings on $\hat{X}$. To construct the variables $t_i = \frac{w_i(u)}{w_0(u)}$ we need the solutions to the PF equations which are linear in logarithms of the variables. Following our discussion in section five we write

$w_i(u)$

$$= \frac{1}{2\pi i}\frac{\partial}{\partial\rho_i}w_0(u, \rho)$$

$$= \frac{1}{2\pi i}\frac{\partial}{\partial\rho_i}\sum_{n,m}\frac{\Gamma(4(n + \rho_1) + 1)}{\Gamma^3(n+\rho_1+1)\Gamma^2(m+\rho_2+1)\Gamma(n-2m+\rho_1-2\rho_2+1)}u^{n+\rho_1}v^{m+\rho_2}\Big|_{\rho_1=\rho_2=0}$$

$$= w_0(u)\log u_i + \tilde{w}_i(u)$$

$$(7.8)$$

where

$$\tilde{w}_i(u) = \frac{1}{2\pi i}\sum_{m,n}d_i(n, m)u^nv^m \qquad (7.9)$$

with ($\psi = \Gamma'/\Gamma$)

$$d_1(n,m) = \{4\psi(4n+1) - 3\psi(n+1) - \psi(n-2m+1)\}c(n,m)$$
$$d_2(n,m) = \{-2\psi(m+1) + 2\psi(n-2m+1)\}c(n,m) \tag{7.10}$$

We can also write down the remaining solutions of the PF equations. The ideal $\mathcal{I}$ is generated by $I_1 = \Theta_u^2(\Theta_u - 2\Theta_v)$ and $I_2 = \Theta_v^2$, so that a basis of $\mathcal{I}^\perp$ is $\{1, \Theta_u, \Theta_v, \Theta_u^2, \Theta_u\Theta_v, 2\Theta_u^3 + \Theta_u^2\Theta_v\}$. The elements at degrees zero and one lead, via (5.13), to the periods w_0 and $w_{1,2}$ already given above. The elements at degrees two and three give the remaining solutions, with up to two and three powers of logarithms, respectively.

The topological triple couplings, i.e. the infinite radius limit of the $\langle 27^3 \rangle$ couplings on $\hat{X}$, are, using the rules given in [16] and an obvious notation, $\kappa^0 = 8J^3 - 8JD^2 - 16D^3$. Here J and D refer to the divisors which correspond to the two lattice points ν_0 and ν_6, respectively with the associated moduli parameters a_0 and a_6. To transform to the divisors whose associated moduli parameters are u and v, we have to take linear combinations (c.f. (6.11)) and define $D_1 = J$ and $D_2 = \frac{1}{2}(J - D)$. In this basis the triple intersection numbers are $\kappa^0 = 8D_1^3 + 4D_1^2D_2$. They will be used to normalize the couplings $\langle 27^3 \rangle$. Ratios of the couplings in the latter basis can also be read off from the elements of $\mathcal{I}^\perp$: $D_i^{(2)} = \frac{1}{2}\kappa^0_{ijk}\partial_{\rho_j}\partial_{\rho_k}$ and $D^{(3)} = -\frac{1}{6}\kappa^0_{ijk}\partial_{\rho_i}\partial_{\rho_j}\partial_{\rho_k}$.

Now we are almost done. What is left to do is to make the coordinate transformation from the complex structure moduli space of $\hat{X}^*$ with coordinates u, v to the Kähler structure moduli space of $\hat{X}$ with coordinates t_1, t_2 and to go to the gauge $w_0 = 1$, i.e. to divide the Yukawa couplings by $(w_0(z(t)))^2$ and express them in terms of $q_i = e^{2\pi i t_i}$ (this involves inversion of power series). At lowest orders this leads to the expansions

$$\kappa_{111} = \kappa^0_{111} + N(1,0)\,q_1 + (N(1,0) + 8N(2,0))\,q_1^2 + N(1,1)\,q_1q_2 + O(q^3)$$
$$= 8 + 640\,q_1 + 80896\,q_1^2 + 640\,q_1q_2 + O(q^3)$$
$$\kappa_{112} = \kappa^0_{112} + N(1,1)\,q_1q_2 + O(q^3) = 4 + 640\,q_1q_2 + O(q^3)$$
$$\kappa_{122} = \kappa^0_{122} + N(1,1)\,q_1q_2 + O(q^3) = 640\,q_1q_2 + O(q^3)$$
$$\kappa_{222} = \kappa^0_{222} + N(0,1)q_2 + (N(0,1) + 4N(0,2))\,q_2^2 + N(1,1)\,q_1q_2 + O(q^3)$$
$$= 4\,q_2 + 4\,q_2^2 + 640\,q_1q_2 + O(q^3)$$

$$\tag{7.11}$$

from which we can read off the numbers of instantons at lowest degrees. Results at higher degrees can be found in refs.[15] and [16][34].

This completes our discussion of this model, which served as a demonstration of the techniques outlined in earlier sections. These techniques have been applied for models with up to three moduli in [16] and [17] and can easily be extended to even more moduli. The hard part seems to be to set up the PF equations. It is not always as easy as in the example above (cf.[16]).

---

[34] Note that in ref. [16], the degrees of the rational curve are defined with respect to the basis h_J, h_D as $n_J = \int_C h_J$ and $n_D = \int_C h_D$ whereas here we define them with respect to the basis h_{D_1}, h_{D_2} as $(n_1, n_2) = (\int_C h_{D_1}, \int_C h_{D_2})$.

8   Conclusions and Outlook

We have tried to convey an idea of the main concepts necessary to understand recent developments of Calabi-Yau compactification of string theory. One of the main advances in the past few years has been the use of mirror symmetry to compute Yukawa couplings. The information necessary to get the Kähler potential, which is of course essential in order to normalize the fields and hence the Yukawa couplings, is also contained in the Picard-Fuchs equations. This has been done explicitly for models with $h^{1,1} = 1$ in refs.[12] and [13], and for a few models with $h^{1,1} = 2$ in [15]. We have conjectured that, at least in the vicinity of the large radius limit, we have constructed quite generally the correct prepotential from which one can get the Kähler metric. The analysis presented here was restricted to the region in moduli space close to the large complex structure and large radius. In the references just cited, this has been extended to the whole moduli space. Since one expects the internal dimensions, or, equivalently, the vacuum expectation values of the moduli, to be of order one (in units of $1/\alpha'$), one needs expressions for the Yukawa couplings which are valid in this range.

Even though there has been considerable progress towards the computation of phenomenologically relevant couplings in strings on Calabi-Yau manifolds, there is still a lot of work left to do. The computation of the Yukawa couplings involving the E_6 singlets is one of them. Also, as long as we have no information on the value of the moduli, the Yukawa couplings are not yet fixed. A potential for the moduli might be generated by non-perturbative string effects. At present there is no hope to compute them. Some information of their possible form can be obtained from studying functions of the moduli with the correct transformation behaviour under duality transformations. However, for general CY compactifications the duality groups are not known, and even if so, one still has to face the task to construct functions of the moduli which are candidates for a non-perturbative potential. One would expect that in general this will not lead to a unique answer. Also, most of the things said here seem to be restricted to the symmetric (2,2) theories. An open problem is the treatment of more general string vacua. We hope to be able, in the not so far future, to report some progress on some of these issues, may be even at the same occasion.

Acknowledgement: S.T. would like the organizers of the school for the opportunity to present our (and others') results and for the very enjoyable week in Helsinki. We thank S.T. Yau for his collaboration on most of the matters discussed here.

References

[1] For introductions, see e.g. M. Green, J. Schwarz and E. Witten, *Superstring Theory*, Vols. I and II, Cambridge University Press, 1986; D. Lüst and S.

Theisen, *Lectures on String Theory*, Springer Lecture Notes in Physics, Vol. 346, 1990; M. Kaku, *Introduction to Superstrings*, Springer 1988

[2] E. Verlinde, H.Verlinde, *Lectures on String Perturbation Theory*, in *Superstrings 88*, Trieste Spring School, Ed. Greene et al. World Scientific (1989) 189

[3] C. Callan, D. Friedan, E. Martinec and M. Perry, Nucl. Phys. B262 (1985) 593, A. Sen, Phys. Rev. D32 (1985) 2102 and Phys. Rev. Lett. 55 (1985) 1846

[4] D. Nemeschanski and A. Sen, Phys. Lett. 178B (1986) 365

[5] For review, see A. Giveon, M. Porrati and E. Rabinovici, *Target Space Duality in String Theory*, hep-th 9401139, to be published in Phys. Rep.

[6] L. Dixon in *Proc. of the 1987 ICTP Summer Workshop in High Energy Physics and Cosmology*, Trieste, ed. G. Furlan et al., World Scientific

[7] W. Lerche, C. Vafa and N. Warner, Nucl. Phys. B324 (1989) 427

[8] B. Greene and M. Plesser, Nucl. Phys. B338 (1990) 15

[9] *Essays on Mirror Manifolds* (ed. S.-T. Yau), Int. Press, Hong Kong, 1992

[10] R. Dijkgraaf, E. Verlinde and H. Verlinde, Utrecht preprint THU-87/30; A. Shapere and F. Wilczek Nucl. Phys. B320 (1989) 669; A. Giveon, E. Rabinovici, G. Veneziano, Nucl. Phys. B322 (1989) 169

[11] S. Ferrara and S. Theisen, *Moduli Spaces, Effective Actions and Duality Symmetry in String Compactifications*, in Proc. of the Third Hellenic School on Elementary Particle Physics, Corfu 1989, Argyres et al.(eds), World Scientific 1990

[12] P. Candelas, X. De la Ossa, P. Green and L. Parkes, Nucl. Phys. B359 (1991) 21

[13] D.Morrison, *Picard-Fuchs Equations and Mirror Maps for Hypersurfaces*, in *Essays on Mirror Manifolds* (ed. S.-T. Yau), Int. Press, Hong Kong, 1992; A. Klemm and S. Theisen, Nucl. Phys. B389 (1993) 153; A. Font, Nucl. Phys. B391 (1993) 358

[14] A. Libgober and J. Teitelbaum, *Duke Math. Journ., Int. Res. Notices* 1 (1993) 29; A. Klemm and S. Theisen, *Mirror Maps and Instanton Sums for Complete Intersections in Weighted Projective Space*, preprint LMU-TPW 93-08

[15] P. Candelas, X. de la Ossa, A. Font, S. Katz and D. Morrison, *Mirror Symmetry for Two Parameter Models I*, preprint CERN-TH.6884/93

[16] S. Hosono, A. Klemm, S. Theisen and S.-T. Yau, *Mirror Symmetry, Mirror Map and Applications to Calabi-Yau Hypersurfaces*, to appear in Comm. Math. Phys.

[17] S. Hosono, A. Klemm, S. Theisen and S.-T. Yau, *Mirror Symmetry, Mirror Map and Applications to Complete Intersection Calabi-Yau Spaces*, preprint LMU-TPW-94-03 (to appear)

[18] P. Aspinwall, B. Greene and D. Morrison, *Space-Time Topology Change: the Physics of Calabi-Yau Moduli Space*, preprint IASSNS-HEP-93-81; *Calabi-Yau Moduli Space, Mirror Manifolds and Space-Time Topology Change in String Theory*, preprint IASSNS-HEP-93-38; *The Monomial-Divisor Mirror Map*, preprint IASSNS-HEP-93-43

[19] P. Berglund, P. Candelas, X. de la Ossa, A. Font, T. Hübsch, D. Jancic and F. Quevedo, *Periods for Calabi-Yau and Landau-Ginsburg Vacua*, preprint CERN-TH.6865/93; P. Berglund, E. Derrich, T. Hübsch and D. Jancic, *On Periods for String Compactification*, preprint HUPAPP-93-6

[20] P. Berglund and S. Katz, *Mirror Symmetry for Hypersurfaces in Weighted Projective Space and Topological Couplings*, preprint IASSNS-HEP-93-65

[21] P. Candelas, G. Horowitz, A. Strominger and E. Witten, Nucl. Phys. B258 (1985) 46

[22] Easily accessible introductions to Calabi-Yau manifolds are G. Horowitz, *What is a Calabi-Yau Space?* in *Unified String Theories*, M. Green and D. Gross, editors, World Scientific 1986 and P. Candelas, *Introduction to Complex Manifolds*, Lectures at the 1987 Trieste Spring School, published in the proceedings. For a rigorous mathematical treatment we refer to A.L. Bessis, *Einstein Manifolds*, Springer 1987; P. Griffiths and J. Harris, *Principles of Algebraic Geometry*, John Wiley & Sons 1978; K. Kodaira, *Complex Manifolds and Deformation of Complex Structures*, Springer 1986

[23] A good and detailed introduction to Calabi-Yau manifolds is the book by T. Hübsch, *Calabi-Yau Manifolds*, World Scientific 1991

[24] S.T. Yau, Proc. Natl. Acad. Sci. USA 74 (1977), 1798; Comm. Pure Appl. Math. 31 (1978) 339

[25] K. Kodaira, L. Nirenberg and D. C. Spencer, Ann. Math. 68 (1958) 450

[26] G. Tian, in *Mathematical Aspects of String Theory*, ed. S. T. Yau, World Scientific, Singapore (1987)

[27] A. Strominger, Phys. Rev. Lett. 55 (1985) 2547, and in *Unified String Theory*, eds. M. Green and D. Gross, World Scientific 1986; A. Strominger and E. Witten, Comm. Math. Phys. 101 (1985) 341

[28] L. Dixon, D. Friedan, E. Martinec and S, Shenker, Nucl. Phys. B282 (1987) 13; S. Hamidi and C. Vafa, Nucl. Phys. B279 (1987) 465; J. Lauer, J. Mas and H. P. Nilles Nucl. Phys. B351 (1991) 353; S. Stieberger, D. Jungnickel, J. Lauer and M. Spalinski, Mod. Phys. Lett. A7 (1992) 3859; J. Erler, D. Jungnickel, M. Spalinski and S. Stieberger, Nucl. Phys. B397 (1993) 379

[29] D. Gepner, Phys. Lett. 199B (1987) 380 and Nucl. Phys. B311 (1988) 191

[30] J. Distler and B. Greene, Nucl. Phys. B309 (1988) 295

[31] M. Dine, N. Seiberg, X-G. Wen and E. Witten, Nucl. Phys. B278 (1987) 769, Nucl. Phys. B289 (1987) 319

[32] S. Ferrara, D. Lüst, A. Shapere and S. Theisen, Phys. Lett. 225B (1989) 363

[33] P. Aspinwall and D. Morrison, Comm. Math. Phys. 151 (1993) 245

[34] E. Witten, Comm. Math. Phys. 118 (1988) 411

[35] T. Banks, L. Dixon, D. Friedan and E. Martinec, Nucl. Phys. B284 (1988) 613

[36] M. Ademollo et al., Nucl. Phys. B11 (1976) 77 and Nucl. Phys. B114 (1976) 297 and Phys. Lett. 62B (1976) 105

[37] A. Schwimmer and N. Seiberg, Phys. Lett. 184B (1987) 191

[38] M. Dine and N. Seiberg, Nucl. Phys. B301 (1988) 357

[39] E. Witten, Nucl. Phys. B202 (1982) 253; R. Rohm and E Witten, Ann. Phys. **170** (1986) 454

[40] P. Candelas, E. Derrick and L. Parkes, Nucl. Phys. B407 (1993) 115

[41] R. Schimmrigk, Phys. Rev. Lett. 70 (1993) 3688 and *Kähler manifolds with positive first Chern class and mirrors of rigid Calabi-Yau manifolds*, preprint BONN-HE 93-47

[42] N. Seiberg, Nucl. Phys. B303 (1988) 286

[43] S. Cecotti, S. Ferrara and L. Girardello, Int. J. Mod. Phys. A4 (1989) 2475; S. Ferrara, Nucl. Phys. (Proc. Suppl.) 11 (1989) 342

[44] P. Candelas, P. Green and T. Hübsch, *Connected Calabi-Yau compactifications*, in Strings '88, J. Gates et al. (eds), World Scientific 1989 and Nucl. Phys. B330 (1990) 49; P. Candelas, T. Hübsch and R. Schimmrigk, Nucl. Phys. B329 (1990) 582

[45] L. Dixon, V. Kaplunovsky and J. Louis, Nucl. Phys. B329 (1990) 27

[46] B. de Wit, P.G. Lauwers, R. Philippe, S.Q. Su and A. van Proeyen, Phys. Lett. 134B (1984) 37; B. de Wit and A. van Proeyen, Nucl. Phys. B245 (1984) 89; J.P. Derendinger, S. Ferrara, A. Masiero and A. van Proeyen, Phys. Lett. 140B (1984) 307; B. de Wit, P.G. Lauwers and A. van Proeyen, Nucl. Phys. B255 (1985) 569; E. Cremmer, C. Kounnas, A. van Proeyen, J.P. Derendinger, S. Ferrara, B. de Wit and L. Girardello, Nucl. Phys. B250 (1985) 385

[47] M. Kreuzer and H. Skarke, Nucl. Phys. B388 (1993) 113; A. Klemm and R. Schimmrigk, Nucl. Phys. B411 (1994) 559

[48] P. Candelas, A. Dale, A. Lütken and R. Schimmrigk, Nucl. Phys. B298 (1988) 493; P. Candelas, A. Lütken and R. Schimmrigk, Nucl. Phys. B306 (1989) 105

[49] V. Batyrev and D. van Straten, *Generalized Hypergeometric Functions and Rational Curves on Calabi-Yau Complete Intersections in Toric Varieties*, preprint 1992

[50] I. Dolgachev, *Weighted Projective Varieties* in Lecture Notes in Mathematics 956 Springer-Verlag (1992) 36

[51] A. R. Fletcher, *Working with Weighted Complete Intersections*, Max-Planck-Institut Series, No. 35 (1989) Bonn

[52] D. G. Markushevich, M. A. Olshanetsky and A. M. Perelomov, Comm. Math. Phys. 111 (1987) 247

[53] J. Erler and A. Klemm, Comm. Math. Phys. 153 (1993) 57

[54] L. Dixon, J. Harvey, C. Vafa and E. Witten, Nucl. Phys. B261 (1985) 678 and Nucl. Phys. B274 (1986) 285

[55] S.S. Roan and S. T. Yau, Acta Math. Sinica (NS) 3 (1987) 256; S.S. Roan, J. Diff. Geom. 30 (1989) 523

[56] M. Kreuzer and H. Skarke, Nucl. Phys. B405 (1993) 305; M. Kreuzer, Phys. Lett. 314B (1993) 31; A. Niemeyer, Diplom Thesis, TU-München, 1993

[57] T. Oda, *Convex Bodies and Algebraic Geometry: an Introduction to the Theory of Toric Varieties*, Ergebnisse der Mathematik und ihrer Grenzge-

biete, 3. Folge, Bd. 15, Springer Verlag 1988; V.I. Danilov, Russian Math. Surveys, 33 (1978) 97; M. Audin, *The Topology of Torus Actions on Symplectic Manifolds*, Progress in Mathematics, Birkhäuser 1991; W. Fulton, *Introduction to Toric Varieties*, Annals of Mathematics Studies, Princeton University Press 1993; V.I. Arnold, S. M. Gusein-Zade and A.N. Varchenko, *Singularities of Differntial Maps*, Birkhäuser 1985, Vol. II, Chapter 8

[58] V. Batyrev, Duke Math. Journal 69 (1993) 349

[59] V. Batyrev, Journal Alg. Geom., to be published

[60] V. Batyrev, *Quantum Cohomology Rings of Toric Manifolds*, preprint 1992

[61] V. Batyrev and D. Cox, *On the Hodge Structure of Projective Hypersurfaces in Toric Varieties*, preprint 1993

[62] F. Hirzebruch, Math. Annalen 124 (1951) 77

[63] M. Beltrametti and L. Robbiano, Expo. Math. 4 (1986) 11.

[64] D.Aspinwall, B.Greene and D.Morrison, Phys.Lett. 303B (1993) 249 and *The Monomial-Divisor Mirror Map*, preprint IASSNS-HEP-93/43

[65] S.-S. Roan: *Topological Couplings of Calabi-Yau Orbifolds*, Max-Planck-Institut Serie No. 22 (1992), to appear in *J. of Group Theory in Physics*

[66] E.L. Ince, *Ordinary Differential Equations*, Dover 1956

[67] M. Yoshida, *Fuchsian Differential Equations*, Braunschweig (1987) Vieweg

[68] P. A. Griffiths, Bull. Amer. Math. Soc. 76 (1970) 228

[69] P. Griffiths, Ann. of Math. 90 (1969) 460

[70] P. Candelas, Nucl. Phys. B298 (1988) 458

[71] I.M.Gel'fand, A.V.Zelevinsky and M.M.Kapranov, Func. Anal. Appl. 28 (1989) 12 and Adv. Math. 84 (1990) 255

[72] D. Morrison, *Where is the Large Radius Limit*, preprint IASSNS-HEP-93/68 and *Compactifications of Moduli Spaces Inspired by Mirror Symmetry*, preprint DUK-M-93-06

[73] M. Reid, *Decomposition of Toric Morphisms*, in Progress in Mathematics 36, M. Artin and J. Tate, eds, Birkhäuser 1983

[74] S. Ferrara and A. Strominger, in the Proceedings of the Texas A & M Strings '89 Workshop; ed. R. Arnowitt et al., World Scientific 1990

[75] P. Candelas, P. Green and T. Hübsch Nucl. Phys. B330 (1990) 49; P. Candelas and X.C. de la Ossa, Nucl. Phys. B355 (1991) 455

[76] R. Bryant and P. Griffiths, *Some observations on the Infinitesimal Period Relations for Regular Threefolds with Trivial Canonical Bundle*, in Progress in Mathematics 36, M. Artin and J. Tate, eds; Birkhäuser, 1983

[77] A. Strominger, Comm. Math. Phys. 133 (1990) 163

[78] L. Castellani, R. D'Auria and S. Ferrara, Phys. Lett. 241B (1990) 57 and Class. Quantum Grav. 7 (1990) 1767; S. Ferrara and J. Louis, Phys. Lett. 278B (1992) 240

[79] A. Ceresole, R. D'Auria, S. Ferrara, W. Lerche and J. Louis, Int. J. Mod. Phys. A8 (1993) 79

[80] X.-G. Wen and E. Witten, Phys. Lett. 166B (1986) 397

Lecture Notes in Physics

For information about Vols. 1–399
please contact your bookseller or Springer-Verlag

New Series m: Monographs

Springer-Verlag and the Environment

We at Springer-Verlag firmly believe that an international science publisher has a special obligation to the environment, and our corporate policies consistently reflect this conviction.

We also expect our business partners – paper mills, printers, packaging manufacturers, etc. – to commit themselves to using environmentally friendly materials and production processes.

The paper in this book is made from low- or no-chlorine pulp and is acid free, in conformance with international standards for paper permanency.